PLANNING FOR SUSTAINABILITY

How can human communities sustain a long-term existence on a small planet? This challenge grows ever more urgent as the threat of global warming increases.

Planning for Sustainability, Second Edition presents a wide-ranging, intellectually well-grounded and accessible introduction to the concept of planning for more sustainable and livable communities. The text explores topics such as how more compact and walkable cities and towns might be created, how local ecosystems can be restored, how social inequalities might be reduced, how greenhouse gas emissions might be lowered, and how more sustainable forms of economic development can be brought about.

The second edition has been extensively revised and updated throughout, including an improved structure with chapters now organized under three sections: the nature of sustainability planning, issues central to sustainability planning, and scales of planning. New material includes greater discussion of climate change, urban food systems, the relationships between public health and the urban environment, and international development.

Building on past schools of planning theory, *Planning for Sustainability* lays out a sustainability planning framework that pays special attention to the rapidly evolving institutions and power structures of a globalizing world. By considering in turn each scale of planning—international, national, regional, municipal, neighborhood, and site and building—the book illustrates how sustainability initiatives at different levels can interrelate. Only by weaving together planning initiatives and institutions at different scales, and by integrating efforts across disciplines, can we move towards long-term human and ecological well-being.

Stephen M. Wheeler, PhD, AICP is Associate Professor in the Landscape Architecture Program, Department of Human Ecology, at University of California, Davis. His areas of interest include sustainable development, planning for climate change, urban design, and built landscapes of metropolitan regions.

PLANNING FOR SUSTAINABILITY

Creating livable, equitable and ecological communities

Second edition

Stephen M. Wheeler

Routledge
Taylor & Francis Group

LONDON AND NEW YORK

First edition published 2004
Second edition published in 2013
by Routledge
2 Park Square, Milton Park, Abingdon, Oxon OX14 4RN

Simultaneously published in the USA and Canada
by Routledge
711 Third Avenue, New York, NY 10017

Routledge is an imprint of the Taylor & Francis Group, an informa business

© 2004, 2013 Stephen M. Wheeler

British Library Cataloguing in Publication Data
A catalogue record for this book is available from the British Library

Library of Congress Cataloging in Publication Data
Wheeler, Stephen (Stephen Maxwell)
Planning for sustainability / Stephen M. Wheeler. – 2nd ed.
p. cm.
Includes bibliographical references and index.
1. City planning–Environmental aspects. 2. City planning–Social aspects.
3. Urbanization–Environmental aspects. 4. Community development, Urban.
5. Regional planning–Environmental aspects. I. Title.
HT166.W484 2013
307.1'216--dc23   2012045164

ISBN: 978-0-415-80988-7 (hbk)
ISBN: 978-0-415-80989-4 (pbk)
ISBN: 978-0-203-13455-9 (ebk)

Typeset in Joanna MT
by Cenveo Publisher Services

RECYCLED
Paper made from
recycled material
FSC
www.fsc.org   FSC® C013604   Printed and bound by CPI Group (UK) Ltd, Croydon, CR0 4YY

CONTENTS

LIST OF BOXES

LIST OF FIGURES

PREFACE TO THE SECOND EDITION

A lot has changed since I wrote the first edition of this book nearly a decade ago. "Sustainability" has become better established as a central goal of many professions and communities. Academic interest has continued to grow, and many universities including my own have launched majors and minors related to sustainability. Cities, towns, regions, states, provinces, and entire nations have prepared sustainability plans. Subjects such as food systems, financial markets, and global warming have become more central concerns of sustainability advocates.

More worrisomely, our global situation has changed as well. We are much more aware now that global warming is an overwhelming threat to humans and other species on the planet, and that the climate is changing in profound ways. Though I've followed this topic myself for more than 30 years, I will admit to treating it until the mid-2000s as one of a number of comparable challenges facing us. Global warming is more than that: it is the pre-eminent sustainability crisis that may bring down industrial civilization. Consequently, I've woven discussion of the climate challenge into many of the new edition's chapters.

Another major change in conditions has to do with financial systems, equity, and governance. The rich have always been with us, and at many, perhaps most times throughout history have sought to control societies toward their own ends. But within the last decade we have developed new understandings about how the wealthy and the corporations they control have manipulated political systems and public consciousness to the detriment of sustainability, especially in North America but also in many other nations around the world. Addressing this situation is a core need of sustainability planning.

Worsening inequality and control of social systems by the wealthy have had a number of effects. One has been the loosening of financial regulations, resulting in speculation-fueled economic disasters such as the late-2000s housing collapse in the US, the early-2010s financial turmoil in Europe, and collapse of a housing bubble in China. Another result has been the undermining of democracy in many nations, both by direct influence on elections and by the more subtle shaping of public opinion against the very idea that the governments should actively ensure long-term social and environmental welfare. In the United States particularly we have seen politicians try to literally starve government of money through anti-tax crusades and endless budget-cutting (though the military and corporate subsidies are usually exempted). Through movements such as the Tea Party, neo-conservative rhetoric has advanced narrow concepts of individual rights at the expense of a sense of collective responsibility for our future. Right-wing interests have also sought to deny basic scientific understandings of climate change and other ecological problems. This reactionary politics is a major barrier in the way of progress toward sustainability.

Other new issues have emerged as well. An obesity epidemic has spread in many countries, leading to many unfortunate public health impacts and rising expenditure to deal with them. Partly as a result, in the early twenty-first century activists have turned more attention toward the question of how to promote healthier and more sustainable food systems, as well as how to enable pedestrian and bicycle travel and healthy living generally.

As it has depleted traditional, easily exploitable sources of oil and gas, the fossil fuel industry has turned to riskier and dirtier production methods including deep sea drilling, tar sands mining, and hydraulic fracturing ("fracking") of shale rock formations. These methods have led to much opposition from environmentalists, and have become a new front in the battle against fossil fuels.

Lastly, the emergence of China, India, and other developing nations as global economic powerhouses has affected sustainability issues around the world. China, for example, has eclipsed the United States as the world's largest greenhouse gas emitter, and is rapidly constructing entire new cities that will affect that nation's sustainability prospects for generations to come. Although it is very difficult for any of us to be aware of what is going on in all parts of the world, adopting a global approach to sustainability challenges is increasingly important.

Though much of the news during the past decade has been worrisome, there have been encouraging developments as well. Population growth is slowing in many societies, to some extent alleviating past fears of a population explosion (more on that later). Poverty, though it remains very much urgent and present, is lessening somewhat as more of the world's population enters the middle class. Social tolerance is on the increase in many parts of the world, and initiatives such as allowing gays and lesbians to marry that would have been unthinkable a generation ago are now becoming commonplace. Protests in the Arab world have inspired millions toward more open and democratic societies, though this is very much a work in progress. The Occupy Movement in the early 2010s helped call attention to enormous underlying inequalities within industrial societies, as well as to the extent of corporate influence over politics. Within such movements social media have proven a powerful force for progressive change, raising the possibility that networks of citizens, nongovernmental organizations (NGOs), and local governments may become a political force able to counterbalance the private sector and often-ineffectual higher levels of government.

Though overall our progress toward sustainability has been lamentably slow, these trends are deeply hopeful.

This thoroughly rewritten and expanded edition of *Planning for Sustainability* tries to take such changes into account. It adds new material to chapters, updates information, and includes new examples and graphics. Most notably, whereas the previous version organized material primarily by scale in order to give a sense of which sustainability initiatives might be tackled at which level of organization from global to local, this edition includes a new section examining key sustainability strategies by planning issue. Separate chapters on topics such as climate change planning, energy, transportation, land use, urban design, housing, social equity, economic development, and population thus present information in ways corresponding more closely to the traditional treatment of issues within planning and other professions. This part of the book also includes a chapter on governance, since it is becoming increasingly clear that without more functional political systems we cannot plan effectively for sustainable societies. Although it has not been seen so historically, effective governance is a planning task as much as anything else, and attention in that direction is urgently needed.

This edition has some slightly different emphases than the previous one. Since it is clear that traditional forms of political mobilization—such as lobbying and working through the major political parties—aren't working well in many countries, I now place more emphasis on new forms of creative action both inside and outside of political systems. This change arises from my growing conviction that we will need to "think outside the box" to bring about sustainability. Frequently this will mean using tools such as social media, the arts, and the organizations of civil society, as well as creating small-scale demonstrations of different ways of living, building, and working. Traditional methods of social change will still be necessary as well, but they alone do not seem enough.

Also, since it is also becoming increasingly clear that green technology alone can't save us from global warming and other crises, I emphasize the need for more fundamental changes in social ecology, individual cognition, and lifestyle. Our current situation in other words is the result of a crisis of values, spirit, and social organization that reaches its most extreme manifestations in the behavior of Wall Street traders, conservative politicians, and the advertising industry, but is much more widely present within industrial cultures. Only by confronting this crisis directly can we get beyond it. I am not alone in this view; Donella Meadows, the *Limits to Growth* lead author who first developed the concept of sustainability in 1972, believed strongly that a change in paradigms or worldviews was essential to bring about more sustainable societies. Many other writers such as Joanna Macy, Thomas Berry, Wendell Berry, Fritjof Capra, David Korten, Hazel Henderson, Vandana Shiva, and the Dalai Lama have written along these lines as well.

In short, I have taken the opportunity of a second edition as a chance to rethink the material in light of the past decade's events. Rather than updating a few bits here and there, I've revised content throughout to try to keep this volume on the cutting edge of its subject.

Planning for Sustainability in its current edition is a bit different than many other texts on sustainability. Rather than providing an upbeat account of one green strategy after another, it tries to balance positive opportunities with real concern about our future and inquiry into the fundamental root causes of unsustainability. I also try to address some topics that often aren't touched by more traditional approaches, including population, consumption, lifestyle, and ethics. Many environmental writers, for example, have been reluctant to discuss population for fear of being seen as anti-immigrant or insensitive to cultures in which large families are the norm. However, I feel strongly that such subjects must be brought into the picture, and have tried to do so here.

Moving toward sustainability will be a long process, no matter how quickly we act in the near term—the "great work" of the next several generations, as Thomas Berry puts it.[1] It's important to understand this challenge as deeply and fully as possible, being realistic about our progress or lack thereof, but also hopeful that in the long run we can in fact help our species evolve towards a more sustainable and humane global civilization. In that spirit I've tried to explore as many different dimensions of sustainability as I can in this book, and hope that the result is useful to those like myself seeking to understand how we might in fact move towards a "livable, equitable, and ecological" future.

ACKNOWLEDGMENTS

As with the first edition, I wish to thank all those who have inspired me to work in the field of urban sustainability, especially the late Donella Meadows and David Brower, and other colleagues and teachers including Timothy Beatley, Manuel Castells, Ernest Callenbach, Elizabeth Deakin, Allan Jacobs, Clare Cooper Marcus, Carolyn Merchant, and Richard Norgaard. Thanks also to my former colleagues at Urban Ecology and Friends of the Earth, and the many authors who have inspired me over the years, including Christopher Alexander, Robert Bellah, Thomas Berry, Kenneth Boulding, Lester Brown, Murray Bookchin, Fritjof Capra, Barry Commoner, Judy Corbett, Michael Corbett, Herman Daly, John Forester, Herbert Girardet, David Harvey, Paul Hawken, Dolores Hayden, Randy Hester, Jane Jacobs, David Korten, James Howard Kunstler, Aldo Leopold, Amory Lovins, Hunter Lovins, John Tillman Lyle, Kevin Lynch, Lewis Mumford, Robert Putnam, Theodore Roszak, E.F. Schumacher, Sim Van der Ryn, and William H. Whyte. Anyone wishing historical background in the urban sustainability related issues could do worse than start with this list.

Eight anonymous reviewers provided very useful comments on either the first or second edition of this book. My colleagues and students at the University of California, Berkeley, the University of New Mexico, and the University of California, Davis, have been invaluable sounding boards for various ideas.

The excellent staff at Routledge shepherded this edition through its development and production. Thanks especially to Andrew Mould, Faye Leerink, and Lisa Salonen, and also to Chris Shaw, the copy editor, and Christine James, the proofreader.

I am deeply grateful to my parents for their constant love and encouragement. My daughter, Maddy, has been an inspiration and a joy during the writing process. Most of all, thanks to my wife, Mimi, for her love, humor, and insight from personal experience as an editor into the long process of book production.

The author and publishers would like to thank the following for granting permission to reproduce copyright material: Robert Thayer for Figure 5.1; Contra Costa County Redevelopment Agency for Figures 6.1 and 6.2; Sustainable Seattle for Box 6.1; Metro Resource Data Center for Figure 10.5; Tom Jones for Figure 13.1; Classic Communities for Figure 13.2; Princeton Architectural Press for Figure 21.1d; The Brookings Institution for Figure 22.2; Town of Markham for Figure 24.5; City of Albany, California and Freedman, Tung, and Bottomley for Figure 24.7; Calthorpe Associates for Figures 24.8 to 24.10;

Steve Price, Urban Advantage (www.urban-advantage.com) for Figures 24.11 and 24.12; and University of California Press for Figure 24.13.

Every effort has been made to contact copyright holders for their permission to reprint material in this book. The publishers would be grateful to hear from any copyright holder who is not here acknowledged and will undertake to rectify any errors or omissions in future editions of this book.

1

INTRODUCTION

The past century has seen the rise of an urban and suburban landscape that is profoundly different from anything created before. From the office parks, malls, freeways, residential tracts, and abandoned inner cities of many affluent countries to enormous megacities in developing countries, human communities have taken on dramatically new forms and characteristics, with radically new lifestyles and economies. Cities that once occupied a few square miles now cover thousands; populations that once walked most places are now dependent on motor vehicles. People are surrounded by vastly larger quantities of material goods, but often feel the lack of social connection, meaning, or ability to control their world.

Although recent patterns of development have brought many benefits, they are unsustainable in that they cannot be continued in the same ways for very much longer. Today's development practices—both economic and physical forms of development—consume enormous amounts of land and natural resources, damage ecosystems, produce a wide variety of pollutants and toxic chemicals, create ever-growing inequities between groups of people, fuel global warming, and undermine local community, economies, and quality of life. Since the changes are incremental, it is hard to appreciate how rapidly our world is being transformed and how fundamentally these processes affect our lives and the choices available to us.

One of the main challenges of the twenty-first century will be to bring about more sustainable human communities. The broad and diverse field that has come to be known as "planning" can play a central role in meeting this goal, in that it both establishes overall visions of desired futures and also deals with the nuts-and-bolts of how communities, regions, and countries are built and run, including how they relate to natural ecosystems.

Since the origins of formal urban and regional planning activities more than 100 years ago, results have been decidedly mixed. Much good has been done in terms of improving human welfare, but unfortunately planners themselves have led many unsustainable development practices. They have issued the building permits for suburban sprawl, programmed the monies for ever-expanding freeway systems, set up urban renewal programs that at times have bulldozed vibrant neighborhoods, assisted with the rise of an economy run by global corporations, and, most importantly, failed to be as creative as they might at developing alternative development visions. Planning must do better. It can and should reorient itself in the twenty-first century to focus on the challenge of creating more sustainable communities. This role will acknowledge existing politics and traditions but seek every opportunity to bring about creative change. Though it will not be easy, such sustainability planning can be an exhilarating and meaningful path for new generations of planners, architects, landscape architects, engineers, political leaders, progressive developers, and community activists.

The purpose of this book is to provide a systematic background to the subject of sustainability planning as it cuts across many different specialties and scales. The chapters in Part One examine how the sustainability concept has developed and assumed center stage in global debates, its general implications for planning, how it relates to planning theory, some tools useful for sustainability planning, and some main ways that sustainability might be pursued at various scales of planning. Part Two, new to this edition, investigates particular issues such as climate change, energy and materials use, urban growth, transportation, housing, social equity, and population. The aim is to highlight key issues within these fields that are important to sustainability. Part Three examines international, national, state and provincial, regional, municipal, neighborhood, and site- and building-scale planning in turn. Sustainability efforts of different types are possible at different scales and one main point of this book is that action at all of these levels is necessary, and must become better integrated into a coherent, mutually reinforcing framework if sustainable development is to come about.

Although much of the focus of the book is on North American planning, it attempts to take a global perspective, bringing in examples from Britain, the European Union, Asia, and elsewhere. Similar sustainability problems exist in many parts of the world, not least because the same technologies and modes of development are spreading everywhere. Many of the potential solutions to current problems are similar as well. Every part of the world can learn from the others, and the most rapid and creative change will come about if we do learn in this way.

Though many existing books discuss sustainable development in one way or another, the intent here is to more fully develop sustainability planning theory in a way that is broadly accessible and useful to students, practitioners, and lay readers alike. I would also like to invite readers personally to become more actively involved in planning for sustainability, whether through projects in their homes, yards, neighborhoods, and local ecosystems, or within larger-scale political, economic, and social systems. All of us can play some role in bringing about a more sustainable society, and doing so may be one of the most satisfying and meaningful activities we can undertake.

THE NEED FOR CHANGE

To start with, we need to take note of the nature and variety of current sustainability problems. We have around us an ongoing disaster that—partly because it takes place in slow and incremental fashion, and at scales and in places seemingly removed from our daily lives—is little discussed by the media, politicians, or most citizens. In part our development crisis is one of environmental damage, leading to phenomena such as global warming, resource depletion, and the loss of species that are difficult or impossible to reverse. In part this crisis is one of misguided urbanization, leaving us with the fragmented landscape of subdivisions, freeways, malls, commercial strips, and office parks that cover vast quantities of land and form the daily reality for many of us. In part it is one of transportation, having to do with rising traffic congestion and overdependency on motor vehicles. In part it is a problem of economic development, having to do with the worldwide promotion of high levels of material production and consumption, the growing power of global corporations, and the decline of smaller-scale businesses that could form a more locally oriented, socially responsible economy. In part we have a challenge of housing, which is frequently scarce, unaffordable, or inappropriately designed and located. In part we face a crisis of poverty and growing inequality, which leaves enormous numbers of people worldwide without access to decent-paying work, good schools, health care, or other necessities of life, while promoting overconsumption among the insular and politically powerful rich. And in part we have a crisis of spirit in which our values, empathy, and methods of understanding both individually and collectively are not what is needed to create a more sustainable world.

All of these problems are interrelated. Though much effort is underway to address them, such crises are not being acknowledged adequately by current political and professional leaders, and must be tackled in far more comprehensive ways if we are to heal the damage of the past and head in more positive directions in the future.

One of the best ways to appreciate the unsustainability of current development patterns is to observe them firsthand. For example, my wife and I recently traveled up Highway 101 in California, a freeway running much of the length of the state. While we were still more than 30 miles away from the main Bay Area cities we began to see the outposts of suburban and exurban sprawl. "For Sale" signs appeared on either side of the road, and many fields were no longer being farmed—an indication that speculators were holding them waiting for prices to rise. As we traveled onwards through what was once some of the richest farming country in the world, the road soon became surrounded by a varying mixture of ranchettes, housing subdivisions, warehouses, office parks, strip commercial developments, big-box stores, golf courses, and motor vehicle dealerships. A chaotic, low-density exurban world, in other words, dominated by large-scale corporate activity and with little sense of place, sensitivity to the environment, or social cohesion.

Crossing a row of hills we descended toward San Jose and neighboring towns—the area known as "Silicon Valley" and envied the world over as a prototype of successful economic development. Whereas the great cities of the past created attractive squares, boulevards, and residential districts, Silicon Valley is organized around tacky commercial strips, crowded freeways, and generic office parks. There is very little public space in this landscape. Housing is astronomically expensive, roads are congested, and air and water pollution are both significant problems for the region. Workers routinely must commute 60 miles or more to their jobs. Many work for low wages in poor conditions. Rather than being the world's success symbol, by many tokens the development of this area has been a failure.

As we reached the southern border of Oakland it became apparent that the region's development boom had not touched that city's lower-income, African American neighborhoods. The blocks we passed by were characterized by empty lots, dilapidated houses, and boarded-up factories. This land, urbanized 50 to 100 years ago, has now been abandoned by many employers and wealthier residents, leaving these impoverished areas to cope with a depleted tax base, toxic contamination, declining schools, crime, and a lack of jobs.

We finally reached home after traveling in heavy traffic for two hours through an urbanized landscape that showed few signs of the beauty, culture, livability, and ecological richness that attracted many people to the Bay Area in the first place. Rather than using its great wealth to create livable, equitable, and ecological communities, our society had done much the opposite. Similar patterns of unsustainable urban development are occurring the world over, though they take somewhat different forms in different places and times. Radically different alternatives are needed.

In contrast, a growing number of communities offer at least partial examples of development that is much more socially and ecologically healthy. The downtown revitalization of Portland, Oregon, the transit systems of Paris, Toronto, and Curitiba, Brazil, the pedestrian districts of Copenhagen and most other European cities, the ecological wastewater treatment marsh of Arcata, California, the democratic budgeting process of Puerto Alegre, Brazil, the relative social equality of northern European countries, the wind farms of China and Spain, and many other examples offer hope for the future. Many such cases will be described later in this book. These examples suggest that different ways of developing our world are possible, and that the action of dedicated individuals can make a difference.

To better understand the overall situation, let us look at some key implications of current development trends more closely.

Climate change

In the early twenty-first century it has become increasingly clear that global warming is the most serious sustainability problem we face. By the early 2010s measurements were beginning to show atmospheric CO_2 concentrations above 400 parts per million (ppm), up from around 285 ppm before the Industrial Revolution. Moreover, the rate of increase in humanity's greenhouse gas (GHG) emissions was accelerating, not decreasing. With more heat, water vapor, and energy in the atmosphere, extreme weather events are becoming more common worldwide. Computer models show likely temperature rises of at least 2° Celsius (3.6° Fahrenheit) by 2050 even if GHG emissions are reduced substantially and heating of between 2 and 5.5° C (3.6 to 10° F) by 2100 under various scenarios.[1] Paleoclimatic data suggests the strong possibility that even hotter temperatures, up to 16° C (29° F) or more, could be reached after that.[2] Corresponding sea level rise is likely to be between one and two meters this century. Effects on crops, water supplies, and coastal cities would be bad enough, but an even worse impact is likely to be acidification of the oceans (which absorb about the half the carbon dioxide we emit). By killing much life in the seas, we may sow the seeds of our species' destruction.[3] Clearly, heading off dangerous climate change is now our top sustainability priority. Doing so, however, means change in almost every other dimension of social and economic development.

Land use and suburban sprawl

Another main though less apocalyptic set of problems concerns land use. Although most pronounced in motor vehicle-dependent communities of the United States, Canada, and Australia, suburban sprawl is advancing almost everywhere else as well. The problem is not just that this style of development tends to be low density—some sprawl may occur at moderate to high densities—but that it possesses many other characteristics that increase GHG emissions and work against the evolution of livable, walkable communities. Sprawl development is often fragmented (the landscape becomes a mosaic of inwardly oriented projects that don't relate to one another or are separated by oversized roads), discontiguous (developments leapfrog out into the countryside), homogeneous (each project contains only housing, offices, or stores), poorly connected (street networks are characterized by cul-de-sacs, loop roads, or other poorly connected patterns), and ecologically destructive (new development fails to take into account natural landscape features and helps generate pollution and excessive resource use). All of these characteristics undermine livability and sustainability. Though certain forms of suburban sprawl have been known throughout history, by far the largest amount has been built since the Second World War. Factors leading to this rapid suburban and exurban expansion include a continual rise in motor vehicle ownership, government road-building, subsidies for suburban homeownership, race and class prejudice, and ideals of country living based on the English tradition of the picturesque country estate.[4] As researchers such as Harvey L. Molotch, John R. Logan, Mark Gottdiener, and Marc Weiss have shown, sprawl has also been facilitated by "growth coalitions" of development interests and politicians which have orchestrated suburban development.[5] In the early 1920s Sinclair Lewis's fictional Babbitt epitomized this boosterish growth mentality, seeking suburban expansion with very little sense of the true social and environmental costs.[6] A century later, attitudes are much the same in many communities throughout the world.

Sprawl consumes enormous amounts of land. The US Department of Agriculture and American Farmland Trust estimate that some 1.5 million acres of open space was lost to urbanization annually in the United States during much of the 2000s.[7] Poorly planned suburban development also degrades or destroys wildlife habitat and natural ecosystems,[8] leads to dependency on motor vehicles, and often segregates different demographic communities from one another. Suburban expansion tends to take place at the expense of older urban areas, which are then plagued by abandoned factories and empty storefronts. Older industrial cities such as Detroit and St Louis have lost up to half the population they had in their heyday. British urban areas such as Manchester, Newcastle, Liverpool, and Glasgow have also lost population, although not at such a rapid rate. In the 1980s and 1990s suburban populations grew ten times faster than central-city populations in the largest US metropolitan areas,[9] as many businesses and residents left older urban areas for the suburbs. Despite significant central city revitalization in the 2000s, the amount of population growth in the core of large metropolitan areas was still less than 10 percent that of suburban and exurban areas.[10] Similar versions of the suburbanization phenomenon are occurring elsewhere in the world as well.

Although rapid suburban expansion is sometimes seen as necessary to handle regional population increase, most of the physical growth of urban areas stems from inefficient land use rather than population needs. According to the US Environmental Protection Agency (EPA), the amount of land consumed by urbanization in 34 large US metropolitan areas grew 2.65 times faster than the rate of population growth during four decades in the late twentieth century.[11] Sprawl occurred even in some Northeastern and Midwest metropolitan areas that lost population. The problem, in other words, is not so much population growth as the ways we use land (see Figures 1.1. to 1.5).

Figure 1.1 Vanishing open space. The Tassajara Valley southeast of Oakland, California, has been targeted for thousands of homes

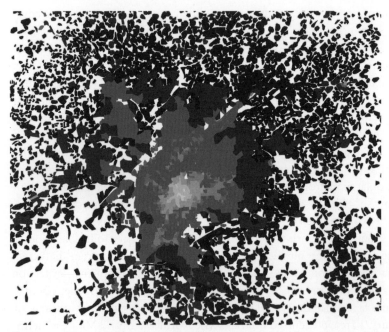

Figure 1.2 The growth of a metropolitan area: Atlanta. Each darker increment represents 20 years of urban expansion, up to 2000. The frame is about 50 miles across

Figure 1.3 The American dream? A 2,000-square-foot house with a three-car garage and sport utility vehicle (SUV) has become the way of life in many newer American suburbs

Figure 1.4 Lake in the desert. Private lakes such as this one in Las Vegas have been created by depleting groundwater

Figure 1.5 The malling of the countryside. Regional malls such as this English example are appearing in Europe as well as in North America, undercutting local shops and economies

Transportation and motor vehicle dependence

Growing motor vehicle use is one of the foremost crises of urban development in many countries. "Vehicle miles traveled" per capita in the United States was increasing at around 3.8 percent a year at the beginning of the millennium, meaning that the average person drove twice as much as he or she did 25 years before.[12] The 2000s saw some slowdown in vehicular travel, in part due to the 2008 recession and changing behaviors among young adults, but the 2008–2012 decline was only 3 percent from the previous high level.[13] In the United Kingdom, passenger car travel rose ten times between 1952 and the turn of the century.[14] There has been some indication that motor vehicle travel and car ownership rates are stabilizing and even declining slightly in some industrialized countries,[15] but this trend is offset by the rapid spread of motor vehicles in developing countries.

The share of travel undertaken by motor vehicles in preference to other modes is growing steadily in most countries. Some 91 percent of Americans are driving to work rather than walking, biking, or taking public transportation, according to the 2009 census data. The percentage of US residents who travel anywhere by foot is negligible, in part because our communities are not designed for pedestrians. These same trends are present elsewhere; for example in Shanghai the combined biking and walking share of travel fell from 72 percent in 1986 to 54 percent in 2004.[16] The negative effects of growing motor vehicle use include declining quality of life in neighborhoods impacted by traffic, excessive use of nonrenewable resources, air pollution, global warming, public health problems ranging from obesity to childhood asthma, and tens of thousands of deaths each year due to traffic accidents and pollution (see Figures 1.6 and 1.7).

Figure 1.6 Growing motor vehicle use. Growth in miles traveled annually per capita (US)
Source: US Department of Energy

Energy and resource use

Due in large part to the nature of recent development and lifestyles, our resource use remains at unsustainable levels. Depletion of petroleum reserves is one area of concern—global oil production is expected to peak and then start declining early in this century. But other resources such as water are also being depleted worldwide, not just in desert locales such as

Figure 1.7 The vehicle-dominated world. Many people's main experience of urban areas is now the landscape seen through car windshields, as here in Beijing

Las Vegas and Southern California, but in metropolitan areas such as Atlanta and São Paolo as well. The concept of recycling water remains little explored even in arid regions. Permeable paving and natural drainage swales—techniques which allow rainfall to recharge the local water table rather than running off rapidly, often causing erosion or floods—are still rare.

Most resource use in our society still takes place in linear fashion, with raw resources being acquired, used once, and then discarded as waste. British households still let 59 percent of their waste be trucked to landfills and incinerators in 2011. Although this was a marked improvement over a figure of 85 percent in 2000, it is still short of the European Union (EU) goal of 50 percent by 2020.[17] The overall US recycling rate was only 34 percent in 2012, although particular cities achieved much higher rates, such as San Francisco at 78 percent.[18] Recycling rates for materials such as plastics are extremely low, ranging from 13 percent in the US to 31 percent in the EU and 35 percent in Australia.[19] We are still a long way from achieving "closed loop" resource cycles, in other words where materials recirculate extensively, waste is eliminated, and new inputs come from renewable sources.

Pollution and environment

One consequence of our current resource use patterns is the generation of a great deal of pollution and waste, much of it toxic. Between 50,000 and 100,000 synthetic chemicals are in commercial production,[20] and "persistent organic pollutants" such as dioxins and PCBs are now found in most parts of the world. Some $31 billion dollars worth of pesticides are sold worldwide every year, many for use in urban areas by homeowners or building maintenance workers. One study of trees at 90 sites in both tropical and temperate countries found no locations that were free from contamination by DDT, chlordane, and dieldrin.[21]

Urban areas are now littered with "brownfield" sites—contaminated land formerly used by industrial, commercial, or military-related enterprises. The firms that created such problems have often moved on or gone out of business, and in many cases new owners or public agencies are left to coordinate and pay for cleanup. A lot of these sites remain for years as vacant land, dissuading many people from locating near them.

Every few years new forms of pollution emerge into the public spotlight. Rachel Carson's (1962) book Silent Spring awoke millions to the dangers of pesticides and other toxic chemicals in the environment. Damage to the Earth's ozone layer caused by chlorofluorocarbons became a cause for worldwide concern in the 1980s. Anthropogenic global warming had been recognized to be a possibility in the late nineteenth century, but was not widely covered by the media until the 1990s. In the early 2000s much attention was focused on the damage pollution and harmful fishing techniques have caused in the Earth's oceans. New crises continue to emerge. The basic problem is that our industrial economy does not take the full effects of its resource use and waste production into account, a prerequisite for a more sustainable society. Many of these wastes and pollutants also have serious implications for public health, and have been found to increase rates of illnesses ranging from childhood asthma to cancer.

Economic development, inequality, and poverty

Sustainability problems worldwide have been driven in large part by forms of economic development that promote overconsumption within affluent societies, exploit workers and natural resources, concentrate wealth and power in a few hands, and leave millions of other people in impoverished circumstances. Critiques of prevailing forms of economic development are not new, but take on new urgency in an age of economic globalization, growing inequality, and severe ecological challenges. A sizable political movement world-wide is calling attention to the need to rethink capitalism in particular. Though not often covered in the mainstream media, this movement has resulted in a resurgence of locally oriented businesses, protests such as the 2011 Occupy events, and "people's summits" coordinated by nongovernmental organizations (NGOs).

Large and growing inequities both within and between societies have driven much of the recent criticism. The top 1 percent of Americans owned about 36 percent of the nation's wealth in 2009, while the bottom 80 percent owned about 13 percent, a level of inequal-ity almost as great as before the Great Depression (see Figure 1.8).[22] Globally, even worse inequalities exist: the world's richest countries, including oil kingdoms on the Persian Gulf, the US, Norway, Switzerland, and the Netherlands, have per capita gross domestic products (GDP) 40–80 times as large as the world's poorest countries, even after the dif-ferent buying power of currencies is taken into account.[23] Such disparities mean that many societies have little money with which to improve human welfare or protect their environ-ments, and that billions of the world's people are concerned simply with daily survival.

Within many countries increasingly affluent and often gated enclaves exist just a short distance from impoverished central cities or squatter settlements. In developing countries 50 percent or more of the population may be living illegally in informal housing con-structed of cardboard, plywood, or other scavenged materials. Many also eke out livings in the informal economy, while globalization produces enormous wealth and an upper class working in modern skyscrapers. Even in affluent countries a large fraction of the population feels constant economic insecurity, which frequently then leads to racial and ethnic ten-sions as residents worry that other groups within society are poised to take their livelihood.

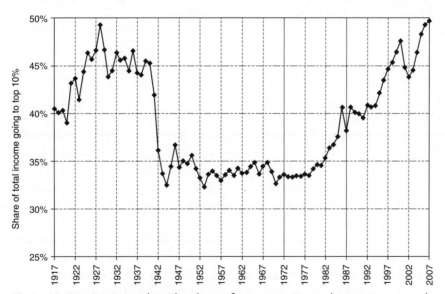

Figure 1.8 Growing inequality. The share of income going to the top 10 percent has climbed greatly in the US since the 1970s

Source: Atkinson et al., 2011

Sense of community

Declining sense of community is another concern that is far less tangible but no less serious. Most urban and suburban development during the past 50 years has been relatively generic, with little sense of place, history, or cultural distinctiveness. Every place looks like everywhere else, a "geography of nowhere," as author James Howard Kunstler has put it.[24] Many of us now live in suburbs without any real downtown or walkable public spaces other than the mall. Parks, greenways, public squares, and sidewalks are often missing. The same generic chain stores and restaurants appear everywhere. Branches of multinational corporations displace locally owned businesses. Driving everywhere, suburban residents spend much of their time alone in their cars. Women, children, teenagers, the disabled, and the elderly are often isolated in their homes, especially if they are unable to drive. Partly as a result of all these factors, political scientists such as Robert Putnam have documented a long-term decline in the extent to which citizens participate in community groups and social institutions.[25] This trend has troubling implications for the health of democratic government. Putnam believes that this decline of "social capital" is at least partly related to the physical nature of our cities and towns.

One could go on at length about the problems of current development patterns. But most of us already know that something is seriously wrong. A long line of literary works has also covered many of these issues, including the writings of Lewis Mumford in the early and mid-twentieth century, Jane Jacobs' *Death and Life of Great American Cities* (1961), Rachel Carson's *Silent Spring* (1961), Ian McHarg's *Design with Nature* (1969), Dolores Hayden's *Redesigning the American Dream* (1984). James Howard Kunstler's *The Geography of Nowhere* and *Home from Nowhere* (1993, 1996), Robert Putnam's *Bowling Alone* (2000), Timothy Beatley's *Native to Nowhere* (2004), Julian Agyeman's *Sustainable Communities and the Challenge of Environmental Justice* (2005), and recent works by New Urbanist designers such as Peter Calthorpe, Andres Duany, and Elizabeth Plater-Zyberk.

In many ways, what is being enacted is a "tragedy of the commons," to use the term that biologist Garrett Hardin made famous in the 1960s. Developers, corporations, and ordinary citizens have maximized private benefit at the expense of shared public goods—including the quality of our cities and towns, not to mention the environment. Individualism has triumphed over collective responsibility. In a conservative era dominated by free market philosophies, there has been little political interest in recognizing that someone, usually the public sector, must stand up for the common good. Developers build on every available piece of land at the urban fringe, landowners feel it is their right to seek maximum economic return from "their" property, owners of factories and motor vehicles each contribute their pollution to the metropolitan airshed, and developers often produce buildings with little thought to the streetscape and public environment that is being created.

Government, which exists in large part to protect and advocate for public interests, is routinely attacked by those who would like to privatize everything. Individualistic attitudes have been institutionalized through law, government policies, corporate practice, advertising, the media, and many other structural elements of our society. These structural forces also reinforce individualistic, consumption-oriented values, types of behavior, and modes of thought within individuals.

Put together, these forces create societies that have great difficulty in planning for sustainability. They also lead to a world that is dangerously unstable. At any time it may produce social, political, financial, or environmental crises. Drama aside, on a daily basis such forces cause an enormous amount of needless suffering and environmental devastation. Developing more functional social ecologies is one of the central tasks of the twenty-first century. The field of planning will be crucial to this challenge, and politicians, professional planners, architects, environmentalists, developers, and citizens of all sorts will have roles to play in meeting it.

A NEW ERA OF SUSTAINABILITY PLANNING

Moving towards a better, more sustainable future is above all a planning challenge, in the broadest, most creative possible meaning of the term. It is a challenge of envisioning future directions, carefully evaluating and studying them, and then working in a variety of creative ways to get from here to there.

In common usage "planning" refers to a wide range of systematic activities designed to ensure that desired goals are achieved in the future. These goals could include environmental protection, urban revitalization, economic development, social justice, and many other ideals. As a formal profession, planning is most often concerned with managing land development at the urban and regional scales, but the field has broadened enormously since its origins, and now can be said to encompass the act of planning for desired future conditions at all scales of endeavor, within the three main spheres of social life: the public sector (i.e. government), the private sector (i.e. business), and civil society (i.e. the myriad nonprofit organizations and networks that increasingly serve to counterbalance the first two sectors). We are entering an era in which increasingly the goals of planning relate to sustainability in one way or another—specifically, to the long-term improvement of human and ecological welfare.

The roots of planning as an organized profession go back to the late nineteenth century, when the extremely rapid growth of cities following the Industrial Revolution led to a profusion of urban problems such as inadequate sanitation, water supply, transportation, and housing. In cities throughout Europe and North America, hundreds of thousands of industrial workers and their families were crammed into tenement housing without adequate light, water, or sewer facilities. Planning historian Sir Peter Hall has labeled this late-nineteenth-century industrial city the "city of dreadful night."[26] Many professions responded to the urban crises of this period. Engineers devised large-scale sewer and water systems. Architects and public health workers proposed housing regulation to ensure access to light and air. Landscape architects formed the backbone of the mid-nineteenth-century Parks Movement, and joined with architects to promote turn-of-the-century City Beautiful ideas. Meanwhile, early planning visionaries such as Ebenezer Howard, Patrick Geddes, and Catalan engineer Ildefons Cerda applied broad and holistic styles of thought to urban problems, coming up with concepts such as Howard's "garden city."

After early development of zoning in German cities such as Frankfurt in the 1890s, city planning as a profession became formally established in Britain and North America during the 1910s and 1920s when municipalities first created planning commissions, planning staffs, and zoning laws. Development projects began to come under greater public scrutiny.

Rapid construction of roads and bridges took place to accommodate urban growth and a new transportation technology: the automobile. Harvard established the first US academic city-planning course in 1909, and many other universities followed suit. Anxious to establish the legitimacy of their field, the new planning professionals soon left behind the idealistic visions of Howard and Geddes in favor of more pragmatic approaches. Quantitative measurement of urban data became preferred to designerly visions or normative advocacy. Unfortunately, city planners also often found themselves with little power within municipal political systems and grew used to a disempowered role as statisticians, bureaucrats, and overseers of the zoning code.

During the twentieth century, new planning specialties arose in response to particular problems of the growing metropolis, such as transportation planning, community development planning, and environmental planning (see Figure 1.9). A quantitative, scientifically oriented form of regional economic analysis also became widespread in the 1950s and 1960s. Seduced by positivistic science and the worldview known as modernism, planners assumed the stance of objective experts using scientific method, modern technology, and economic analysis to determine paths of urban development. This approach is typified by Walter Isard's 1975 statement that

> A regional scientist is not an *activist* planner. ... The typical regional scientist wants to surround himself with research assistants and a computer for a long time in order to collect all the relevant information about the problem, analyze it carefully, try out some hypotheses, and finally reach some conclusions and perhaps recommendations. His findings are then passed on to key decisionmakers.[27]

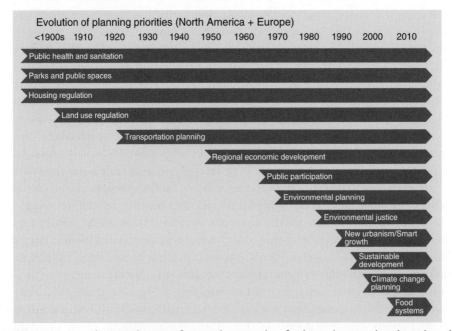

Figure 1.9 Broadening planning focus. The agenda of urban planning has broadened greatly over the past century

Much of this approach still lingers. One cause of today's urban problems is that planners in the past often abandoned holistic understanding of urban environments and active involvement in problem-solving in favor of detached quantitative analysis. Marxist critics of the field beginning in the 1960s and 1970s also argued that planners had ignored the realities of political and economic power within society, and were simply facilitating capitalist development rather than dealing with pressing needs for equity and improvement in social conditions. Environmentalists, historic preservationists, and neighborhood activists fought against some of the worst excesses of modernist planning, such as urban renewal in which entire neighborhoods of older housing were bulldozed to create sterile new apartment blocks. As modernist planning became increasingly discredited in the late twentieth century, the stage was set for something different.

During the second half of the twentieth century planning was a rapidly growing profession, both in its traditional urban context and in many non-traditional contexts as well. Virtually every city or town in industrialized countries created a planning department, often composed in the US of a "current planning" division that oversees development permitting and the zoning code, and an "advanced planning" division responsible for developing longer-range plans. But planners also came to work within many other types of government agencies—regional agencies, state and national governments, international development agencies, transportation authorities, park districts, and public utilities among them—as well as for private corporations, development companies, consulting firms, community development corporations, and various types of nonprofit nongovernmental organizations (NGOs). Planning often overlapped with policy analysis, though it tended to be more action-oriented, and with other specialties such as urban design, engineering, law, and political organizing. People in many different professional roles did what we might call "planning." But it is the activity that matters, more than the label, and how forward-thinking, strategic approaches can help promote sustainable development in whatever the context at hand.

Overall, twentieth-century urban planners were successful at meeting some goals but not others. Cities in the industrialized world now have decent sanitation, water, roads, and zoning and building regulation designed to ensure public safety and protect property values. Planning was also fairly successful in organizing society to accommodate new technologies such as the motor vehicle—although without sufficient reflection on the long-term impacts. Health care, housing, social welfare functions, and environmental protection have usually improved as well. Those gains should not be forgotten. But progress hasn't been enough in many other areas. And we now face a different set of problems than existed a century or more ago, including climate change, resource depletion, excessive reliance on motor vehicles, and persistent, often worsening inequality.

To address these new challenges, we are now entering the era of sustainability planning. For decades many visionary writers, planners, architects, landscape architects, developers, politicians, and local activists have been exploring steps to create more sustainable, livable, and equitable communities. Governments at many levels have also been creating new planning documents that go by names such as sustainability plans, climate change plans, bicycle and pedestrian plans, or green plans. Sustainability goals have been woven into traditional planning documents, policy reports, and institutional names. The profession's central organization in the US, the American Planning Association, has sought ways to support sustainability planning,[28] as have organizations in other countries such as the Canadian Institute of Planners and the Royal Town Planning Institute and Town and Country Planning Association in Britain.

However, the process of redefinition is just beginning, and faces huge hurdles because of politics, economics, and public attitudes. Much work remains to be done to figure out how to best nurture and speed sustainability planning.

In redefining the field we need to recognize that planning has limits—not every aspect of urban, regional, or global development can or should be planned out. History has supplied many examples of grand visions by architects or social theorists that proved too rigid or inappropriate in practice. The misplaced faith of twentieth-century planners in scientific knowledge and research is also a cautionary tale. Some theories of urban development such as the organic theory of urban form developed by Christopher Alexander in books such as *A Pattern Language* (1977) hold that cities are best developed incrementally, learning from experience in the process, rather than according to some overarching master plan. Towns and villages often developed in this slow, incremental way historically, before the advent of large-scale city-building in industrialized societies, and many of these places are among the most delightful urban communities currently. However, even if more incremental and community-based development is desired it helps to have a framework into which such development can fit, ensuring that it meets certain basic criteria and perhaps prohibiting inappropriate actions. The trick will be to develop sustainability planning visions with a large degree of humility regarding planners' ability to predict the future, and with a great openness to learning from experience and collaborating with the wide range of constituencies affected by any initiative.

LEVELS OF GOVERNMENT AND SCALES OF PLANNING

Insofar as the public thinks about it at all, planning in North America is often associated with the process by which local governments approve land development and build roads or other infrastructure. But an enormous variety of planning activities occur at many levels of government and within the private sector as well. Understanding how these different scales of planning interrelate and of how action at some levels can reinforce efforts at other scales is essential. Each level of planning will be discussed in greater detail in subsequent chapters, but a brief introduction is appropriate here.

At the very broadest level, a growing number of global agreements and programs constitute an international scale of planning that is relatively recent, depending in part on the existence of international institutions that are themselves recent. Chief among these institutional actors are the United Nations agencies, the World Bank, the International Monetary Fund, many bilateral development agencies, and nonprofit organizations such as the ICLEI: Local Governments for Sustainability, which helps local governments around the world develop and implement sustainability-related programs. Most large international development agencies were established in the period after the Second World War, when the industrialized countries sought to assist in global reconstruction, ensure the success of their own model of development, or promote their own political hegemony, depending on one's point of view. The nonprofit sector activities are even more recent, with many groups

established after the 1960s. ICLEI, for example, was established in 1990. With consciousness of common global problems such as climate change spreading, broad frameworks of
goals and policies at the international level are likely to be expanded in the years ahead. For
those issues that are less than global, a related area of opportunity lies in regional agreements between several countries with common borders or issues. An example is the series
of agreements between Canada and the United States to protect environmental quality in
the Great Lakes.

Nation states represent the next lower scale at which planning typically takes place.
Although in countries such as the United States and Canada federal governments have
relatively little power or interest in directly regulating the spatial development of cities and
urban regions, they do plan in many other ways for goals such as social welfare, environmental protection, and energy provision. The British government has prepared a very
extensive set of planning policy guidance documents for local and regional agencies, and
at times has taken a strong role in structuring and overseeing local government institutions. National governments in Sweden, France, the Netherlands, Japan, and other countries have also been heavily involved in urban and regional development policy. Moreover,
in countries such as the US that have little explicit planning policy, federal laws and programs related to transportation, taxation, housing, civil rights, and environmental protection profoundly affect the sustainability of local and regional communities without
explicitly acknowledging that they are doing so.

Planning by states or provinces in those countries large enough to contain them is a
further level of large-scale planning. In Canada, the provinces have great power to review
land use decisions, allocate infrastructure, and revise local institutions. The province of
British Columbia has played a strong role in park and land use planning, for example,
while the province of Ontario completely restructured local governments in the Toronto
area in 1998. In Australia, the six state governments also have a high level of control over
urban planning issues with the power to override or restructure local governments. In the
United States, a number of states have adopted statewide growth management planning
frameworks, as well as housing programs, environmental review requirements, pollution
regulation, and waste reduction policies. The planning and construction of large-scale
transportation projects is also usually handled at the state level.

Regional planning has often been sought by those wanting to address problems such as
air and water pollution, watershed planning, transportation, affordable housing development, and equity. These problems typically cross the jurisdictional boundaries of local city
and county governments. Such planning might occur at the level of a metropolitan region
(consisting of one or more central cities along with dozens of outlying suburbs and counties), or on the scale of watersheds, airsheds, commute sheds, bioregions, economic
regions, or cultural regions. Unfortunately, regional planning is usually the scale least represented by government institutions. There are few strong regional agencies in the US, and
local, state, and national governments typically feel threatened by the possibility of regional
agencies usurping some of their power.

Local government—including cities, towns, counties, and a wide variety of special agencies set up for specific purposes such as managing school districts or park systems—is the
level at which the most detailed urban planning is usually done. Most municipal governments

have a full-time planning staff whose responsibility is to process development applications and to prepare longer-range plans and policies. Many consultants, nonprofit organizations, neighborhood associations, and business groups are active participants in planning at the local level as well. In addition to citywide general or comprehensive plans, local governments in turn prepare plans for smaller scales of development, including large areas within cities, smaller neighborhoods, and, occasionally, specific sites of special importance. Economic and community development topics can be very much a part of such plans in addition to physical or spatial planning questions. Municipalities may also prepare "functional" plans to deal with specific topics not covered in their comprehensive plans.

Neighborhood and site planning is typically done by developers, urban designers, architects, and landscape architects, operating within the context of municipal zoning codes, plans, and design guidelines (if these exist). The character of particular places, their relation to local ecosystems, economic development initiatives, and the nature of community facilities are typically determined at this level. Neighbors and community groups frequently play a significant role in affecting plans at this scale.

In theory goals and policies set at each of the higher levels establish the context within which planning is carried out at smaller scales. The reality, of course, is not so clear-cut. In the absence of specific incentives or penalties, broad global, national, state, or regional goals may be overlooked at lower levels. Decision-makers at any given level are also often not used to keeping broader perspectives in mind, and may have little personal stake in doing so, especially if they stand to benefit politically or financially by particular courses of action that consider only the local interest. Some US states require different scales of plans adopted by local and regional governments to be consistent with one another, but many do not. Integrating different scales of planning, then, is a task of raising awareness at every level of the overall context into which specific actions fit, and structuring incentives, mandates, and educational processes that help encourage such thinking.

RECENT TRENDS IN PLANNING AND DESIGN

Although political situations often seem stacked against sustainability initiatives, recent developments in the planning field offer hope that constructive approaches to current problems can be found. In the past couple of decades a number of new planning and urban design movements have sprung up to challenge twentieth-century development directions, incorporating a new awareness of social and ecological issues. Taken together, these efforts help lay the foundation for sustainability planning.

Within architecture and urban design the movement known as the "New Urbanism" appeared in the early 1990s and has become a strong force for re-evaluating the physical layout of communities. Originating with a number of architects promoting neotraditional community planning—based in large part on the traditional American small town—the movement held its first annual congress in 1993 and published a charter in 1996. Founders include Peter Calthorpe, the husband-and-wife team of Andres Duany and Elizabeth Plater-Zyberk, Daniel Solomon, Stefanos Polyzoides, Elizabeth Moule, and Douglas Kelbaugh. In Britain, similar design principles have been championed by the Prince of Wales's Foundation for Building Community. On the continent, architects Leon and Rob Krier have designed similar new communities in a denser, more traditional European form.

In large part a reaction to the placeless, unwalkable landscape of suburban sprawl, the New Urbanism calls for improved design at building, neighborhood, city, and regional levels to create more walkable, livable communities.[29] New Urbanists for example call for narrowing streets, adding sidewalks, placing porches on the front of houses, tucking garages behind them out of the way, creating street grids or other connecting street patterns instead of cul-de-sacs, and organizing neighborhoods around mixed-use centers and attractive public spaces. Most of these design elements are similar to those used in communities before the age of the automobile. The movement is open to criticism on a number of fronts—in particular for being focused on better-designed suburban development, often for relatively affluent residents, rather than the creation of truly "urban" places, and for not incorporating green building design and landscaping. However, it nonetheless represents a much-needed attempt to establish a new model of neighborhood design that has many sustainability advantages.

Just a few years later, in the mid-1990s, a movement for "Smart Growth" gathered steam. With its origin in state efforts to control metropolitan growth, Smart Growth focused especially on mechanisms to promote more compact, economically efficient urban development, in particular through preserving open space, limiting the outward expansion of cities, promoting infill development, and redesigning communities in ways similar to those proposed by the New Urbanism. Many of the same individuals have been active in both movements. One of the main motivations in the case of Smart Growth was to reduce infrastructure costs for local governments. Although states such as Oregon, Vermont, New Jersey, and Florida had been pursuing growth management programs since the 1960s in some cases, Maryland took particular leadership in the new round of Smart Growth policy through its 1997 legislation. This policy framework tied state infrastructure funding to the local establishment of priority growth zones, channeled new state facilities into existing town centers, and made additional funding available for open space acquisition. Other states such as Minnesota and Washington followed suit. Actual implementation of Smart Growth policies has been limited in most places due to opposition from development interests, the unwillingness of local governments to cede authority to higher levels, and conflicted views within the public, many of whom like large-lot suburban or exurban living.[30] However, the concept remains influential.

A third, somewhat more diffuse movement called for "Livable Communities." A series of annual conferences on this theme has been held in Carmel, California, since the 1970s, and organizations such as Livable Oregon have promoted this theme within state and local planning. Walkability, diverse and mixed-use development, and the establishment of a wide range of urban amenities to make cities more livable have been main themes of this movement.[31] Livability advocates build on the previous writings of twentieth-century humanist urban critics such as Mumford, Jacobs, William H. Whyte, and Bernard Rudofsky.

A related, likewise diffuse assortment of professionals, in large part coming from the field of public health, has sought to promote "healthy cities." This term appears to have been coined at a Canadian conference in 1985, and was used in the 1990s to refer to a wide range of issues related to pollution, toxic chemicals, safety, homelessness, education, community, and urban quality of life. But in the 2000s both public health and planning professions focused more specifically on the epidemic of obesity and the need for communities to promote "active living" and healthier food systems.[32] Their efforts have resulted in programs to make cities more bicycle and pedestrian friendly, and in new strategies to develop urban agriculture and limit unhealthy food.

Beginning in the 1980s, advocates for "Environmental Justice" called attention to the ways in which society saddled minority and low-income neighborhoods with the negative side-effects of urban development—especially exposure to toxic chemicals and industrial pollution.[33] Environmental Justice leaders challenged mainstream environmental groups in the US to broaden their agenda and staff diversity, most notably in a series of letters to the "Gang of 10" heading the major national organizations. In the 1990s and 2000s this movement looked in addition at the inequities involved in other areas such as transportation and climate change. Meanwhile, the Occupy Movement of the early 2010s has focused attention on economic inequities, such as the divide between the top 1 percent of the population in terms of wealth and the remaining 99 percent.

Interest in ecological restoration has grown steadily in recent decades. One conceptual foundation of this has been the field of landscape ecology, based on the work of Richard Forman and others, which developed a new way of looking at landscapes in urbanizing or rural areas, emphasizing "patches" of wildlife habitat, "corridors," "edges," and so forth. The result is a set of terms and concepts that makes possible more systematic approaches to preserving and restoring natural landscape elements within cities and towns.[34] Landscape architects and planners such as Michael Hough, Anne Whiston Spirn, John Tillman Lyle, Rutherford H. Platt, Timothy Beatley, and Marina Alberti also explored the relation between cities and natural landscapes during this period, arguing that the two are profoundly interrelated.[35] Meanwhile, more specific environmental planning movements for creek and wetland restoration, use of native species in landscaping, ecological sewage treatment, green streets, and green building have contributed to more environmentally oriented development practices.[36]

Lastly, many efforts to increase public participation in urban development decisions have arisen over the past four decades. "Participatory planning" has become a prime objective in local government, indeed has been required in many cases, leading to an expanded emphasis on public meetings, workshops, design charettes, and consensus-based goal-setting. Realizing that much of what planners do is mediate between various interests, planning theorists have established the perspective known as "communicative planning," which emphasizes processes of communication, participation, and consensus-building.[37] Related movements for "Community Design" and "Community Based Planning" have called for planners and urban designers to collaborate with local residents in addressing neighborhood problems and designing new buildings, public spaces, and community facilities.[38] In conjunction with these efforts for public participation, some observers have called for a more activist role on the part of planners in advocating for underserved constituencies within local government beginning with movements for equity planning and advocacy planning established in the 1960s.[39]

In short, in recent years many related planning movements have been converging toward a similar end: more environmentally, socially, and economically sustainable communities.

Although the concept of sustainability itself was first applied to other fields such as forestry, agriculture, energy use, fisheries management, and international development, by the late 1980s and 1990s the profession began to take it up, and "sustainable city" programs emerged in a few pioneering locales. Some resulted from grassroots activism, some were based on municipal initiative, some benefited from the support of national governments, and some were facilitated by multilateral entities such as the European Community and United Nations agencies. Main areas of activity were the west coast of North America, where Seattle, Portland, Olympia, San Francisco, San Jose, and Santa Monica established programs, Canadian cities such as Toronto and Vancouver, and Britain and continental Europe, where "Local Agenda 21" programs inspired by the Rio Earth Summit proliferated. Most of these municipal initiatives were small and preliminary, often focusing on energy and resource use. But the concept of sustainability took hold and eventually began addressing other planning topics as well, including transportation, urban design, and architecture. As the 1990s progressed urban planning academics started to explore the subject, conferences and books adopted it as a theme, and universities began to offer related courses.

In the 2000s interest in sustainability expanded further, driven in particular by the need to plan to reduce GHG emissions and adapt to climate change. Many jurisdictions developed "climate action plans" of one sort or another, which dovetailed with other sustainability initiatives. New issues such as urban food systems and active living emerged. Clearly interest in sustainability as a goal of planning processes at all levels has grown enormously over several decades, but just as clearly the obstacles to actually bringing about sustainable urban development are huge. Consequently there is a need to better define the subject, figure out strategies for implementation, and develop political and institutional mechanisms to bring it about.

This book aims to address those needs by exploring how planning initiatives at different scales can fit together to produce more sustainable cities and towns. The ideas presented here are not just for professional planners, but for citizens, local activists, students, elected officials, developers, architects, engineers, and others concerned with how cities and towns can become more sustainable, livable places. Whatever our roles, each of us can help bring about a more sustainable future, sharing strategies, working together, and supporting one another in this endeavor.

Part One

The nature of sustainability planning

2

SUSTAINABLE DEVELOPMENT

The concept of "sustainability" in its modern sense emerged in the early 1970s in response to growing understanding that modern development practices were leading to worldwide environmental and social crises. The term "sustainable development" quickly became a catchword for alternative development approaches that could be envisioned as continuing far into the future.

The verb "sustain" has been used in the English language since 1290 or before and comes from the Latin roots "sub" + "tenere," meaning "to uphold" or "to keep." The Oxford English Dictionary traces the adjective "sustenable" to around 1400 and the modern form "sustainable" to 1611. However, the word appears to have been used mainly in legal contexts until recently, as in "the Defendant has taken several technical objections to the order, none of which ... are sustainable" (1884). The phrase "sustainable development" appears to have been first used in 1972 by Donella Meadows and other authors of *The Limits to Growth*, and by Edward Goldsmith and the other British authors of *Blueprint for Survival* in that same year. Once in use, the term became one of those inevitable expressions that so neatly encapsulates what many people are thinking that it quickly becomes ubiquitous.[1] Yet the conceptual roots of the term "sustainability" go far deeper, and have to do with the evolution of human attitudes toward the environment within Western culture.

ROOTS OF THE CONCEPT

Environmental issues have almost always been concerns of human societies. For millennia people have had to develop their communities and livelihoods within the context of pre-existing ecosystems and geographies, and there has been frequently an uneasy balance between human and non-human worlds. Parts of the Mediterranean were extensively deforested during antiquity, and environmental collapse may have contributed to the decline of the Mayan cultures. Many societies have had elaborate rituals and institutions devoted to maintaining what we might call "sustainable" resource use, in particular norms and taboos around the use of "common pool resources" such as fisheries, forests, water sources, and grazing land.[2] Ecosystems or individual species have often been prominently represented within many spiritual traditions, especially those of indigenous peoples living close to the land.

As early civilizations developed they profoundly changed pre-existing natural environments. Indigenous cultures cleared forests, managed landscapes by setting fires, domesticated or transplanted species, and hunted or drove into extinction many non-human forms of life. Few landscapes anywhere were undisturbed by human contact. Writers beginning with Aristotle put forth terms such as "second nature" to refer to elements of the natural world that had been influenced by interaction with humans. So in some ways current sustainability debates are the modern version of age-old concerns about how to maintain human societies within the context of natural ecosystems.

However, although many early civilizations bumped up against ecological limits, the coming of the Industrial Revolution in the late eighteenth and nineteenth centuries made the impacts of human actions far more dramatic. Skies in parts of Britain, continental Europe, and North America were blackened with coal smoke. Forests were cut down to produce lumber or charcoal for iron smelting, and rivers and streams were fouled with sewage and industrial wastes. Clearcutting and inappropriate agricultural practices such as plowing across contours often led to erosion and flooding.

One response to the negative effects of industrialization was a strong strain of romantic or transcendentalist philosophy in which nature was asserted as an antidote to industrial civilization. Writers such as Henry David Thoreau and John Muir, and Romantic poets such as William Wordsworth, Percy Shelley, John Keats, and Alfred Lord Tennyson extolled nature as a spiritually rejuvenating alternative to industrial society. As Muir put it, natural things were "the terrestrial manifestations of God." This spiritual perspective underlaid the "preservationist" strand of the early environmental movement.

Another response to the pressures of industrialization was more pragmatic: to study environmental impacts and find alternative strategies that avoided them. In 1864 George Perkins Marsh published the first systematic consideration of how humankind was altering the natural landscape in his book *Man and Nature*, based on detailed observation of environmental changes in southeastern France and New England. Marsh focused in particular on deforestation, which he saw as leading to increased runoff, floods, landslides, and other ecological disasters. Warning that humans were upsetting the balance of nature, he prophesied that long-term ecological decline would lead to decline in human populations.[3]

A few decades later, German foresters developed "sustained yield" techniques of forest management.[4] Applied particularly to the Black Forest in southwestern Germany, these concepts influenced Americans who trained at continental forestry schools in the late nineteenth century, including Gifford Pinchot, who later became President Theodore Roosevelt's chief forester. Pinchot and others imported European concepts of sustained yield resource management back into the United States, where they influenced the growing conservationist movement. In contrast to the preservationist viewpoint, which ascribed intrinsic value to nature, this sustained yield approach was relatively utilitarian, concerned with preserving natural resources for future human use. Such anthropocentric attitudes, in which ecosystem elements are viewed as valuable mainly in terms of their potential human use, are still prevalent today.

A mixture of preservationist and conservationist sentiments is contained in Aldo Leopold's mid-twentieth-century notion of a "land ethic"—a human responsibility to care for particular lands and ecosystems, discussed most fully in *A Sand County Almanac* (1948). Although trained in utilitarian forest management perspectives, Leopold came to believe that humans had an ethical responsibility to steward and safeguard natural ecosystems, and that these had intrinsic value apart from human use. This more ecocentric perspective helped lay the groundwork for the rise of deep ecology in the 1970s and 1980s, an even more radical philosophy which sought to put the well-being of the global environment first, with human priorities revised to reflect their role as just one small element of the global system. These varying perspectives on the relation between humans and ecological systems helped lay the foundation for late-twentieth-century sustainability debates.

Public concern about the relation between industrial development, urban expansion, and the environment grew steadily after the Second World War. War production had stimulated a huge expansion of petrochemical industries that in the postwar period created many new pollution, toxic materials, and resource use problems. Works such as William Vogt's best-selling *Road to Survival* (1948) and Fairfield Osborn's *Our Plundered Planet* (1948) helped tie the rise of ecological problems to this growth in industrial development. Other writers called attention to the alienation and conformity of 1950s industrial society in bestselling works such as David Riesman's *The Lonely Crowd* (1953) and William H. Whyte's *The Organization Man* (1956). Meanwhile, in many books between the 1920s and the 1970s the great urban planning critic Lewis Mumford linked large-scale urbanization, technology, and warfare, warning of the dangers of the "technopolis," in which anti-humanistic technology was the primary value.[5] In books such as *The Culture of Cities* (1938), *The City in History* (1961), and *The Urban Prospect* (1968), Mumford advanced instead an ideal of the city as an organic community, designed on a human scale, oriented towards human needs, fueled by a life-enhancing economy, surrounded by undeveloped lands, and with streets filled with people instead of automobiles. This vision is remarkably similar to recent sustainable city ideals.

Modern environmentalism—in which advocates became far better organized, adopted an increasingly broad agenda, and brought about a wave of environmental legislation—is generally dated to the late 1960s and early 1970s.[6] During this time social critics, futurists, feminists, peace activists, and environmentalists critiqued existing notions of development and proposed alternative paradigms emphasizing the spiritual, the natural, and the human over values of profit and economic progress. Particularly significant were Rachel Carson's book *Silent Spring* (1962), which first called attention to the dangers of pesticides and other toxic chemicals in the environment, Kenneth Boulding's *The Meaning of the Twentieth Century* (1964), Barry Commoner's *The Closing Circle* (1971), and Theodore Roszak's *Where the Wasteland Ends* (1972).[7] Barbara Ward and Rene Dubos' book *Only One Earth* (1972) was also influential. The report of an unofficial commission set up by Maurice Strong, Secretary-General of the 1972 United Nations Conference on the Human Environment, this widely distributed volume warned about threats to global survival and included an explicit description of the greenhouse effect[8] and warnings about "unsustainable" growth in automobile usage.[9]

Current events also helped change public consciousness. Views of the Earth from space, first taken by astronauts on their way to the moon in the 1960s, helped people conceptualize the planet as a whole for the first time. The debacle of the Vietnam War helped throw into question the prevailing pattern of US economic and political control over developing countries, and exposed the underside of "the military–industrial complex" that Dwight

Eisenhower had warned against in his farewell address. Earth Day splashed environmental problems onto front pages and magazine covers in 1970. The 1972 UN Conference on Environment and Development, held in Stockholm, for the first time brought together public officials and NGOs from around the world and gave them a forum to share ideas and strategies. The 1973 energy crisis hit the pocketbooks of millions of people, many of whom suddenly realized that their fossil fuel use could not continue to expand forever.

At a more philosophical level, in the late 1960s humanistic psychologists such as Abraham Maslow and Carl Rogers pointed out ways in which human potential is shaped by the surrounding social and cultural environment, and ways in which human nature can perhaps be shaped in healthier directions in the future. Their work helped counter pessimistic views of human nature as warlike and competitive, which had been reinforced by violent events earlier in the twentieth century. The implication of this optimistic humanism, endorsed as well by spiritual philosophers such as Pierre Teilhard de Chardin[10] and later by feminist "stage theories" of psychological development,[11] is that people and perhaps entire societies can evolve towards more conscious, compassionate, and sustainable modes of existence, given the right conditions.

By the 1970s the ground had been laid for new perspectives on global development.[12] In Limits to Growth (1972) Meadows and other MIT researchers modeled trends in global population, resource consumption, and pollution, and found that regardless of the range of assumptions they entered the model showed the human system crashing in the mid-twenty-first century. But they argued that "it is possible to alter these growth trends and to establish a condition of ecological and economic stability that is sustainable far into the future."[13] The sooner such efforts began, Meadows and her coauthors believed, the greater the likelihood of their succeeding. This conclusion was opposed by conservative economists such as Julian Simon, who argued that economic mechanisms would naturally take care of resource problems by reducing consumption or substituting other resources for those depleted.[14] In some cases, this market-based process clearly did happen. However, both 20 years later and 30 years later Meadows and her colleagues revisited their model and found its basic predictions still accurate. Indeed, they warned that the human population had reached a situation of "overshoot" in terms of resource limits, and would need to take strong action to correct unsustainable trends.[15]

Meanwhile, in their own 1972 book Blueprint for Survival Goldsmith and other editors of the British journal The Ecologist drew on the work of the Limits to Growth group as well as nineteenth-century British economist John Stuart Mill in calling for the creation of a stable global society.[16] More synthetic and polemical than the Meadows group, Goldsmith et al. began with a sweeping critique of industrial society, stating that

> The principal defect of the industrial way of life with its ethos of expansion is that it is not sustainable. ... Radical change is both necessary and inevitable because the present increases in human numbers and per capita consumption, by disrupting ecosystems and depleting resources, are undermining the very foundations of survival.[17]

In ways that presaged much later sustainability literature, they then systematically reviewed various strategies for resource management, agriculture, and social and political reform. As they quite eloquently put it,

> Our task is to create a society that is sustainable and will give the fullest possible satisfa
> tion to its members. Such a society by definition would depend not on expansion but on
> stability. This does not mean that it would be stagnant; indeed, it could well afford more
> variety than does the state of uniformity that at present is being imposed by the pursuit
> of technological efficiency. We believe that the stable society … is much more likely than
> the present one to bring the peace and fulfillment that hitherto have been regarded, sadly,
> as utopian.[18]

Once introduced, the concept of sustainable development diffused rapidly not just through the networks of environmental activists but also among economists, ethicists, and spiritual leaders concerned about the course of global development. A 1974 conference of the World Council of Churches issued a call for a "sustainable society," and the earliest book with the word "sustainability" in the title appeared in 1976, a volume entitled *The Sustainable Society: Ethics and Economic Growth* by Lutheran theologian Robert L. Stivers. Herman Daly's writings about a "steady state economy," discussed further in the next chapter, were also influential at this time. The sustainability literature got one of its strongest pushes from Lester Brown and others at the Worldwatch Institute, a Washington, DC-based organization which in the late 1970s began publishing an extensive series of papers and books related to global sustainability, including the Worldwatch Papers and annual *State of the World* reports.[19] The tide of literature on sustainability swelled in the 1980s with the International Union for the Conservation of Nature (IUCN)'s influential *World Conservation Strategy* (1980), the President's Council on Environmental Quality's *Global 2000 Report* (1981), and above all the 1987 report of the World Commission on Environment and Development, chaired by Norwegian Prime Minister Gro Harlem Brundtland. These reports documented the growth of global environmental problems and critiqued notions of "development," although generally accepting the desirability of continued economic growth. Even fiction writers contributed to the re-evaluation of development trends. Ursula Le Guin's science fiction novels explored alternative societies wrestling with problems of overconsumption, inequality, and environmental destruction. Ernest Callenbach's *Ecotopia* books laid out a vision of a harmonious and sustainable society created when local activists in Northern California secede from the United States.[20]

With the release of the Brundtland Commission report *Our Common Future* in 1987 and the United Nations Rio de Janeiro "Earth Summit" conference in 1991, calls for sustainable development entered the mainstream internationally. The influence of the IUCN and Brundtland reports in particular flowed from the broad participation of mainstream governmental officials within these bodies, which gave their findings an air of authority going beyond the "alarmist" reports of the *Limits to Growth* researchers, *Global 2000*, or the Worldwatch Institute. The Brundtland Commission in particular received input from literally thousands of individuals and organizations from around the world. Initiated at the request of the United Nations Secretary-General, it followed in the footsteps of two other highly respected UN-sponsored commissions, the Brandt Commission on North-South Issues and the Palme Commission on Security and Disarmament Issues. A more authoritative body to explore the topic would have been hard to find. Following Brundtland and the Rio Earth Summit, national reports such as the *Sustainable America* report of the President's Council on Sustainable Development (PCSD) in 1996 attempted to establish sustainable

ections for particular countries. The 1996 United Nations Habitat II "City
 ıbul also took slow but significant steps towards establishing global con-
 the sustainability agenda can be applied to urban planning. The tide of
 professional literature related to sustainability grew steadily during the
 rly 2000s, and although some initially expected the subject to be a passing
 ın no sign of diminishing in the early decades of the new century.

DEFINITIONS AND PERSPECTIVES

Despite several decades of discussion, no perfect definition of sustainable development has emerged. The most widely used is that of the Brundtland Commission: "development that meets the needs of the present without compromising the ability of future generations to meet their own needs."[21] However, this formulation is open to criticism for being anthropocentric and for raising the difficult-to-define concept of needs. (Does every household really need two cars? A VCR? A 2000-square-foot house on a 5000-square-foot lot? What happens if every household worldwide has these things?) Many groups have also criticized the Brundtland Commission's approach for being too accommodating to the interests of the industrialized countries and for not questioning the desirability of continued economic growth.[22]

Other definitions include that given by the World Conservation Union in 1991: "improving the quality of human life while living within the carrying capacity of supporting ecosystems." This version raises the problematic notion of "carrying capacity," which is useful to think about for educational purposes but extremely hard to pin down in practice. It is one thing to say that the carrying capacity of a given watershed is a certain number of deer; deer populations can be counted and analyzed over time, and are relatively rooted to a particular place. It is far more difficult to say that a given region or the planet as a whole can support a certain number of human beings, when humans readily transport themselves and the resources they use over vast distances, and can substitute some resources for others if these become scarce.

Still other writers prefer to define sustainability in terms of preserving existing stocks of "ecological capital" and "social capital." This approach builds on the economic wisdom of living on the interest of an investment—in this case the Earth's stock of natural resources—rather than the principal. For example, British economist David Pearce argues that sustainable development "is based on the requirement that the natural capital stock should not decrease over time."[23] Although conceptually appealing, this approach likewise has an anthropocentric flavor, and involves difficult questions of measurement and whether resource substitution should be allowed.

Most sustainability advocates throw up their hands when faced with the definitional question and fall back on the Brundtland formulation. My own preference is to use a relatively simple, process-oriented definition emphasizing long-term welfare: "Sustainable development is development that improves the long-term health of human and ecological systems." This definition avoids fruitless debates over "carrying capacity," "needs," or sustainable end states, while emphasizing the process of continually moving towards healthier human and natural communities. In theory at least the directions of this process can be agreed on through participatory processes in which all relevant stakeholders are represented, and progress can be measured by means of various performance indicators (see Box 2.1).

Box 2.1 SOME DEFINITIONS OF SUSTAINABLE DEVELOPMENT

Theme: meeting the needs of future generations
"Sustainable development is development that meets the needs of the present without compromising the ability of future generations to meet their own needs."
—*Brundtland Commission (1987)*

Theme: carrying capacity of ecosystems
Sustainable development means "improving the quality of human life while living within the carrying capacity of supporting ecosystems."
—*World Conservation Union (1991)*

Theme: maintain natural capital
"Sustainability requires at least a constant stock of natural capital, construed as the set of all environmental assets."
—*David Pearce (1988)*

Theme: maintenance and improvement of systems
"Sustainability ... implies that the overall level of diversity and overall productivity of components and relations in systems are maintained or enhanced."
—*Richard Norgaard (1988)*

Theme: not making things worse
Sustainable development is "any form of positive change which does not erode the ecological, social, or political systems upon which society is dependent."
—*William Rees (1988)*

Theme: sustaining human livelihood
Sustainability is "the ability of a system to sustain the livelihood of the people who depend on that system for an indefinite period."
—*Otto Soemarwoto (1991)*

Theme: protecting and restoring the environment
"Sustainability equals conservation plus stewardship plus restoration."
—*Sim Van der Ryn (1994)*

Theme: oppose exponential growth
"Sustainability is the fundamental root metaphor that can oppose the notion of continued exponential material growth."
—*Ernest Callenbach (1992)*

Theme: grabbag approach
"Sustainable development seeks ... to respond to five broad requirements: (1) integration of conservation and development, (2) satisfaction of basic human needs, (3) achievement of equity and social justice, (4) provision of social self-determination and cultural diversity, and (5) maintenance of ecological integrity."
—*International Union for the Conservation of Nature (1986)*

Although writers on sustainability share the same basic concerns about the directions of global development, there are also several recurrent debates between them. One main rift is between those who maintain a faith in technology, scientific rationality, and economic growth and those who don't. The former approach often fits well with the mainstream conservation movement within industrialized countries and with large international development agencies and research institutes that are used to engaging in detailed scientific, economic, and policy analysis. The aim becomes to achieve ecological goals by quantifying environmental impacts, analyzing economic policy options, fine-tuning regulation of private industry, and adjusting market incentives. In contrast, others believe that sustainable development is fundamentally incompatible with current capitalist economic structures, attitudes, and lifestyles. For example, Australian sociologist Ted Trainer argues that "a sustainable society must be based on non-affluent living standards, on highly self-sufficient and small-scale local economics, and on zero economic growth."[24] This camp has definitely been in the minority in official circles, but finds considerable support at the grassroots level.

A second main division is between those who focus on ecological crises and those who emphasize social needs and equity. Deep ecologists and mainstream environmentalists in the industrialized countries tend to fall into the first camp, while social ecologists and grassroots activists in developing countries take the latter perspective. Activists in the so-called "developing countries" often see First World concern about the global environment as a way to deny them the advantages that industrialized countries already enjoy, and criticize sustainability advocates in North America and Europe for not focusing sufficiently on the problem of First World overconsumption. Some also criticize the Brundtland Commission's work for embracing conventional concepts of economic growth without paying attention to overconsumption and exploitation in developing countries. However, many others recognize "the intimate connection between the ecological crisis and the broader issues of social and economic justice," as *Ecologist* co-editor Nicholas Hildyard puts it,[25] and have sought to conceptualize "sustainable development" in a way that takes both environmental and equity needs into account.

A third area of contention concerns the extent to which indigenous peoples should be used as models of sustainability. On the one hand, many deep ecologists and social activists agree with Helena Norberg-Hodge and Peter Goering that "traditional societies are the only tested models of truly sustainable development."[26] Writers such as Jerry Mander point to the wisdom of native cultures that have learned to live relatively harmoniously with the land, and argue that such cultures illustrate a quality of spirit that is a necessary antidote to Western materialism.[27] On the other hand, others dismiss this viewpoint as romanticism, and argue that indigenous peoples frequently behaved in unsustainable ways themselves. The Plains Indians, for example, reportedly stampeded large herds of buffalo off cliffs, and Paleolithic hunters may have caused the mass extinction of many species. There is probably something to be said for both points of view, though on the whole traditional peoples seem to have lived with a reverence for land and nature that industrial society would do well to learn from.

A final area of potential confusion concerns changes within ecological science itself, in particular the move away from the notion that ecosystems naturally reach a point of balance or harmony, towards a more process-oriented view that acknowledges the somewhat chaotic, unpredictable, constantly changing nature of natural systems. The former viewpoint, developed following the traditional ecological theories of Eugene P. Odum and others, might imply a search for steady-state conditions of human development. The latter perspective would allow for more continual change as long as it headed in directions that nurtured human and ecological well-being.

As the preceding history suggests, advocates of sustainable development have brought a number of different perspectives to the table. A good starting place is to look at four main groupings of writers: environmentalists, economists, equity advocates, and spiritually and ethically oriented writers. Environmentalists tend to be motivated by the threat of ecological crises; they range from environmental managers working within large corporations (adopting a more-or-less utilitarian attitude toward the environment) to deep ecologists and Earth First! sympathizers (adopting more ecocentric attitudes). Economists use the language and tools of economics, a quasi-science that emphasizes monetary valuation of things and the goal of efficiency. The tendency of economic writers is to bring environmental and social issues into an economic framework of analysis, for example by viewing sustainable development as a process of maintaining natural capital, or by seeking market-based mechanisms for cleaning up environmental pollution. Equity advocates often focus on inequality, exploitation, and First World overconsumption, and develop detailed analyses of how concentrations of political and economic power lead to exploitation. Such individuals and groups often mobilize politically against economic globalization and to regain local control over economic activity. Spiritual writers and ethicists dwell on the need for a transformation of values and mind-sets as a precondition to sustainable development. By reconnecting with the Earth, each other, and our own relation to the universe, this viewpoint suggests, humans will become better able to coexist with one another and the planet. Ecofeminist critiques of development follow a similar path, arguing that specifically male values, mind-sets and institutions are much of the problem.

Such a categorization is simply a useful way of organizing the sustainability literature, and parallels the "Three Es"—environment, economy, and equity—that are often seen as the goals of sustainable development. It should be stressed that many writers combine more than one approach. Box 2.2 provides a general overview of how some authors may be viewed in relation to these groupings, with lists of names arranged in rough order of chronology.

MODERNIST, POSTMODERNIST, AND ECOLOGICAL WORLDVIEWS

Any pivotal concept like "sustainable development" must be seen against a backdrop of the slow, massive shifts in outlook that shape history at particular times. In this case, the sustainable development movement can be seen as part of a larger reaction against the modernist worldview that dominated global development during the twentieth century and that continues its influence today, although often under a postmodern guise. (Whether postmodernism should be viewed separately from modernism is an ongoing debate, as will be discussed further in a later chapter.) In contrast, sustainability can be seen as a key goal of an ecological worldview that has been slowly gaining adherents for many decades, and that represents a potential alternative to both these others.

Box 2.2 PERSPECTIVES ON SUSTAINABLE DEVELOPMENT

Environmentalists	Economists	Equity advocates	Spiritual writers and ethicists
Environmental concerns paramount; ranges from "environmental management" to "deep ecology"	*Economics as the focus and language of choice; emphasis on incorporating environmental concerns into an economic framework*	*Structural inequality, exploitation, and First World overconsumption as primary concerns; emphasis on resisting economic globalization, reclaiming the commons and local control over development*	*Focus on a transformation of values and mind-sets; reconnection with the Earth and each other; search for an alternate paradigm to 20th-century modernity*
Predecessors Malthus Thoreau 19th-century German forestry	*Predecessors* John Stuart Mill Kenneth Boulding E.F. Schumacher	*Predecessors* Marxist, Socialist, Anarchist critiques of capitalism	*Predecessors* Teilhard de Chardin Gregory Bateson Paul Goodman Ivan Illich
Conservationists Gifford Pinchot	*Steady state economics* Herman Daly	*Social ecologists* Murray Bookchin	*New paradigm writers* Ervin Laszlo Fritjof Capra
Preservationists John Muir	*Environmental economics* David Pearce Michael Redclift	*Dependency theory* Andre Gunder Frank	
20th-century natural resource scientists Aldo Leopold Rachael Carson Barbara Ward Rene Dubos	*Ecological economics* Robert Repetto Robert Costanza Kerry Turner Johan Holmberg Richard Norgaard	*Development critics* Edward Goldsmith Nicholas Hildyard/ *The Ecologist* magazine Frances Moore Lappe Helena Norberg-Hodge Arturo Escobar Anti-WTO Activists	*Environmental ethicists* Baird Callicott Timothy Beatley
Global environmentalism Donella Meadows Lester Brown/WW Institute World Resources Institute Brundtland Report Earth Summit/Agenda 21 President's Council on Sustainable Development	*Restorative economics* Paul Hawken *Local self-reliance* David Morris *Ecological footprint analysis* William Rees		*Ecopsychology* Theodore Roszak *Green politics/ ecofeminists* Charlene Spretnak Petra Kelly Carolyn Merchant
Deep ecologists Arne Naess Bill Devall/George Sessions	*Economic democracy* Martin Carnoy Derek Shearer	*Third World activists* Vandana Shiva Martin Khor	*Spiritual writers* Gary Snyder Thomas Berry Matthew Fox Thich Nat Hahn Dalai Lama
Bioregionalism Kirkpatrick Sale	*Socially responsible investment* The CERES principles	*Environmental justice* Robert Bullard Carl Anthony	
Environmental management ISO 14000			

The modernist worldview has taken on different manifestations at different times in the visual arts, in literature, in architecture, in science, and in philosophy. However, it is based on a number of core elements:

- a desire to leave traditional forms behind and to create a new, "modern" world often oriented around technology;
- a faith in science, rationality, and an objective viewpoint;
- a search for universals often connected with science;
- methodological approaches that break problems down into their constituent parts and that tend to view the world atomistically and mechanically; and
- a frequent discomfort with normative statements and value-based discourse.[28]

Between the 1920s and 1970s modernist architects cast aside traditional or classical forms and experimented with sleek new designs that often used new materials such as glass, steel, and concrete. The modernist movement in architecture was represented by the Congrès International des Arts Modernes (CIAM) and the 1938 Charter of Athens, authored in large part by the most famous modernist architect, Le Corbusier. Although many modernists endorsed a humanistic political philosophy with laudable social goals, their design aesthetic emphasized forms of development—in particular the slab-like "towers in a park" scheme emulated by low-budget US public housing and urban renewal—that were later seen as anti-human. The style and works of many modernists also exhibited an arrogance that led quite understandably to a backlash.[29]

In the urban planning field, modernists moved away from the ecological holism of Geddes and Mumford to embrace the social sciences and highly quantitative forms of analysis. The ideal of the planner as a detached, objective expert took over. At the same time, planners adopted an unquestioning faith in material progress and economic development. Towards the end of the twentieth century these goals came into question owing to the bleakness, ecological degradation, inequities, and questionable livability of the resulting urban environments.

Within international development, modernist attitudes meshed well with the rise of post-Second World War development practices relying on large-scale infrastructure and technology. The "Green Revolution"—through which Western countries convinced developing countries to substitute fertilizers, pesticides, and hybrid seeds for indigenous agricultural practices—is a classic example. Biotechnology may represent a more recent version of this approach, which relies on science, technology, and large inputs of nonrenewable resources and capital to increase agricultural yields. Worldwide, developing countries also rushed to emulate modernist First World urban development by building automobile infrastructure, huge industrial plants, and North American-style suburbs, often with disastrous results.

To some extent the postmodernist viewpoint represents a rethinking of the values and assumptions of modernism. The ideal of universal development principles or design ideas has been shattered—these have been shown to often create sterile and monotonous communities. Instead the postmodern perspective acknowledges the value of many different cultures and viewpoints. "Anything goes" might be the mantra. Within architecture, postmodernism is characterized by a mixing of styles and forms within a single building. Buildings often become playful, borrowing from here and there, as in Philip Johnson's famous AT&T building in New York, which emulates a piece of Chippendale furniture. The results are often a welcome relief to bland, faceless modernist design. But the motivations underlying postmodern design are far more than just playfulness. Whereas modernism followed Le Corbusier's dictum "Form follows function," Nan Ellin points out that postmodernism might be said to follow a number of new principles with less commendable motives: "form follows fiction" (Disney World, Las Vegas), "form follows fear" (gated communities, sanitized semi-public spaces such as malls), "form follows finesse" (projects designed by egotistical architects trying to carve out niches for themselves), and "form follows finance" (urban landscapes most fundamentally shaped by flows of capital).[30]

As geographer Michael Dear notes in his study of Southern California, urban regions have been fragmented into a postmodern melange of edge cities, gated communities, and social groups more connected to global electronic networks than to particular places.[31] The relatively simple model of a central city and suburbs that prevailed until recently is fading as a wide variety of different spaces and cultures emerge within the postmodern urban environment. However, urban geographer David Harvey has argued that postmodernism may not be a radically new state—the underlying logic of capitalist production has not changed in his view, simply some of its surface manifestations.[32] Shiny new suburban office towers, regional malls, and gated communities are just new window-dressing for the same dynamics of economic power that fueled modernism.

The main problem with postmodernism as a philosophical framework lies in establishing grounds for ethical and moral judgments—that is, for action of any sort that might seek goals such as a sustainable society. For many, the result of the postmodernist outlook is a nihilistic relativism that denies the existence of any shared values or grounds for social change. If anything goes, is there any point in trying to build cities one way as opposed to another? Are there grounds for adopting certain planning policies, economic development strategies, or design guidelines as opposed to others?

The ecological worldview, in contrast, acknowledges cultural diversity but seeks to ground the development of society in fundamental values that we all share by virtue of being human and sharing a small planet. This perspective emphasizes interdependence, based in part on scientific understandings of the radical interconnectedness of the "web of life." It views the world in terms of overlapping complex systems and organic unity, rather than as an atomistic collection of people and material things, as in positivistic science and neoclassical economics. It emphasizes flexible, evolving systems that can learn and adapt. Unlike postmodernism, the ecological perspective holds the possibility of justifying ethical belief and action, in that these are necessary to sustain social and ecological systems.[33] This ecological worldview—and the challenge of sustainable development in particular—can be seen as a grand narrative replacing the modernist ideals of technological and material progress.[34]

Differences between the modernist, postmodernist, and ecological viewpoints are summarized in Box 2.3.

Modernism advances a very strong value set, one that places priority on scientific and technological tools and methods. Within planning and urban development, modernist

Box 2.3 MODERNIST, POSTMODERNIST, AND ECOLOGICAL WORLDVIEWS

	Modernist worldview	Postmodernist worldview	Ecological worldview
Values	Universal values based on modern science	Pluralistic values based on cultural and cognitive traditions	Acknowledges pluralism but also a shared core value set based on common problems
Cognitive approach	Atomistic (break problems down into constituent parts; view world as collection of individual elements)	Acknowledges pluralistic ways of viewing the world	Emphasizes interrelationships, networks, systems
Core influences	Newtonian physics; neoclassical economics	Twentieth-century physics (relativity, uncertainty principle)	Ecological science; chaos theory; systems theory
Political implications	Reinforces centralized political authority	Undermines centralized political authority	Emphasizes flexible and evolving relationships between different political institutions
Preferred planning modes	Rational, comprehensive planning	Decentralized local planning to meet pluralistic community needs; communication to gain consensus on directions	Emphasizes communication and education to help evolve public understanding; advocacy planning to achieve shared goals; evolving incentives and mandates between different levels of government

outlooks have underlain the expert-driven, technocratic planning common during the mid-twentieth century, within which planners determined urban problems through abstract quantitative analysis and saw themselves as impartial analysts and researchers. Neoclassical economics, with its even stronger value set oriented around economic efficiency, growth, and material progress, went hand-in-hand with this mind-set. The modernist approach forbade planners from acting in any normative fashion, even while it advanced such strong values of its own.

Postmodernism works against value-based planning for a different reason—all viewpoints are seen as equally valid. Since truths are seen to be relative to culture and the existence of any universal beliefs is questioned, no rationale remains for choosing a certain path of development over others. But still, as Harvey points out, the values of capitalist economics underlie the postmodern perspective. Radical pluralism itself can also be seen as a value. These give postmodernism a strong though unacknowledged normative bias.

The ecological viewpoint respects different cultural perspectives, and it values maintaining this diversity. However, it also calls for common values and rules that are fundamental to survival on a small planet. Thus without being backed by modernist science (although supported by more recent scientific findings showing a radically interrelated universe), universals can be reached. Many of these points of global agreement have been expressed since the 1940s in United Nations conventions and declarations, in particular the UN's Universal Declaration of Human Rights in 1948, and have been expressed more recently through the Agenda 21 agreement emerging from the 1992 Earth Summit and the 2000 Earth Charter.

Sustainability, then, can be seen as one of the core values and goals of an emerging ecological worldview that weaves together recent developments in physics, ecology, and psychology along with core elements of many of the world's great spiritual traditions (which support the importance of ethical action within an interdependent world). This cognitive outlook sees the world in terms of interdependence and coevolving complex systems, and supports values, ethics, and actions that likewise emphasize interdependence.

Environmental, economic, equity, and spiritual or ethical perspectives on sustainability can all fit with this worldview. "Sustainability" itself is a code word for other values—principally the sustaining and nurturing of life on the planet—that become a starting point for action in urban planning as in other fields. Acknowledging this normative foundation implies a conscious direction to future action that is very much needed (see Figure 2.1).

THE ROLE OF VALUES AND INSTITUTIONS

Values are priorities that people adopt—consciously or otherwise—based on their worldviews and assumptions about reality. These priorities then motivate behavior that follows from these more general cognitive outlooks. If they were being logically consistent someone subscribing to a worldview based on free-market economics might value competition, entrepreneurship, and individual freedom. Someone subscribing to an ecocentric view of

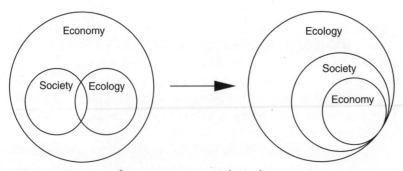

Figure 2.1 Transition from economic to ecological perspective

the world might value the integrity of natural ecosystems, closeness to nature, and a low impact lifestyle. Someone with a strongly feminist perspective might value equality, civil rights, and social welfare policies aimed at caring for women, children, and the elderly.

Values either can be explicitly developed as the basis for action, or they can be adopted unconsciously as a set of de facto guidelines for how an individual or society leads life. Societies have often set out a few basic values as representing their core beliefs. For example "life, liberty, and pursuit of happiness" is perhaps the most basic statement of American values, while the Canadian constitution refers instead to "peace, order, and good government." However, in practice daily social or individual life may be based on a hodgepodge of different and frequently conflicting values.

Talk of values has unfortunately often been co-opted by "family values" conservatives in the United States and by similar reactionary groups elsewhere. These groups often use "values" language in hypocritical ways—valuing "family" may mean advocating for a traditional, patriarchal model of the family with the wife staying at home, rather than endorsing good child care, education, health care, and parental leave policies to support today's working parents; valuing "life" may simply mean opposing abortion, rather than actively nurturing human welfare. (One common joke is that for "pro-life" protestors life begins at conception and ends at birth.)

However, defining sets of progressive values can be extremely useful in bringing about social change—and in planning activities of all sorts—in that stating values helps groups clarify their goals and then move on to look at what politics and programs might achieve them. For example, the Global Green Party, building on the work begun by the German Green Party in the early 1980s, adopted a list of six core political values at a 2001 conference in Canberra that includes ecological wisdom, social justice, participatory democracy, non-violence, sustainability, and respect for diversity. For its part the Green Party USA has adopted a somewhat broader set of ten key values that adds decentralization, community-based economics, feminism, and personal and global responsibility.

The "Three Es" of sustainable development can be considered to represent a rather condensed value set as a basis for change. Some sustainability advocates have sought to expand this list to include concepts such as empowerment, education, and the like. Others might come up with much larger sets of values such as those of the Green Party USA. In the end the exact formulation is not as important as the fact that sustainability-oriented politics is based on *something*—some set of core beliefs and priorities that can then be planned around, and that reflects global needs for healthy societies and ecosystems. Developing such values explicitly helps eliminate the deep gulf that often occurs between stated and de facto values in a society, producing a politics of hypocrisy in which constructive change is difficult. Arguably the United States is in such a situation, professing to value democracy while tolerating a weak and corrupted version, claiming to value peace while waging wars, and touting the virtues of free markets while actually subsidizing large quasi-monopolistic corporations.

Within societies values are in turn propagated and shaped by a wide variety of institutions—social, political, cultural, and economic structures and traditions that to a large extent determine the ways we see the world and live our lives. These institutions include systems of laws, courts, and government; corporations, advertising, and the media; and a large number of informal rules and codes of behavior. A substantial literature, led in part by the "structuration theory" of British sociologist Anthony Giddens and the shared resource theory explorations of Elinor Ostrom,[35] has grown to examine how such institutions structure values and behavior within society.

Changing institutions, then, is a way to change values, and vice versa. Working to reform institutions—for example, election processes, government agencies, planning codes and procedures, and tax structures—can be seen as central to establishing a context in which more sustainable development can come about. The role of institutions will therefore be a recurrent theme in later chapters. Institutions, values, and worldviews all form part of the context in which people develop their individual approaches to sustainable development.

3

THEORY OF SUSTAINABILITY PLANNING

The concept of sustainable development relies on a number of underlying themes that help determine its nature and application. As we have seen these viewpoints arise from the environmental movement and the "small planet" consciousness of the late 1960s and early 1970s, and also from the much longer-term move away from the modernist perspectives of the mid-twentieth century and the Cartesian mind-set that gained sway during the Enlightenment (and which has been reinforced by the ideologies and institutional forces of capitalism). Rooted in the ecological worldview discussed in the previous chapter, interest in sustainability has emerged as individuals and organizations seek to address the shortcomings of the modernist worldview and its models of development.

The more specific characteristics that make sustainability planning different from business-as-usual in the profession include a long-term approach to decision-making, a holistic outlook integrating various disciplines, interests, and analytic approaches, a questioning of traditional models of growth and acceptance that limits to these exist, a new appreciation of the importance of place, and proactive involvement in healing societies and ecosystems. These approaches can help reorient planning debates in constructive ways to address current development challenges. They encourage people to conceptualize problems in different ways than in past mainstream practice, "providing a different kind of dialogic space in which particular conceptions of the good might be fostered," as Susan Owens and Richard Cowell have put it.[1]

ELEMENTS OF THE SUSTAINABILITY PLANNING APPROACH

A long-term perspective

Quite obviously, sustainability planning seeks to bring about a society that will not only exist but thrive far into the future. This expanded time horizon is implicit in the word "sustain." It is also why the 1972 publication of *Limits to Growth* was so earthshaking, in that for the first time researchers used scientific methods to attempt to predict the future of human civilization 100 years or more in advance. Their finding that global society might well crash halfway through the twenty-first century—as well as subsequent information about resource depletion, overpopulation, global warming, and species loss—has caused millions of people worldwide to question prevailing development directions.

Operating mainly within local government, which tends to operate from a relatively short-term and small-scale perspective, the planning profession as traditionally conceived has had difficulty adopting a long-term viewpoint. Typically, planning documents address a 5–20-year time horizon, though occasionally plans covering up to 50 years are appearing these days. Local politicians often consider only a 1–4-year time frame—until the next election. Cost–benefit analyses for large infrastructure projects consider at most a 20-year time frame. Indeed, as will be discussed further in Chapter 4, economic analysis is fundamentally incapable of considering costs and benefits more than about 30 years into the future, owing to the existence of discount rates which assume that the economic consequences of particular actions are worth more in the present than similar effects in the future. All of these factors reinforce a relatively short-term viewpoint within local government planning.

Much longer perspectives are needed that take into account human and ecological well-being 50, 100, or 200 years from now—during future generations, in other words. To do this, planners may need to more specifically assess how near-term actions can lead to long-term goals. They may need to figure out new ways to illustrate for the public how particular buildings, transportation systems, patterns of land use, or economic development programs help or hinder sustainability far into the future. In the area of land use, for example, local officials may need to address questions such as who will clean up industrial land after it is used, how future options for parks, open spaces, or recreational corridors can be preserved within new development, and how mechanisms can be put in place to encourage the eventual recycling of large shopping centers and office parks into more balanced and intensive communities. In each case, near-term modifications to project design or public policy may be able to improve the future flexibility and sustainability of development. Tools such as sustainability indicators and ecological footprints, discussed later in this book, are also ways to encourage attention to the long-term impacts of present development.

In order to understand the history and evolution of urban environments and how current problems arose, time horizons need to be expanded not just into the future, but into the past as well. It is crucial to understand how urban places have coevolved with environmental, economic, social, political, or technological factors, and how specific planning actions have influenced this evolution. For example, land subdivision and development permits lock in patterns of land use that will persist for hundreds of years. Road systems are very difficult to change once established; their long-term implications need to be considered up front. The disconnected fragments of recent suburban sprawl cannot be easily rearranged. Zoning systems and land use designations have frequently frozen in place development concepts that then acquire fierce political constituencies if anyone attempts to change them. Understanding how such mechanisms have worked in the past is the prelude to better planning in the future.

The relatively new discipline of environmental history examines this sort of coevolution between places, cultures, environments, and modes of production, although not necessarily with a focus on planning. In his genre-defining book *Nature's Metropolis* (1991), William Cronon, for example, presented a history of Chicago that shows how the location of the city, its growth, and its current character stem from factors such as the location of natural resources in the upper Midwest, decisions by railroads and other industries to locate or build infrastructure in certain places, and the environmental characteristics of the often swampy prairie south of Lake Michigan.[2] Similar analyses can be developed for any city or town.

A long-term perspective also means being able to look at small, incremental changes in the present and to see how they can interrelate and reinforce one another to build a more sustainable society in the future. Lack of this understanding leads to a fragmented and disjointed situation in which individual actions do not reinforce one another, but cultivating it can help planners organize otherwise isolated projects and initiatives so that they add up to a transformative whole.

To take a physical planning example, if several new buildings are added piecemeal to a moribund downtown district they may not do much to help revitalize the area. Each developer may design an inwardly oriented building that barricades itself from the street and does little to create attractive public spaces. But if developers and planners adopt a strategic vision in which individual building projects work jointly with many other physical planning and economic development initiatives to create a vibrant, pedestrian-oriented new downtown community, then an entirely different picture emerges. Planners and developers may identify other sites for infill housing or cleanup. They may redesign the streetscape to make it more pedestrian friendly, with wider sidewalks, more street trees, cafe seating, and bulb-outs on the street corners to slow traffic and improve pedestrian safety. They may identify a site for a new mini-park or public square, or where a creek can be restored as an urban centerpiece. They may adopt a set of design guidelines so that all new downtown buildings contain attractive retail spaces along the sidewalk, with parking behind the structures rather than in front. They may take steps to preserve and renovate historic buildings. They may adopt an economic development strategy to encourage certain types of locally owned small businesses. Along with these initiatives, planners may take parallel steps to implement an urban growth boundary outside of town and to prohibit big-box retail that would kill smaller downtown businesses.

Each of these steps is small and incremental in itself, and citizens may not initially see them as significant or related. But together they form part of a long-term strategy for something very exciting. In the arena of social planning, similar steps can fit together as well. A society can strengthen its educational system and emphasize critical thinking skills. It can adopt a progressive tax structure and living wage requirements to minimize inequalities of wealth. It can provide and perhaps require national service programs through which young people learn to help communities different than their own. It can ensure decent housing, food, and health care. It can promote small, locally owned businesses rather than give subsidies to multinational corporations. It can provide citizens with meaningful opportunities to be involved in local government decision-making. Many such steps, each significant in its own right, can reinforce one another in reaching toward the goal of a healthier social ecology.

A holistic outlook

As the above examples suggest, a second main characteristic of sustainability planning is a holistic outlook that emphasizes the relationships between different elements of human and natural systems, embodying an ecological understanding of the world. In practice this means two things for planning. First, it means linking the different planning specialties that have historically been compartmentalized, such as planning for housing, transportation, land use, environmental quality, and economic development, as well as integrating more general goals of planning such as the Three Es. Such integration counteracts the separations between disciplines that frequently occurred during the modernist era, in part owing to the emphasis on technical specialization. Second, a holistic perspective means linking different scales of planning, searching for ways that planning efforts at international, national, regional, local, neighborhood, and site scales can reinforce one another. This likewise has often not been considered by planners and governmental officials, who understandably have frequently been preoccupied with their own immediate level of government.

In the late nineteenth and early twentieth centuries many of the first planning theorists and practitioners adopted such a broad and holistic viewpoint. Figures such as Ebenezer Howard, Patrick Geddes, Lewis Mumford, Daniel Patrick Burnham, Charles Eliot, and Frederick Law Olmsted addressed many different aspects of the urban context, including natural resources, parks, transportation systems, building and community design, regional planning, and social dynamics. Howard's classic 1898 *Garden Cities of Tomorrow*, for example, was not just a physical prescription for decentralizing the overcrowded nineteenth-century industrial city into a constellation of new towns, but a social and economic vision featuring collective property ownership and development of industries that could support local residents. Daniel Burnham's *Plan of Chicago* (1909) included proposals for grand avenues and public spaces, a regional park system, transportation corridors, and recreational facilities along the waterfront. For 50 years Mumford was perhaps the ultimate synthesizer of diverse issues, writing passionately about architecture, social issues, urban development, regional planning, economic systems, the natural landscape, and the impact of technology. His two monumental histories of urbanization, *The Culture of Cities* (1938) and *The City in History* (1961), were for years the bibles of urban planning students.

However, as the modernist era gained momentum, planning became increasingly fragmented into technocratic specialties. Transportation planning, housing, land use, urban design, environmental planning, and economic development became separate disciplines within both academia and the professional world. Consulting firms often specialized in only a few of these areas. Separate professional associations were set up for many of these specialties. Urban designers became almost a separate species from policy analysts and economists, each with very different ways of seeing the world. The result was the loss of a perspective on how different aspects of urban environments fit together into a whole.

The task now is to weave different perspectives and specialties back together and to reinforce a sense of how all urban development actions relate to one another. What is needed are individuals who can see the whole and integrate the different components of urban development. Urban design must be integrated with public policy and economics, transportation planning with land use and housing, quantitative methods with qualitative analysis. Other disciplines such as architecture, landscape architecture, sociology, and environmental science must be brought in as well.

Attempts to work across disciplines are often referred to as "interdisciplinary." Such an approach is a good starting place, but often doesn't go deep enough. People tend to stay within their disciplines except for occasional joint meetings or acknowledgments of each other's perspectives. Such efforts may be better referred to as "multidisciplinary" since they simply combine existing, isolated specialties without a strong integrating vision. In contrast, "transdisciplinary" approaches seek to actively link theories and concepts across specialties to address contemporary issues, operating outside the traditional disciplinary silos. And it may be possible to have free-floating "conceptual thinking" that generates ideas not rooted in past disciplines at all (though drawing upon many past traditions and points of view).[3] Much of the challenge of sustainability planning is to operate within multidisciplinary or conceptual thinking approaches while still maintaining the ability to work with others who are more bound to traditional disciplines and ways of viewing the world.

It is particularly essential to weave together planning goals such as "environment, economy, and equity"—the "Three Es" of sustainable development, which will be explored at greater depth in Chapter 4. Remedial focus may be needed on two of the three "Es" that are generally under-represented within our current economic and political decision-making structures: environment and equity. For these to be considered equally with economic objectives, professionals of all sorts will have to play a very active role, calling these issues to the attention of decision-makers, involving advocacy groups who can speak to these issues within public debates, and nurturing urban social movements that can counterbalance entrenched economic interests. A minority of the planning profession has sought to do this historically, for example, under the "equity planning" banner, but much more is needed.

The Three Es can be seen as symbolizing the horizontal integration of goals at a single scale of planning. But also needed is a vertical integration of objectives across different scales of planning—site, neighborhood, local, regional, state/provincial, national, and global scales. Actions at each level must be seen in terms of their impacts on other levels. In practice, much the reverse has usually happened. In the United States, for example, the federal government has maintained the fiction that land use is entirely a local concern, and has been blind to many of the ways that its actions affect local development. (For example, federal policy-makers have refused to consider the role that federal funding of highways and home mortgage loan insurance has played in stimulating suburbanization.) Most states have likewise refused to establish planning or growth management programs, leaving such considerations up to cities and counties. Meanwhile, local governments have vigorously resisted considering the effects of their planning and development policies on regional, state, or global levels, often being preoccupied instead with securing local economic development and tax base.

Understanding the interrelationships between different scales of planning takes practice, training, and the experience of different contexts. Many of our governmental and legal institutions are set up in ways that discourage such coordination of scales. For example, having many taxes collected by local government gives local leaders every incentive to increase the municipal tax base at the expense of broader interests, competing with neighboring municipalities for limited tax revenues by zoning for sales tax-producing regional malls and office parks rather than affordable housing, which typically consumes more in government services than it produces in tax revenues. This phenomenon is known as "fiscalization of land use." Likewise, having so much authority vested in national governments—rather than international treaties and institutions such as the United Nations—discourages global thinking and strong collective action to deal with problems like global warming.

At this point in history institutions are much stronger at some scales of planning than others. International institutions are weak and still relatively recent. It is only since the establishment of the United Nations in 1948, other institutions such as the World Court, the UN Charter on Human Rights and other treaties, and the major UN conferences of the 1970s and 1990s that a global level of democratic coordination has begun to emerge. Regional planning institutions are also weak in many countries. In the United States, for example, planning at a metropolitan level is usually overseen by relatively weak Councils of Governments (COGs)—voluntary associations of local government that have no statutory authority over most aspects of land use and urban development. As a result, most local governments ignore these regional agencies and their guidelines.

On a personal level it is understandable that many people focus on their homes, neighborhoods, and daily lives without a sense of how these affect regional or global sustainability. The impacts of personal actions on the daily environment are immediate; those affecting larger scales are more remote and may occur far in the future. Cultures often do not encourage connecting the global with the local, or other forms of critical thinking. Increasing this holistic understanding at the individual level is perhaps the most fundamental task before us.

A variety of initiatives can help integrate different scales of planning (many of these will be discussed more later, as well). Action by higher levels of government to set general goals for planning—with local and regional governments performing the implementation—is one way to begin to structure a framework for sustainable development. Various United Nations conferences and organizations have begun to establish such principles at a global level. Many national governments have also adopted a variety of land use, environmental, and civil rights policies, while some state governments have adopted detailed land-use planning goals. Higher-level governments can also offer incentives and mandates to lower government to encourage them to implement such goals. Technical support and coordination between different levels of government are extremely useful as well. Educational events and conferences can increase understanding between planners and other activists working at various scales. These and other mechanisms can boost intergovernmental understanding and coordination, better integrating the different scales of planning. This sort of flexible, evolving, mutually supportive framework between different levels of government often goes under the name of "governance," as opposed to more rigid models of government institutions in the past.

Active involvement in problem-solving

A final key element of the sustainability planning perspective concerns the role of professionals, politicians, and ordinary citizens, and the need for all of these groups to participate more actively and constructively in problem-solving. Sustainable development requires enormous changes in the ways that things are currently done, and will only come about through the dedicated work of many people.

Although many of us must fill professional roles as technical experts, mediators, and facilitators who have to be dispassionate and removed from advocacy, sustainable development requires a proactive, normative, and personally involved stance whenever possible. This balance is not easy. Many planners and other professionals are profoundly uncomfortable with the notion that they should be seen as advocates of any particular viewpoint. Yet it is a myth of positivist science that such a thing as a totally detached, "objective" role exists. By adopting a passive, technocratic attitude planners may inadvertently facilitate unsustainable forms of development.

Planners, politicians, and citizen activists may often need to go out of their way to express the needs of under-represented groups, stand up for the environment and future generations, and try to compensate for unequal power dynamics within governmental decision-making. With every given situation, planners can actively look for ways to create a more sustainable situation either immediately or 5, 10, 20, or more years down the line. This is not to say that they should force their beliefs on others. But they do need to actively call the attention of the public, politicians, and community interests to the long-term perspective, to the need to balance economic, environmental, and equity objectives, to the interconnections between issues, to the interests of under-represented groups, to visions of a more sustainable society, and to the implications of decisions at each level of government for long-term social and ecological welfare. And they need to figure out how to actually get things done, in particular how to navigate through often gridlocked institutional bureaucracies in order to bring about positive change. This alone often requires a highly active, entrepreneurial approach.

Particular action is usually needed to give voice to under-represented points of view— such as those of lower-income communities, future generations, people on other parts of the globe, other species, and natural ecosystems. Planners and other government officials may want to invite spokespersons or advocacy groups to meetings, inform them about planning processes, bring under-represented perspectives to the attention of others within planning debates, and frame issues in such a way that all constituencies are heard. Planners and other professionals can also help nurture grassroots movements for equity, environmental protection, or other sustainability-related goals through technical assistance, donations and support grants, strategic advice, and service on organizational boards or staffs.

The Code of Ethics and Professional Conduct adopted by the American Institute of Certified Planners (AICP) and revised in 2009 requires planners to take an active role that is generally consistent with sustainability planning.[4] The Code of Practice of the Canadian Institute of Planners is similarly forward-looking. Unfortunately the Code of Professional Conduct of the British Royal Town Planning Institute is less idealistic. AICP ethical requirements vis-à-vis the public interest are listed in Box 3.1. Under this Code, if planners cannot play an active and constructive role in promoting sustainability within a particular position, then it may be unethical for them to continue. In choosing employment, planners and other professionals should avoid situations in which they are likely to be placed in such ethical conflicts, and if stuck in such a position they must be willing to consider moving to another job. Increasingly, many people with professional degrees work for non-profit housing developers, community development corporations, nonprofit advocacy groups, regional agencies, and private foundations—groups in which they can take on

---

Box 3.1 PLANNERS' RESPONSIBILITY TO THE PUBLIC INTEREST

Our primary obligation is to serve the public interest and we, therefore, owe our allegiance to a consciously attained concept of the public interest that is formulated through continuous and open debate. We shall achieve high standards of professional integrity, proficiency, and knowledge. To comply with our obligation to the public, we aspire to the following principles:

- We shall always be conscious of the rights of others.
- We shall have special concern for the long-range consequences of present actions.
- We shall pay special attention to the interrelatedness of decisions.
- We shall strive to provide timely, adequate, clear, and accurate information on planning issues to all affected persons and governmental decision-makers.
- We shall give people the opportunity to have a meaningful impact on the development of plans and programs that may affect them. Participation should be broad enough to include those who lack formal organization or influence.
- We shall seek social justice by working to expand choice and opportunity for all persons, recognizing a special responsibility to plan for the needs of the disadvantaged and to promote racial and economic integration. We shall urge the alteration of policies, institutions and decisions that oppose such needs.
- We shall promote excellence of design and endeavor to conserve and preserve the integrity and heritage of the natural and built environment.
- We shall deal fairly with all participants in the planning process. Those of us who are public officials or employees shall also deal evenhandedly with all planning process participants.

(From the American Institute of Certified Planners (AICP) Code of Ethics and Professional Conduct, as amended in 2009)

---

advocacy roles that may not be available within local government. These employment sources may prove better opportunities to work actively on behalf of sustainable communities and to reconcile personal ethics with the need to make a living.

An important part of activist planning is to structure processes so that positive long-term change can come about. This requires proactive vision and strategic thinking. Good communications skills, a sense of humor, ability to work with different constituencies, and savvy combined with an entrepreneurial approach are also extremely useful. These skills can be learned and honed to allow sustainability professionals to be effective in working for change in difficult situations. Such efforts often mean cooperation between a wide variety of professionals—planners, architects, landscape architects, engineers, policy analysts, environmental scientists, lawyers, and politicians. They may also require any given individual to play different roles at different times, for example serving as an expert resource, a facilitator, a networker, an organizer, or an advocate depending on the need. Professionals may also find themselves working in different capacities at different times in their careers. But, throughout, sustainability planning depends on the ability to identify and seize opportunities for constructive change.

Acceptance of limits

The notion of "limits to growth" has long been at the core of sustainable development debates, and indeed was the title of the 1972 book that first used the term. In a society based in large part on a belief in "progress" as quantitative growth in production and consumption, this concept is truly revolutionary. The possibilities of applying this theme to planning are numerous. First and perhaps most fundamental is the challenge of planning for economic development that produces qualitative improvement in human and environmental conditions, rather than continual quantitative growth in consumption and material use. This task requires prioritizing forms of economic activity that produce goods and services that people really need, such as affordable housing, education, health care, healthy and nutritious food, cultural and recreational opportunities, and the like. It also means avoiding quick and easy forms of economic development that may produce immediate tax revenues but little long-term sustainability benefit—retail strips, auto malls, petrochemical industries, and large manufacturing complexes aimed at producing vast quantities of low-end consumer products.

Living within limits also has clear implications for land use planning, in that it encourages a compact city form and the prevention of sprawl. Planning tools such as urban growth boundaries (UGBs) may be particularly useful in this regard because they literally set limits to urban expansion that can be seen and understood by the general public. The notion of limits applies as well to water use, especially within arid or semi-arid areas, and may mean planning for extensive water conservation and recycling. In fact, such steps are already bringing about a transformation of landscaping practices in parts of the southwestern US, where concepts such as xeriscaping—landscaping using drought tolerant, often native vegetation—are taking hold. The fields of energy and resource use provide other challenges to live within limits, especially in terms of reducing and eventually eliminating fossil fuel consumption.

Lastly and perhaps most challenging of all will be the task of planning for a stable global population, so that the species can live within planetary limits. Planners have not traditionally addressed population issues, which are often highly controversial. But population growth profoundly affects local growth management efforts and quality of life, as well as broader sustainability topics such as global resource use. Planning for communities with excellent health, education, and family planning services can help reduce the population growth rate, especially if these resources are available to women. At national and international levels, planning for forms of development that improve the condition of the world's least affluent countries is extremely important, both to reduce internal population growth rates and to reduce the immigration pressures that challenge already-developed countries.

A focus on place

Though it may seem less significant at first, a focus on "place" has been a key principle of many sustainable city advocates as well as other recent movements such as the New Urbanism. Such an emphasis means nurturing the health and distinctiveness of specific, geographical locations. Focusing on place is particularly necessary right now because development throughout much of the last century has done exactly the opposite—creating an aspatial, global realm of homogeneous, interchangeable communities with little connection to local landscapes, ecosystems, history, culture, or community. This process is facilitated by economic globalization, which tends to produce standardized products and generic urban environments with little authentic connection to past cultures or ways of life. One result of this non-place-oriented approach has been that it has facilitated the exploitation of natural ecosystems and local human communities—since the exploiters tend to feel little connection with the people and landscapes being exploited—and to undermine the ethic of stewardship called for by Leopold and others.

The anti-spatial trend of the modernist era was epitomized by urban planning theorist Mel Webber's concept of "community without propinquity," advanced in the 1960s as the era of automobiles and unlimited personal mobility seemed to promise a new technological utopia without the need for local community.[5] In Webber's vision, technology such as the motor vehicle, telephone, and television would allow people to maintain social ties without regard to distance. There is, to be sure, some value in this ideal, and during the early twenty-first century social media in particular have moved us rapidly in this direction. Yet the costs of ignoring local places, local communities, and face-to-face interactions are clearly very high as well.

Many other elements of current urban development also tend to diminish attachment to place. Housing is increasingly mass-produced by large-scale homebuilding corporations, which often produce a generic product without any relation to local landscapes, climate, history, or culture. Office parks and malls likewise follow generic models. Long-distance commuting for work or shopping tends to further detach people from the places where they live. Unique, locally owned businesses are displaced by regional malls full of chain stores. And so on. Within the field of planning itself, abstract economic or social science methods of analysis have displaced the more place-oriented strategies of urban design. The latter were a cornerstone of early-twentieth-century planning, but were then seen as unscientific by many mid- and late-twentieth-century planning experts. Even today many professionals too often sit poring over their computers, analyzing problems abstractly rather than going out and looking at real places, talking to real people, and gaining appreciation of the problems and complexities of actual communities.

A reaction against aspatial planning began in the last quarter of the twentieth century. In 1979 planning scholars John Friedmann and Clyde Weaver stated their belief that the next wave of regional planning would have to emphasize "territory" as opposed to "function."[6] Sociologist Manuel Castells analyzed the struggle of local groups to maintain and assert their own identities—frequently place-oriented—in the face of a global network society.[7] Academic geographers argued for new understandings of how "space" is produced by political, economic, social, or cultural forces, partly on the assumption that ignoring these dynamics allows corporations and powerful elite forces to control our physical world unrecognized.[8] New Urbanist architects and planners sought to build communities with a strong, though often quite traditional, sense of place, while Timothy Beatley and others suggested that reasserting the importance of "place" within planning can assist local stewardship efforts and lead to "biophilic cities" integrating natural and built environments.[9] Such reassertions of local place, character, and community can be seen as key to the effort to resist economic globalization, and to support local identities and communities of resistance instead.

Closely related to "sense of place" is the concept of "livable communities," which likewise has become much discussed in recent decades because of a collective realization that many of our communities do not meet this criterion very well. They often lack neighborhood centers, downtowns, parks, local stores, recreational facilities, cultural amenities, or even sidewalks. Increasing traffic makes streets unsafe and unpleasant, while many neighborhoods are so isolated and disconnected from town centers that residents must get in a car to travel anywhere. These and many other factors have decreased the livability of cities and towns so steadily that we often take this situation for granted.

Since we live in environments that have often been very damaged, in ecological, social, cultural, and economic terms, much of the work of sustainable development involves healing specific places and communities. This approach may mean helping to restore urban ecosystems, for example creeks, wetlands, or wildlife habitat. It may mean reducing greenhouse gas emissions and moving towards a zero-net-energy status for new development. It may mean looking for ways to redress inequities or to respond to the racism and deprivation that have led to the impoverishment of particular neighborhoods. It may mean working to restore a locally oriented, socially responsible economy in particular cities or towns. Or it may mean taking advantage of thousands of other opportunities to build a better future in particular places.

A number of writers have proposed additional sustainability planning themes. In an article in the *Journal of the American Planning Association*, Philip R. Berke and Maria Manta Conroy (2000) proposed six principles: harmony with nature, livable built environments, place-based economy, equity, polluters pay, and responsible regionalism. In their book *The Ecology of Place*, Timothy Beatley and Kristy Manning (1997) suggested additional themes for what they viewed as "the new planning paradigm," including fundamental ecological limits to development, reduced consumption of nonrenewable resources, a restorative and regenerative approach to development, quality of life, community, equity, and full-cost accounting. The Sustainability Plan for the 2012 Olympic Games in London included main themes of "climate change, waste, biodiversity, inclusion, and healthy living."[10] Other authors have advanced yet other principles and goals for sustainability planning.

Clearly many different sustainability strategies are important and complement the more general approaches emphasized in this section. However, I have tried to distinguish here themes that most strongly differentiate sustainability planning from planning-as-usual. Among the main reasons why conventional planning has led to unsustainable development is that it often fails to take a long-range perspective, has not sufficiently integrated different disciplines and goals, hasn't considered the interrelation of actions at different scales, hasn't focused on place, and hasn't emphasized constructive action, in particular to inject under-represented viewpoints and strong visions of a sustainable future into policy debates. In all of these ways sustainability planning is different from business-as-usual. The challenge is to figure out ways of applying these sustainability themes within difficult, real-life situations. But with care and dedication, ways to do this can usually be found.

PAST PERSPECTIVES ON PLANNING

To be applied most effectively, sustainability planning must build on and situate itself in relation to past planning theories. In other words it must make use of the past body of wisdom regarding how human communities can and do develop. Accordingly, this section outlines major existing branches of planning theory, and asks how sustainability planning relates to these.

Rational comprehensive planning

Probably the dominant planning strategy during much of the field's history has been the so-called rational comprehensive planning model. In the United States particularly this approach can be seen as flowing from a longstanding identification with the philosophy of pragmatism, seen, according to Hilda Blanco, as an "application of rational process and ... the best knowledge available to address social problems."[11] Under the rational model, planners analyze situations, define goals, identify obstacles that prevent these from being accomplished, develop alternative solutions, compare these, decide on a preferred approach, implement this, and then evaluate its success.[12] The focus of such a straightforward, linear planning process may be a city, a county, a region, a park district or utility, a country, or any other type of organization. Whatever the jurisdiction in question, professionals follow predictable steps to develop a planning document which may contain both written policies and graphics or maps designating specific places for types of development. Planners' actions within this theoretical model may also include collecting and analyzing background data, making projections, evaluating whether existing programs are working, and recommending modifications to these if necessary. The rational planning process draws heavily on social science methods, especially in the initial data collection and analysis of problems, and can be said to be "comprehensive" if it considers a broad spectrum of interrelated issues and policies.

The roots of this model extend back at least to Scottish visionary Patrick Geddes' philosophy of "survey, analysis, plan" in the early years of the twentieth century. (Geddes emphasized survey methods that involved planners getting out into the field and observing environments firsthand as well as utilizing more abstract quantitative data.) Rational comprehensive planning came to the fore beginning in the 1920s as the urban planning field became professionalized and planners sought systematic and credible methods of determining urban planning policy. In the United States, Secretary of Commerce Herbert Hoover prompted the federal government to pass the Standard State Zoning Enabling Act of 1923 and the Standard City Planning Enabling Act of 1927, establishing legal authority for local planning; in Britain a series of Town Planning Acts beginning in 1909 created similar powers. Many US state governments began requiring cities to create municipal master plans (also called general plans or comprehensive plans), and the rational comprehensive model was generally employed to do this.[13] Following on from their master plan and often serving as elements of it, cities also frequently develop area plans (also called specific plans or, at a very detailed level, sector plans or precise plans) for particular neighborhoods or locations within the city, and functional plans to develop detailed policy in particular issue areas. Various versions of the rational comprehensive model are also now used to develop watershed plans, habitat conservation plans, ISO 14000 plans for improved environmental performance of industrial plants, national spatial plans, and Local Agenda 21 plans to implement principles stemming from the 1992 UN Earth Summit. These specific types of planning related to sustainability objectives will be discussed at greater length in later chapters.

The rational comprehensive method has many strengths. It provides a clear, straightforward method of formulating policy and programs, and is useful at many different levels of planning. It also meshes well with the use of indicators to measure sustainability problems and the effectiveness of policies. Rational planning appears logical to many members of the public, tends to be respected by local political leaders, and can be designed to offer opportunities for public involvement.

However, this model has also been criticized on a number of grounds. For one thing, it is often seen as overly expert-driven, based on a mind-set in which detached, "objective" planning analysts determine policy rather than letting public concerns drive the planning process. It has also been criticized for relying on quantitative analysis of data rather than taking into account less tangible, qualitative elements of the urban environment. Because the process can be abstract and expert-driven in this way, it may fail to develop the public and political buy-in necessary for policies to work. Planners may become seduced by the expert or facilitative roles required in the rational comprehensive model, and may simply focus on fulfilling these responsibilities and allowing the model to work rather than taking on more active entrepreneurial or advocacy roles to ensure that plans are implemented, goals are met, and under-represented interests are considered in the process.

The rational model has also often overlooked the realities of political power, meaning that much hard work has gone for naught when political or economic elites simply follow their own agendas instead. Strategies that incorporated advocacy and hard-nosed political organizing might have been necessary to bring about change instead. Lastly, rational comprehensive planning has often failed to take into account important social and environmental issues that were not part of the intellectual mainstream of the time. In particular, planners operating according to the values and methods of modernist science and economics often have not adequately incorporated environmental or equity objectives into their supposedly rational plans. As a result, these plans have often contributed to unsustainable development practices.

In the late 1950s and 1960s several modified versions of the rational model appeared, responding primarily to the misplaced certainty embodied within its more-or-less scientific approach. Charles Lindblom proposed a "disjointed incrementalism" that viewed planning as a process of "muddling through" day-to-day decisions while avoiding overarching large-scale judgments that could not be supported.[14] Amitai Etzioni proposed a method of "mixed scanning" under which planners were supposed to incorporate elements of both broad perspectives and local angles.[15] Under this approach, planners surveyed the broader scene first to get a sense of perspective and then focused on particular issues and strategies. This method in particular might help incorporate the more holistic approach of sustainability planning.

Neo-Marxist planning theory

In the late 1960s a strong theoretical critique of the rational planning model arose from a number of sources. Writers such as geographers David Harvey and Doreen Massey, sociologists such as Henri Lefebvre and Manuel Castells, and, somewhat later, academic planning theorists such as John Friedmann, Norman and Susan Fainstein, Robert Beauregard, and John Forester called attention to the fact that previous planning theorists had paid insufficient attention to power dynamics within the city.[16] The result, as Harvey argued forcefully, was that equity issues were not addressed within planning, and indeed that planning often allied itself with powerful economic or political forces and increased inequity.

Whether or not specifically based in Marxist theory, which had developed its own specific language and set of constructs, such theorists began developing detailed studies and critiques of power dynamics within urban development, often terming their approach "political economy" or, more recently, integrating ecological analysis, "political ecology." Researchers such as John Logan and Harvey Molotch analyzed the role of "growth coalitions" of developers, real estate interests, and local politicians in promoting boosterish local development.[17] Clarence Stone studied the role of elites in directing the rapid expansion of Atlanta.[18] John Mollenkopf produced a detailed study of how development interests dominated local politics on Long Island.[19] John Friedmann argued for a focus on "social mobilization" in order to counter established interests.[20] Brian Stoker and others developed "regime theory" to look at how ruling elites ran cities.[21] And a large number of authors including Amartya Sen, Arturo Escobar, David Korten, Martin Khor, and Vandana Shiva analyzed how economic globalization has undercut local, regional, or national initiatives for environmental protection and social welfare.[22]

Overall, the range of neo-Marxist theories has provided a rich and nuanced understanding of urban dynamics. But it has been less successful at moving beyond critique to show how better forms of planning might come about. Its most useful theoretical contributions in this regard came through the study of urban social movements (discussed later), including informal and insurgent planning, and through what evolved into the "communicative action" school of planning theory.

Participatory and communicative planning

One main response to the shortcomings of the rational comprehensive planning method was the rise of participatory planning beginning in the late 1960s and 1970s. In the United States, new federal programs such as Model Cities required extensive public involvement; this requirement was also written into environmental review legislation beginning with the National Environmental Policy Act (NEPA) in 1972. In the United Kingdom, Parliament amended town and country planning legislation in 1968 to incorporate a statutory right to public consultation. Environmentalists, historic preservationists, and social justice advocates increasingly demanded a voice in local government planning, leading to new procedures for surveys, workshops, focus groups, stakeholder consultation, and public comment. Researchers such as John Forester began to realize that much of what planners do every day is to meet with various constituencies, network, share information, and facilitate communication. A new theoretical perspective emerged which is often referred to as "communicative" planning, in that it emphasizes both public participation (sometimes within a consensus-based format) and ongoing processes of communication between planners, citizens, developers, government officials, and other parties as the main mechanism through which things get done and people learn. To be useful, as Patsy Healey argues, communicative theory must also be based on awareness of how knowledge and patterns of communication are socially constructed;[23] this perspective thus overlaps with the institutionalist and social learning perspectives discussed later.

Many different sorts of public participation are possible, ranging from relatively tokenistic advisory boards or blue-ribbon commissions to actual community direction of the planning process. In a classic 1969 article social worker Sherry Arnstein proposed a "ladder of citizen participation," beginning with relatively nonparticipatory situations in which planners sought to manipulate the public or viewed meetings and educational materials in a paternalistic way as "therapy" for the public, through relatively tokenistic stages of "informing," "consultation," and "placation," to situations in which the public gained real power through "partnership," "delegated power," and "citizen control."[24] Later planning theorists developed more nuanced theories of collaborative planning, based on the work of German philosopher Jurgen Habermas, in which many different constituencies learned from one another over time through communicative action.[25] Habermas believed strongly that for this process to work it must be free from domination and distortion (as when different parties twist information for particular purposes), and otherwise meet the criteria of a "discourse ethics."[26]

As a minimum, public workshops, notices, and hearings have now been included in almost every planning endeavor. Frequently planners go well beyond this to conduct surveys, hold focus groups, create websites, facilitate consensus processes, or conduct urban design "charettes" in which residents map out the preferred form of new development in their neighborhood. The variety of public participation opportunities in planning for any city or town is often simultaneously exciting, exhausting, and—if constructive actions are not possible within a reasonable period of time—frustrating.

Participation is possible at every scale of planning, though often through different methods. At the site scale, one extreme is represented by the cohousing movement in which future residents design and plan their community themselves with the help of technical consultants. At the neighborhood scale, various design and planning processes run by cities or nonprofit organizations involve the public through workshops, charettes, and/or project boards composed of neighborhood residents and constituencies. At the city scale, the same forums may also be held and many different boards and commissions set up to exert citizen control over the planning process. Electoral politics also of course often comes into play. At regional and national scales direct participation is more difficult, but public perspectives are often incorporated by having representatives of diverse constituencies meet repeatedly over months or years to develop consensus, by conducting surveys and workshops, and by relying on local elected officials to express the views of their constituents.

If done well, participatory planning can help develop policies that are responsive to public needs, in particular the needs of constituencies that may not be represented within the political establishment or planning staff, such as lower-income communities and communities of color. Both the goals of the planning process and the recommended policies and programs may be improved through such input. The process of communication may increase understanding of the issues on all sides and build mutual respect for the diversity of viewpoints. It may also strengthen political buy-in and develop political coalitions that can help implement recommendations. All of these factors make participatory planning extremely attractive for sustainable development purposes, especially in improving equity and responding to the needs of our least-privileged communities.

However, participatory or communicative planning theory is also open to criticism. There is no guarantee that consensus-based planning or efforts directed by local communities will produce good results—they may simply reinforce short-sighted, parochial viewpoints. Inserting broader, longer-term perspectives into such debates is crucial.

There are great difficulties in defining what the "community" is or who represents it.[27] Is it the residents who live in a place, others who may use it occasionally, or still others at greater distances who may be affected by particular local actions? Is it just human residents, or other species as well? Is it just current residents or does it include future generations? Is it adequately represented by those voices who speak the loudest, or even those who vote, or do these individuals represent a biased and unreliable subset of community interests?

Consensus processes often produce lowest-common-denominator solutions and vague goals with few specific means to implement them. Avoiding these outcomes may take skilled facilitation by planners, political leaders, and/or community-based organizations. In the ideal situation all stakeholders potentially affected by a decision would be directly represented within the decision-making process. However, if we consider that these stakeholders might include future generations, people on the other side of the world, other species, and entire ecosystems, then this is clearly impossible. Planners may need to inject such unheard viewpoints into the debate themselves, and articulate sustainability options for the future, while also maintaining a facilitating role. Or else they may need to identify and encourage advocates who can adequately represent these viewpoints within the public process. They often must anticipate such potential viewpoints and incorporate them into the dialogue whether or not these have been well expressed by members of the public. Needless to say, facilitating a public process in this way is not easy.

Participatory or consensus-based planning without buy-in from outside political interests and institutions can be a fruitless exercise. Individuals and groups may spend years on collaboration, only to see their recommendations ignored. For example, Amy Helling documents how regional planners and hundreds of citizens in the Atlanta area spent five years and $4.4 million on a collaborative visioning project to develop a Vision 2020 framework. The resulting report was then ignored by the existing power structure and other regional agencies, which had not bought into the process.[28] Luckily in that case the state government and the US Environmental Protection Agency (EPA) eventually stepped in to force better coordination of regional transportation, land use, and air quality planning. But the earlier idea that citizen collaboration could bring about change without buy-in from regional elites proved to be deeply flawed.

Participatory processes also demand enormous time, money, and commitment from both nongovernmental organizations and planners. Groups must be able to send staff to countless meetings over several years, and individuals need to possess the patience to work through complicated issues in slow discussion. Many organizations simply do not have the resources or experience to engage in such processes, or they may believe that there are more efficient ways to work for change. Participants frequently become burned out, and only those with the biggest gains to make or axes to grind persist. Power imbalances between participants can also skew results. For such reasons participation must be carefully designed to produce the best possible representation and real results in a timely and cost-effective manner. Ironically, that does not necessarily mean having hundreds of public workshops or elaborate multi-year consensus processes. Instead, it may be better to have a limited number of well-structured, well-publicized, and well-attended events that are appropriate to the planning task at hand. Along with these, as Diane Warburton points out in a study of community, participation, and sustainable development, a foundation of education, social learning, and consciousness-raising is critically important, similar to the process that Brazilian educator Paolo Freire once called "conscientisation."[29] Only by developing deeper public awareness of issues and commitment to cooperative and constructive work can participatory planning truly be successful.

Advocacy planning

The concept of advocacy planning arose in the 1960s partially to compensate for past decades of nonparticipatory, top-down action by local governments. Pioneered particularly by Paul Davidoff, a lawyer, planner, and professor at Hunter College in New York, supporters of advocacy planning recommended that planners work with particular constituencies such as low-income communities to make sure their viewpoints were effectively represented. Advocacy planners might even be employed by these constituencies, serving in effect as lawyers or technical experts working on their behalf.[30] Such advocacy is often clearly needed if change is to occur on environmental and social issues. Advocacy planning acknowledges that the core dynamic of social change in many cases is a power struggle in which some groups—developers, corporations, well-heeled neighborhoods, and so on—have far greater access to resources, expertise, and political clout than others. Entire urban or suburban landscapes may be shaped by unseen concentrations of political or economic power.[31] Ignoring these powerful forces and hoping that processes of consensus-building by themselves can produce change may be naive, when what is needed is grassroots organizing, advocacy, coalition-building, and creative use of various media to create alternative power centers in society.

At the same time, planners and other professionals have to walk a fine line between advocacy and maintaining their credibility as technical experts and process facilitators. They may need to make clear when they are taking on one role as opposed to another. When serving as an information source, their research needs to be thorough and respectable, not slanted so as to manipulate data for the needs of a particular constituency. It is also often difficult to take on advocacy roles while employed within a city or regional government, or while maintaining academic or professional ties. Professional agencies tend to look down on advocacy.

Nevertheless, even the most careful, scientific work on development issues can be framed in ways that have direct implications for public policy, urban design, and social organization. Planners can also play an enormously useful role by involving advocates of under-represented constituencies in decision-making processes, injecting a full range of viewpoints into debates, and structuring any form of analysis so as to get at the long-term sustainability issues that really matter.

Advocacy planning dovetails with the rise of NGOs as a significant force in urban planning and development around the world. Since the mid-twentieth century a wide variety of nonprofit groups have been founded to undertake many sustainability-related tasks, ranging from advocating for global environmental protection agreements to undertaking local economic development and affordable housing construction.[32] Although most of these groups are still relatively small and weak, they have achieved many successes, and have in many cases become a valuable counterweight to other sectors of society such as business, labor, and government. Indeed, the rise of "civil society" worldwide is one of the most hopeful phenomena of recent decades. But, for this movement to more profoundly affect the development of human communities, NGOs will need to increase their professional expertise in fields such as urban planning and community development. They will need to advocate their viewpoints convincingly to both government and the private sector, undertake large amounts of development themselves, and counter private sector developers' plans with more ecological or equitable alternative scenarios. Professional planners, historically hired mainly by government or consulting firms, thus may find themselves working increasingly for activist NGOs or in research roles for nongovernmental institutes.

Theories of urban social movements

Behind specific advocacy groups are often urban social movements (USMs)—broad upwellings of public concern about particular issues that may be represented by various constellations of NGOs and political leaders. Examples might include historic preservation, neighborhood preservation, civil rights, and environmental movements, as well as campaigns to protect or improve particular places or achieve local policy change.

Such grassroots uprisings serve as a counterweight to the political power represented by elite social and business groups,[33] and are considered by Friedmann to represent a social mobilization tradition within planning. Indeed, Manuel Castells, one of the foremost scholars of such movements, argues that USMs are crucial to establish new, locally oriented senses of identity that can counter the power of global corporate culture and elite networks in the twenty-first century.[34]

Urban social movements have tackled a wide variety of planning issues, such as calming traffic, claiming civil rights for gay residents, protesting against sterile, modernist urban renewal, and fighting for ecological restoration of creeks, rivers, and parkland. In developing world contexts, they can result in "insurgent planning" that seeks to develop grassroots opposition to post-colonial government and corporate forces.[35] But these movements face a wide variety of challenges in building their power, maintaining it, and channeling their efforts in the ways most likely to bring about constructive change. In particular, unless they are institutionalized through the establishment of strong NGOs and political leaders, USMs may quickly dissipate when grassroots leadership and public interest move on.

Not all USMs are constructive and inclusive efforts—Nimby ("not in my backyard") groups in affluent countries opposing affordable housing or other much-needed local services are an example of negative local action—but many represent the sincere efforts of residents to bring about positive change in the local environments over which they have some control. Active support and advice from planners and other professionals can help these advocacy-oriented groundswells of civic attention bear fruit. Even non-advocacy-oriented planners can acknowledge their importance by making sure they are invited to participate in local plan-making and consensus-oriented planning processes.

Institutionalism

Several waves of theory over the past 50 years, coming mainly from sociology and political science, have stressed the role of institutions in structuring the context in which action takes place. These institutions include not just government agencies and large organizations, but the whole panoply of laws, customs, cultural norms, and social traditions that affect how we live, think, and act. This perspective emphasizes how debates and mind-sets are shaped by the prevailing structures of society. The implication is that to plan effectively, we need to look at these institutions and think about how changes to them can reinforce constructive action in the future. The work of Giddens has been pivitol in this regard,[36] as was the pioneering work *The Social Construction of Reality* published in 1966 by sociologists Peter L. Berger and Thomas Luckmann.[37] Institutional theory is consonant with the approach labeled "social learning" by Friedmann,[38] in that it emphasizes an evolutionary approach to change, with Healey's communicative planning approach, which she sees as dependent on institutionalist analysis and action aimed at "building up the *institutional capacity* of a place,"[39] (emphasis in the original) and with other theories of critical thought in which the emphasis is on how various forces organize knowledge, how generative metaphors structure cognition, and so forth.

Writings about "social capital" can be seen as fitting into the institutionalist perspective. In his classic book *Bowling Alone* (2000), political scientist Robert Putnam analyzes how television, the mass media, suburbanization, pressures of time and money, and generational change have led to a decline in civic engagement and social capital in the United States.[40] These factors in large part represent or stem from the institutions of society. The solution, Putnam believes, lies first in naming our problem, and then in working "to create new structures and policies (public and private) to facilitate renewed civic engagement."[41] In his view this means focusing on a long list of initiatives, particularly on the education of young people, promoting community service programs, making workplaces more "family-friendly and community-congenial," rewarding socially responsible economic activity, promoting more integrated and pedestrian-friendly communities with better public spaces, supporting spiritual communities of meaning, fostering forms of electronic entertainment that reinforce community engagement, promoting group activities, and reforming the political system to make it more participatory and democratic.[42] Though it doesn't offer a single easy strategy to bring about change, such a package of mutually reinforcing initiatives can potentially change the institutions and patterns of social learning that from this point of view determine the overall direction of society.

Despite these broad theoretical approaches, day-to-day urban planning perhaps most often resembles Lindblom's "disjointed incrementalism" or "muddling through."[43] The reasons for this are understandable—chaotic local politics, unpredictable events, the importance of seizing transitory opportunities, and bureaucratic obstacles or inertia. Still, the risk is always that in a "muddling through" process planners or other decision-makers lose sight of broader goals and strategies. They may be sidetracked into technocratic or bureaucratic roles that have little to do with actively addressing urban problems. Worse yet, they may facilitate types of development that later will be seen as unsustainable.[44]

Situating planning within a broader theoretical framework can thus help give greater coherence to day-to-day activities, fit them into an overall strategy, and clarify long-term goals.

SITUATING SUSTAINABILITY PLANNING WITHIN PLANNING THEORY

While following the themes mentioned earlier, sustainability planning may need to draw on many different planning theories and strategies. To be most effective it may need to weave together a range of theoretical perspectives, for example at different times paying particular attention to rational planning methods, communicative processes, underlying structural forces of political and economic power, social movements, and the role of institutions. In keeping with its holistic approach, sustainability planning therefore may be best conceptualized as a "meta-theory," situating its particular perspectives and agenda on top of the best possible foundation of existing social and political theory.

The elements of this foundation will surely evolve over time. In recent decades, for example, new theoretical approaches such as environmental history, institutionalism, and communicative action have emerged, all of which shed light on key questions of how human society evolves and how it might relate more sustainably to the Earth's environment.

Movements such as deep ecology, ecofeminism, and environmental justice have also appeared, and their theoretical perspectives deepen our understanding of potential sustainable development directions. Although individual theorists and activists must often specialize in one intellectual approach or another, different theoretical perspectives should not be seen in isolation, since each contributes valuable knowledge to the overall picture. The implications of each for the overall challenge of creating a more sustainable society should be emphasized wherever appropriate, and the connections between theoretical frameworks pointed out, to avoid unnecessary fragmentation of knowledge that will undercut the ultimate objectives.

This holistic approach stems naturally from the ecological forms of cognition underlying the sustainability perspective, which emphasize connections between things, a dynamic view of intellectual as well as natural systems, and a view of systems as a whole. As discussed previously, this outlook on the world is fundamentally different from that of mid-twentieth-century positivistic science, often referred to as the Cartesian worldview, which still greatly influences many disciplines, and which has led partisans to fiercely endorse one theoretical perspective or another, rather than finding broader understanding.

Planning and power

All that being said, some theoretical mechanisms may be particularly important in explaining why unsustainable development has come about, and may be important as well for understanding how more sustainable directions might emerge. In particular the realities of power—along with the institutions and ideologies that support power—must be understood if sustainable development is to come about in the long run, or if specific planning efforts are to be successful in the short run.

Plans are only effective if they are implemented, inspire action, or otherwise help bring about changes in the world. Whether this happens depends in large part on whether those with power buy into planning processes, learn new ways of looking at the world, or come to see their own interests as dependent on change. Politicians, developers, large businesses, unions, neighborhood associations, and many other groups wield power within societies, and planning at any level takes place within the context of these power relationships. Planning typically takes place through the auspices of government, but as has been well documented, certain groups often dominate government at various levels. Local developers or growth coalitions often dominate municipal government, for example. Citizen coalitions, environmentalists, historic preservationists, civic reformers, and other groups may represent opposing power bases. These constituencies themselves may have varying goals and be in conflict with one another. Public agencies and institutions, including planning staffs, also hold considerable power, in that they are able to structure processes, set agendas, and either expedite or stifle particular initiatives within bureaucracies. Planning efforts that ignore the reality of such power dynamics are likely to fail.

No matter what other theoretical approaches are employed, an understanding of the nature and dynamics of power is essential for successful planning at any level. Sustainability planners and nongovernmental activists will often need to work with various groups inside and outside government to ensure that sufficient political backing exists to implement change. Many of the larger power dynamics within our current society are extremely difficult to change, such as the relative influence of monied interests within government. Over many decades these interests have shaped public opinion, the news media, and the culture itself in certain directions, for example in support of materialist cultures and a certain national identity. In the US, these trends have produced an increasingly dysfunctional politics, resulting in political gridlock and wildly distorted perceptions of reality such as when one political party denies the validity of climate change science. It may require enormous time and effort to help this context evolve toward a situation that is more democratic, community-oriented, and open to addressing current environmental and social problems. However, that's the challenge, and in the meanwhile planners can seek opportunities to build constituencies for smaller-scale near-term improvements, all the while working for longer-term structural change and social learning.

Planners' roles

With this background of sustainability planning theory in mind, how do planners and other professionals work towards sustainable development individually? Much of the answer lies in recognizing the different roles that may be useful in different contexts. It has long been recognized that, at whatever level of government, planners may play a number of roles at different times.[45] Often they serve as process organizers—initiating planning efforts, keeping them on schedule, and ensuring that final results are achieved. Within this role they may also serve as facilitators of meetings and workshops. Planners may work to negotiate between various powerful interests, brokering compromise. They may serve as technical experts supplying information to all parties. (This traditional, apparently nonpartisan role is what the public often expects to see.) They may take on a role as political organizers, either to level the playing field so that under-represented constituencies are heard within the process, or to assemble the necessary political constituencies so that plans will actually be implemented. Planners are often educators, ensuring that information is spread to whoever needs it and helping various participants understand issues and contexts. And, last but not least, planners can adopt a motivational role as visionaries, cheerleaders for civic initiatives, or dramatic speakers inspiring action.

All these roles are important at different times. Planners and related professionals may wear many different hats in the course of a single day, and the exact role at any given time may depend on the context and the other constituencies involved. It may be necessary to keep roles distinct and carefully separated, or it may be appropriate to play several roles at once. Hitting the right balance in all cases tends to require substantial experience, diplomacy, initiative, communicative abilities, humor, and understanding of situations.

Sustainability planning implies a different balance between some of these roles than undertaken by the planning profession historically. The technical expert role is likely to be much less than previously, and even when operating as an expert it is important to always be asking to what use technical information can be put. Likewise, although managing public planning processes will still be a central role, it is incumbent on the planner to be more than just a detached manager, but to pay close attention to the results of the process, and to facilitate public understanding and political action.

Meanwhile, more active planning roles assume increasing importance—those of facilitating consensus, supplying vision, educating the public, and serving at times as an advocate and organizer. A shrewd, entrepreneurial stance may be required as planners and other professionals navigate within institutional bureaucracies and weave their way across political minefields, trying at all times to advance debates and achieve the most constructive, real-world results possible. At times this will be a risky and difficult endeavor. But the results are likely to be worthwhile in terms of producing small- or large-scale movement towards sustainable communities.

4

SUSTAINABILITY PLANNING
AND THE "THREE Es"

As perhaps the pre-eminent symbol of their holistic approach, discussions of sustainable development often focus on how to simultaneously meet goals in the areas of environment, economy, and equity, usually referred to as the "Three Es." In the past these objectives have often been separated within urban planning, governmental decision-making, and development discussions of all types. Public debates have pitted "environment against economy" within controversies such as logging in the US Northwest, coal mining in Britain, air pollution controls in China, and mining in Chile, South Africa, and Australia. Construction unions, developers, and road-building interests have also often fought growth management efforts in metropolitan areas on the grounds that these environmentally oriented policies will destroy jobs. Businesses oppose equity initiatives such as "living wage" ordinances on similar grounds. Even environment and equity have come into conflict at times, such as when recycling facilities with their noise and traffic are located in low-income neighborhoods.

Reconciling these seemingly conflictual goals—and developing new decision-making approaches that reconcile the needs of all three perspectives—is a frequent focus of sustainability planning. Some might argue that a construct such as the Three Es is unnecessarily simplistic, or might want to add additional "Es" such as ethics, education, and empowerment. Certainly what is important in the end is not this concept per se but the underlying ability to weave together many different conceptualizations of core values. But the Three Es represent an excellent starting point in this effort, a relatively simple, straightforward set of criteria that are intuitively understandable to most people. Many civic organizations ranging from the Regional Planning Association of New York and New Jersey to the Bay Area Alliance for Sustainable Development have used the Three Es framework. This model also serves the purpose of elevating equity and environmental goals to the level of economic objectives within day-to-day decision-making, which alone would be truly revolutionary if achieved, and an enormous step towards more sustainable development. So we will focus on this approach here as a useful construct for achieving the broader holistic vision of sustainable development.

SUSTAINABILITY AND THE ENVIRONMENT

The first of the Three Es, environment, is what many sustainability advocates have historically focused on. As we have seen, sustainable development has strong roots in the twentieth-century environmental movement with its increasingly broad definition of the "environment" and its focus on integrating global and local environmental issues. A number of writers see "sustainable communities" as representing a third wave of the modern environmental movement, following the environmental legislation and activism of the 1960s and 1970s, and subsequent attempts at more flexible, market-oriented, and negotiated approaches to environmental cleanup.[1]

However, environmental goals have varied enormously over time and between different groups of advocates. It is important to explore different debates, viewpoints, and themes within this movement to understand its implications for sustainable development.

One main theme within environmentalism has been the dramatic broadening and redefinition of the term "environment" in recent decades. The early conservation groups founded in the late nineteenth century, such as the Sierra Club in 1892 and Audubon Society chapters in 1896, focused primarily on wilderness and wildlife issues. These topics remained the movement's main priorities through the first half of the twentieth century, even for progressive thinkers such as Leopold. But in the 1960s the agenda and political power of the environmental movement expanded to the point where a number of authors begin their histories of modern environmentalism with this decade.[2] Air and water quality, pesticides and other toxic chemicals, energy use, nuclear power, nuclear arms control, environmental justice, urban growth, international development, and global climate change became concerns of environmental organizations between the 1960s and 2000. Many people began to see every element of the world around them as part of their "environment," including air quality within their homes, the nature and origins of their food, the amount of traffic on local streets, and the quality of public spaces in their towns and cities. This broadening of perspective has been rooted in the gradual emergence of an ecological worldview that stresses the interconnectedness of all things. Such a holistic perspective is of course also one of the foundations of sustainability planning.

A second, related change in environmentalism—affecting its goals and character—is the partial shift from the anthropocentric and utilitarian attitudes that have underlain industrial society and capitalism towards more ecocentric approaches. The philosophy of Thoreau and Muir, with its emphasis on the intrinsic value of nature, found much public support in the late twentieth century and began to gain ground on the utilitarian conservationist outlook of Pinchot and many others which was dominant within mainstream environmentalism for much of the century. Deep ecology, spiritual ecology, and recent other strains of environmentalism support this fundamental rethinking of the relation between humans and the natural world. Debates such as over logging of ancient forests, for example, reflect the emerging of a point of view that these ecosystems have value in their own right and should not be put to human use at all, even if they could be managed in a "sustained-yield" fashion.

As previously suggested, the notion of "limits" is another main environmental theme underlying sustainability debates. From the time of Thomas Malthus in the early nineteenth century and probably long before, thoughtful observers have wondered if and when human civilization would reach the limits of a small planet. The Limits to Growth debate of the early 1970s, which as we have seen helped give rise to the term "sustainable development," raised serious questions about planetary limits. In contrast is the viewpoint known as "technological optimism," represented by Simon, which maintains faith that human ingenuity, technology, or the hidden hand of economics will avert catastrophe.[3] To a certain extent Simon's viewpoint has been proven correct—resource depletion has occurred at a slower rate than expected by some environmentalists. Simon even won a bet with environmentalist Paul Ehrlich over how much prices of five metals would rise between 1980 and 1990 (the aggregate price fell by almost one half). Yet in a longer time frame "limits to growth" arguments clearly have much merit. They are especially valid when applied to urban landscapes—land is a limited commodity for which there is no substitute, and the loss of open space and agricultural land can be visually confirmed every day.

On a practical level, integrating environmental goals into planning and activism implies developing as much knowledge as possible about local ecosystems and their history, as well as about environmental law and regulation, environmental planning tools such as environmental impact reports, and best practices of ecological planning and restoration. Such understanding makes possible better decisions about how to balance the range of possible environmental goals with economic and equity objectives. Much of this knowledge, if possible, should be based on detailed, firsthand observation of particular places, since this is the best way to fully appreciate the character of ecosystems and to understand how humans have interacted with them.

In terms of urban development, environmentally oriented principles related to sustainability include compact urban form (which saves open space, reduces driving, and often produces more livable, walkable communities), transit-oriented development (which likewise reduces driving and fossil fuel use), closed-loop resource cycles (ensuring that water, metals, wood, paper, and other materials are reused or recycled), environmental justice (integrating environmental and equity concerns), pollution prevention (steps to prevent pollution in the first place rather than clean up later), the "polluter pays" principle, and the restoration of creeks, shorelines, habitats, wildlife corridors, and other ecosystem components within cities and towns. These and other strategies are ways to move towards a radically greener society, one that can coexist with the Earth's limited resources and often-fragile ecosystems in the long run.

SUSTAINABILITY AND ECONOMICS

Environmental goals often seem in stark contrast to those of economics, and at a broader level the ecological worldview seems at odds with the perspectives of many economists, especially those subscribing to free-market philosophies. Understanding ways in which contemporary economics might better fit with sustainable development is a challenging task.

Many pros and cons of market-based capitalist economics from a sustainability point of view are well known. Such an economic system can be very good at regulating supply and demand, allocating resources, and providing incentives for entrepreneurship and innovation. It is also clearly good at generating wealth and a high level of material comfort for many. These benefits are very significant, and mean that we wouldn't necessarily want to scrap market economics in an ideal world even if we could.

However, our current capitalist economics—both in theory and practice—has many flaws from a sustainability perspective. There is the problem of valuation: it is extremely difficult to put a price on social and environmental goods (such as human health, equity, and environmental quality) so as to factor them into economic decision-making. There is the problem of public goods: it is difficult to make meaningful economic decisions regarding things such as clean air, safe streets, or attractive public spaces that everyone uses but nobody pays for. There is the problem of externalities, those enormous social and environmental impacts of production and consumption that are generally not incorporated into economic decision-making.[4] The price of gasoline, for example, does not reflect the externalities of driving, which include air pollution, water pollution, traffic congestion, degradation of urban quality of life, deaths due to traffic accidents, and the costs of maintaining access to petroleum around the world.

Then there is the problem of discounting the future: the existence of interest rates and inflation means that future costs and benefits are less valuable than those in the present, and that it is very difficult to incorporate the long-term effects of actions into economic equations. Most cost–benefit models are literally incapable of considering impacts more than 30 years into the future. Thus, current economic theory is structurally handicapped in adopting the long-term perspective required by sustainability planning.

Other challenges abound. There is the problem that supposedly "free" markets are distorted by subsidies and regulations. There is the problem that demand and human "needs" are manipulated by corporations and advertising, usually to increase the level of material consumption. There is the problem that capitalism tends toward concentration of wealth and monopoly of power, both of which undermine equity. There is the problem that global trade distances consumers from the true costs of their economic decisions, making it difficult for them to understand these costs, while displacing social and environmental harm onto far-distant people and ecosystems. Capitalist economics assumes continuous expansion in consumption of material goods and resources, a phenomenon that conflicts with the environmental notion of "limits." Last, but not least, there is the problem that economic power tends to subvert democratic institutions and shape cultures to meet its own ends, meaning that economic objectives constantly threaten to overwhelm environmental and equity goals (see Box 4.1).[5]

All of these deficiencies of capitalist economics—and of economic analysis generally—undercut sustainability. Yet many economic tools are useful or necessary in the process of moving towards a more sustainable society. A number of alternative strategies have been proposed over the years to restructure capitalist economics to meet environmental and equity as well as economic goals.

One of the most radical proposed reforms is known as steady-state economics. Noted social philosopher John Stuart Mill first raised this concept in the mid-nineteenth century, but the main advocate since the 1970s has been US economist Herman Daly, who has worked at the World Bank as well as teaching at Louisiana State University and the University of Maryland. Under steady-state economics, human population and consumption are held at constant levels with a minimum throughput of resources, while qualitative, technological, and moral evolution occurs instead of quantitative increases in material production. The endless growth in material consumption, in other words, is brought to a halt. Daly has

Box 4.1 EVALUATION OF MARKET-BASED ECONOMICS FROM A SUSTAINABILITY PERSPECTIVE

Virtue or problem	Result for sustainability
Virtue of efficiency	Good at setting prices, regulating supply and demand, allocating resources
Virtue of motivation	Good at providing incentive for entrepreneurship, innovation, and creativity
Virtue of production	Good at producing a large number of material goods and generating high level of material comfort for many
Virtue of flexibility	Good at substituting resources and technologies and adjusting prices to counter resource scarcity (assuming markets are not overly monopolistic or constrained)
Virtue of analysis	Provides an important set of tools to analyze economic costs, benefits, and returns from particular projects
Problem of valuation	Difficult to place an economic value on social and environmental goods
Problem of public goods	Little incentive to incorporate common-pool resources into economic decision-making
Problem of externalities	Many costs of action not included in economic decision-making
Problem of discounting the future	The existence of interest and discount rates means that future costs and benefits are less valuable than near-term ones, making it difficult to incorporate the long-term effects of actions into decision-making
Problem of manufactured demand	Perceived human "needs" are manipulated by corporations
Problem of equity	Capitalism tends toward concentration of wealth and economic power, producing inequality
Problem of the distancing effect of trade	As production moves farther away from consumers, they do not perceive the true costs and externalities of economic actions
Problem of growth	Most economics assumes continuous growth in production and consumption
Problem of market distortions	Subsidies, regulations, and the manipulation of demand by producers mean that there is never such a thing as a "free" market
Problem of democracy	Economic power tends to subvert democratic institutions for its own benefit, usually undercutting non-economic objectives such as equity and environmental protection

proposed progressive resource depletion quotas as a mechanism to nudge the economy towards this steady level of consumption. The government would set a maximum quantity of nonrenewable resources that could be consumed each year, and then adjust the level downwards annually. From that point market mechanisms would do the rest. Prices would rise correspondingly, promoting conservation and substitution of alternative materials. Other economists such as Kenneth Boulding have floated the more radical idea—perhaps partly tongue-in-cheek—of applying the same sort of quota system to managing human population. Each couple would be issued birth permits, and these would be bought and sold like any other commodity. Although such a system may seem far-fetched, it represents a logical extension of economics to the problem of overpopulation.

The notion of a steady-state economy was most widely written about in the 1970s, when Daly published two volumes on the subject. Needless to say, the concept has not caught on in recent years; our current economy has a very entrenched addiction to material growth, and the basic economic indicators repeated on the evening news directly reflect growth in material production rather than overall quality of life. Yet the steady-state economy remains an important theoretical alternative to growth-oriented capitalist economics, one that may re-emerge in somewhat different guise in response to future resource or environmental crises.

A much more pragmatic approach to reconciling environmental and economic goals has been the discipline of environmental economics, which first appeared in the 1970s. Concerned with how best to use economic mechanisms to reduce pollution, resource use, and other environmental impacts of production, this field seeks to revise a range of mainstream economic tools such as cost–benefit analysis to better include the environment.[6] Often the focus is on attaching economic valuations to elements of the environment so as to include them within economic equations. However, this approach doesn't fundamentally challenge any of the basic assumptions of capitalist economics, and can be seen as a somewhat technocratic and reformist response to the challenge of sustainable development.

In contrast, ecological economics is a more fundamental reform movement within economics that also uses the tools and language of neoclassical economics but seeks to locate the human economy within a much larger context of ecological interactions. According to Robert Costanza, Herman E. Daly, and Joy A. Bartholomew, "Ecological economics sees the human economy as part of a larger whole. Its domain is the entire web of interactions between economic and ecological sectors."[7] This valiant effort to reconcile economic and environmental worldviews meets some of the deficiencies of conventional economics, but still succumbs to others. It generally does not challenge the centralization of power within capitalist institutions or the practice of trying to value most things in monetary terms, and still has difficulty trying to fit intangible or qualitative social and economic goods into a quantified economic framework.

Paul Hawken and Amory Lovins have argued for a restorative economics—a "natural capitalism"—that uses the enormous power of markets to bring about environmental restoration rather than exploitation. Mechanisms to assist in this process might include higher prices for nonrenewable resources and waste disposal and green taxes to internalize the environmental externalities of production.[8] Their approach might be seen as more of an advocacy-oriented version of ecological economics, in which specific policy mechanisms are used to integrate economics into a broader framework including all three Es.

Since the 1960s efforts at local self-reliance have at times posed an alternative to conventional, export-oriented global capitalism. Within international development, countries such as India sought import substitution policies in the 1960s especially. These efforts, which sought to promote locally produced products and restrict imports from abroad, were disparaged by advocates of export-oriented capitalism, and often did not work well because of the difficulty of producing a wide spectrum of goods locally in the face of international competition and political pressure. Within North America, proponents of local self-reliance have likewise advocated local economic development strategies that focus on promoting small, locally owned businesses rather than courting multinationals.[9] Such movements have often coalesced around efforts to keep Wal-Mart and other "big box" retailers out of certain towns (the entire state of Vermont has also sought to stave off Wal-Mart). Alternative currency networks such as Ithaca Hours have also attempted to promote local self-reliance through the dramatic strategy of introducing a new currency that can only be used locally, with notes representing one hour of labor instead of dollars. Anyone receiving such a note as payment can then redeem it at other local businesses for other products or services. Such local networks represent the modern version of ancient barter systems.

Also since the 1960s a sporadic movement for economic democracy, championed by American consumer advocates such as Ralph Nader and Mark Green, has sought to exert democratic control over corporations within the US. Such advocates have argued that states originally gave corporations very limited and strict charters, and that the idea of corporations as an independent power base within society with full legal rights and few responsibilities to the public is anti-democratic.[10] They frequently quote Thomas Jefferson, who wrote in 1816:

> I hope we shall take warning from the example [of hereditary aristocracy] and crush in its birth the aristocracy of our monied corporations which dare already to challenge our government to a trial of strength and bid defiance to the laws of our country.[11]

Large-scale efforts at new corporate chartering procedures have so far made little progress. But in an era of growing corporate excess perhaps such initiatives will come. A related effort is the large and growing movement for socially responsible investment, which seeks to use shareholder power and consumer choice to influence corporate activity. Begun in earnest in the late 1970s and early 1980s with the first socially responsible investment funds, this movement has grown so much that in the early 2000s it claimed to represent a quarter of all investment (much of this through large pension funds which have cautiously adopted social investment "screens" on their portfolios). Shareholder campaigns against corporations that pollute or use sweatshop labor, boycotts, and anti-globalization organizing have also helped put exploitative companies on the defensive, and have forced greater sensitivity within businesses to environmental and equity concerns (see Box 4.2).

Other principles such as "full-cost accounting" and the "polluter pays" principle have been developed in an attempt to reform economics. Both of these refer to situations in which the social and environmental costs of public and private decisions are factored into decision-making, particularly the costs of pollution and eventual cleanup of facilities. Decision-makers would then have a strong or economic incentive to adopt sustainable

---

Box 4.2 ALTERNATIVE ECONOMIC APPROACHES
TO PROMOTE SUSTAINABILITY

Alternative approach	Promotes sustainability
Steady-state economics (Herman Daly, John Stuart Mill)	Responds to the problem of growth by seeking to hold population and consumption constant with minimum throughput of resources, and instead seeking qualitative, technological, and moral growth
Environmental economics (David Pearce)	Concerned with reforming economics to better incorporate externalities, future effects, etc.
Ecological economics (Richard Norgaard, Robert Repetto)	More radical position sees human economy as part of larger web of ecological interactions
Restorative economics (Paul Hawken, Amory Lovins)	Aims to harness economic energy for sustainable development, for example through green taxes
Local self-reliance/import substitution (David Morris)	Emphasizes local ownership, production, consumption, resources
Socially responsible investment (Hazel Henderson, Green America)	Activist movement to influence corporate behavior through collective purchasing and pressure

---

development and resource use strategies. A related term, the "precautionary principle," warns corporate and governmental decision-makers that if they cannot fully understand the effects of their actions, they are best advised to take the least harmful and most sustainable approach.

Such concepts and endeavors hold potential for developing forms of economics that are more compatible with sustainable development. But the balance is still uneasy. Fundamental changes in economic values and processes will be necessary to accommodate environmental and equity goals. As Michael Redclift has put it, "Sustainable development, if is to be an alternative to unsustainable development, should imply a break with the linear model of growth and accumulation that ultimately serves to undermine the planet's life support systems."[12] Leading values behind the free-market capitalism that dominate current global development include efficiency, individual choice, growth in consumption, materialism, and privatization of common resources. Although they may have advantages in terms of motivating an efficient growth-oriented economy, these values frequently displace those of environmental protection and social equity, and will need to be put in a much better balance with these other Es.

SUSTAINABILITY AND EQUITY

Equity is the third and by far the least well-developed of the Three Es. To be sure, it has long been a focus of many community activists, labor unions, and social justice organizers.

However, these constituencies often have relatively little power, and equity concerns frequently take a back seat in planning and political discussions. Often there is literally nobody in the room who will speak up for disenfranchised segments of the population. Equity goals are often poorly understood and articulated by decision-makers, unlike concerns for the environment or economic development. There is little organized constituency for equity at most levels of government, and many powerful forces work for inequity, that is, far greater concentration of wealth and power. Yet virtually all policy-makers interested in sustainability have been forced to acknowledge the importance of addressing equity concerns.

In a global context, a diverse group of writers including Doreen Massey, David Harvey, Edward Goldsmith, Vandana Shiva, Martin Khor, L.S. Stavrianos, and Arturo Escobar have called attention to growing inequities in economic power and distribution of resources, and the ways that these inequities are played out spatially for different human communities around the world.[13] Some, beginning with Andre Gunder Frank and others in the 1960s, have charged that processes of economic globalization create a dependency on the First World by developing countries, and lead to situations in which the benefits of development are exported to the North or given to elites in the South rather than benefiting the poorest and most needy.[14] Equity advocates within the developing world also focus on First World overconsumption and argue that the industrialized countries of the North have no right to advise countries of the South on how to develop if they can't rein in their own consumption. They often view First World countries as exporting the risks and externalities of economic production, exploiting low-wage labor internationally, and seeking control of global resources. Attempts by multinationals based in the North to control the genetic and biological resources of the South have met with special resistance in recent years.

Within First World countries, a somewhat different assortment of inequities has become urgent. These include wealth and tax base disparities between rich and poor communities (especially wealthy suburbs and impoverished central cities), concentrations of poverty, inequitable distribution of affordable housing and transportation infrastructure, inequitable representation within decision-making, and environmental justice questions regarding differential exposure to environmental hazards.

To take up the first of these concerns, growing imbalances of resources between rich and poor communities have become worrisome to many. These disparities occur in large part because of the fact that our metropolitan areas are fragmented into many smaller cities that receive wildly varying amounts of money from local tax revenues. Typically suburban jurisdictions have seen their coffers benefit from mall, office park, and upper-income housing development, while central cities have seen their tax bases decline as businesses and affluent residents leave and once-thriving commercial streets become lined with empty storefronts. Meanwhile, their resource needs rise in order to provide social services for less-affluent communities, to repair aging infrastructure, and at times to clean up brownfield lands left over from past industrial development. Essentially the financial benefits of our current sprawl patterns of development accrue to some local governments, while others are left bearing the costs.

These regional disparities are worsening as suburbs expand. (Interestingly, before the rapid suburbanization of the past 60 years central cities were often better off than suburbs, since they had a rich concentration of businesses and affluent residential districts.) A main result of growing regional inequities has been the creation of highly isolated concentrations of poverty in central cities. Myron Orfield, a former Minnesota state legislator who has studied this problem nationally, finds a disturbing growth of "extreme poverty neighborhoods" in which more than 40 percent of residents are below the federal poverty line and "transitional poverty neighborhoods" where 20 to 40 percent of residents are in this category.[15] William Julius Wilson has also written extensively on the plight of the "truly disadvantaged" in central city neighborhoods where decent jobs no longer exist.[16] This sort of concentrated poverty leads to social isolation and a wide variety of problems that reinforce one another, structurally entrenching a situation of inequality.

A related area of inequity in US urban areas has to do with the provision and distribution of affordable housing. Affordable units have been in woefully short supply in many metropolitan areas in recent decades, since for-profit housing developers prefer to build middle- and upper-income housing rather than low-income units. This general deficiency is aggravated by the fact that many local governments, following local prejudice and political sentiment, actively resist accommodating lower-income populations (which are often members of minority racial or ethnic groups). In an old practice called "exclusionary zoning," many cities and towns have zoned their land for large-lot or single-family development, thus ensuring that only relatively pricey housing is created. Meanwhile, such cities resist zoning for multifamily units such as apartments or condominiums, which often provide cheaper rental housing for minorities. Courts have ruled such practices illegal, but many cities still resist zoning to create housing for a diverse population. American housing policy has also strongly favored homeownership over rental housing (which better meets the needs of the truly poor), tends to segregate affordable housing units in a limited range of locations, and often fails to provide the necessary community services, social services, amenities, or transportation to make affordable units become part of functional neighborhoods for residents of all income categories.

Inequities are also perpetuated these days through NIMBYism. Even though many studies have shown that well-designed, scattered-site affordable housing projects do not decrease surrounding property values, existing residents often fight them based on that fear as well as a general dislike of others different from themselves. Such attitudes often accompany, or are camouflaged by, opposition to any building type that represents higher density than single-family homes. Many planners and elected leaders are easily swayed by Nimbys and fail to stick up for affordable housing or other appropriate forms of development. Cities also fail to allocate funds to subsidize housing for the poorest of the poor, and often seek to meet state or regional affordable housing requirements through senior housing (on the theory that low-income senior citizens are acceptable to local neighborhoods).

Unequal distribution of transportation funding is a further source of inequities within urban regions. Federal transportation money is funneled through state and regional agencies that have often favored freeways or commuter rail systems that serve suburbia over forms of public transit that serve central cities. Increasingly advocacy groups have challenged such policies on civil rights grounds. For example, in the mid-1990s the Los Angeles Bus Riders Union sued that region's Metropolitan Transit Commission over its policy of funding enormously expensive subways and commuter rail instead of cheaper bus service serving low-income residents. In a court settlement the agency agreed to increase bus service and reduce fares. Transit riders in New York have waged a similar campaign. In Washington, DC, community activists complained bitterly that Metro's Green Line subway serving predominantly black northeast DC and Anacostia was the last major line to be built. In the San Francisco Bay Area, social justice advocates staged demonstrations in the late 1990s at the Metropolitan Transportation Commission over its policies of fully funding freeway construction but not bus transit serving central city populations. The agency relented and shifted $375 million within its Regional Transportation Plan to meet that purpose.

US federal spending of many sorts throughout the twentieth century was highly skewed in favor of suburban sprawl—favoring relatively well-off groups within society—rather than the preservation and restoration of urban centers. The interstate freeway system represented a massive subsidy for suburbanization, opening up millions of acres of land around cities to sprawl development. The federal tax deduction for home mortgage interest likewise favors suburban homeowners, who typically have larger mortgages and larger incomes from which to deduct the interest. This provision gives nothing to renters, who are primarily middle- and lower-income individuals and often live in older urban areas. Although this subsidy for homeownership has had some advantages in terms of allowing middle- and working-class families to own their own homes, its overall equity implications appear negative, and its impacts in terms of land use have been disastrous. Federal spending on military production and large-scale waterworks likewise have fueled the growth of suburban, Sunbelt areas such as Los Angeles, San Diego, Atlanta, and Phoenix, while older, more urban industrial areas in the Northeast and Midwest have suffered from disinvestment. The 1980s deregulation and subsequent 1990s federal bailout of the savings and loan industry subsidized a massive, unnecessary construction of suburban office buildings and malls; the lack of mortgage derivative regulation in the 2000s helped fuel residential overbuilding in suburbia. And so on. The inequities implicit in such federal policies in the US and their effects on urbanization generally have rarely been acknowledged by decision-makers.

Within most countries inequities have been built into public decision-making processes. Historically many lower-income or minority groups have not been involved in these processes, have not had the skills or knowledge necessary to participate, or may have had more pressing concerns such as earning a living and surviving. Information may be presented in technical language that ordinary citizens have trouble understanding, especially immigrants who speak different languages at home. Public meetings or hearings on development are frequently held during daytime hours, when many working individuals cannot attend. Many advocates or lower-income residents can simply not afford to spend countless unpaid hours trying to affect public decisions. In contrast, developers are in effect paid for their own extensive involvement (through making profits off subsequent development). Upper-income communities frequently have the time, experience, political contacts, ability to litigate, and access to decision-makers necessary to affect policy. They also may have a sense of empowerment that lower-income communities of color do not— they know they can affect the political process, and may have prior experience of doing so.

A final main area of inequity within US urban development is encompassed by the term environmental justice. This movement initially called attention to ways that minority groups or communities of color are disproportionately exposed to toxic chemicals, pollution, and unwanted land uses such as dumps and incinerators. It has since expanded to include other urban planning subjects such as the lack of parks and recreational facilities within lower-income communities, disparities in transportation services, and the need to restore inadequate infrastructure.

The environmental justice movement has its roots in the 1960s, when citizens' groups first publicized problems such as inner-city children eating lead paint chips and Native American communities in the Southwest suffering from radioactive uranium mine tailings. The movement spread in the 1980s as awareness grew of how communities of color were exposed to a wide range of hazards and many activists realized that affluent white constituencies dominated the mainstream environmental movement and the staffs of national environmental organizations. One of the first organized actions came in 1982 when community groups rallied opposition to a proposed PCB landfill near African American neighborhoods in Afton, North Carolina.[17] More than 500 people were arrested, including Dr Benjamin Chavez, the former director of the National Association for the Advancement of Colored People, and Dr Joseph Lowery, of the Southern Christian Leadership Conference.

A national conference in New Orleans on toxic substances and minorities the next year was the first major effort to systematically link discrimination, environment, and social justice. A 1983 General Accounting Office study, done at the request of Rep. Walter Fauntroy (D-DC), found that three out of every four landfills in the southeastern United States were located near predominantly minority communities.[18] In 1987, a study of 415 hazardous waste sites by the United Church of Christ and the Commission for Racial Justice found risks to minorities were nearly double that to whites, further cementing the link between environmental risk and minority communities.[19]

In 1990 a group of non-Anglo activists sent a letter to the Group of Ten CEOs of national environmental organizations alleging "racism" and "whiteness" of the environmental movement. A second letter soon followed from the Southwest Organizing Project, signed by more than 100 activists and community-based groups alleging that people of color were "the chief victims of pollution." At the time it was found that there were no African Americans or Asian Americans and only one Hispanic among 250 Sierra Club staff, and only five persons of color among 140 staff members of the Natural Resources Defense Council (NRDC). These organizations soon took steps to remedy this situation, and two years later the National Wildlife Federation claimed that 23 percent of its staff were members of minority groups.

In 1991 the First National People of Color Environmental Justice Conference brought leaders of the growing movement together in Washington. A second such conference was held in 2002. The Clinton Administration responded to environmental justice advocates by releasing Executive Order 12898 on Environmental Justice in 1994, setting out "Federal actions to address environmental justice in minority populations and low-income populations." This executive order required each federal agency to make environmental justice part of its mission and to identify strategies for achieving it. An additional executive order in 2000 improved access to federally funded or assisted programs for persons with limited English proficiency.

Other steps to improve equitable access to governmental decision-making processes in the US have been taken a number of times since the 1960s. Title VI of the federal Civil Rights Act of 1964 laid the groundwork by stating that "No person in the USA shall be excluded from participation, denied benefits, or subjected to discrimination in any federally-funded program, policy, or activity on the basis of race, color, or national origin." The National Environmental Policy Act of 1969 required public participation in determining environmental impacts and alternatives to proposed action. The Intermodal Surface Transportation Efficiency Act of 1991 (ISTEA), the Transportation Efficiency Act for the 21st Century of 1998 (TEA-21), and other legislation mandated effective public participation within specific local or regional planning processes. Meanwhile, many local governments have enshrined public participation within their general plans or master plans, and have gone to considerable lengths to carry out neighborhood planning, notify citizens of proposed planning decisions, and incorporate feedback. But all of these programs represent only a start towards the difficult goal of involving under-represented communities in urban-planning-related decision-making.

Efforts to improve interjurisdictional equity in terms of tax resources have proven even more difficult. The main device proposed is regional tax-sharing, but to date this has only been implemented in one US metropolitan area, Minneapolis–St Paul, where since 1974 40 percent of the increase in sales tax revenue has been put into a regional pool distributed by population. A similar initiative was debated for the Sacramento region in 2002. Elsewhere tax-sharing has had little political support. Still, it is a device increasingly discussed as a means to level out resource disparities.

Courts have intervened in recent decades to mandate that many state governments to some extent equalize school funding between jurisdictions. These legal decisions have resulted in state "equalization" programs in California, New Jersey, and elsewhere to provide some base level of funding for each pupil. Also, the distribution of "community development block grants" (known as CDBG funds) from the federal government to cities to some extent promotes equity, in that these funds are allocated on the basis of population and are a remnant of larger federal "revenue-sharing" programs first instituted under President Richard Nixon in the early 1970s. In general any strategy that collects taxes at a higher level of government—such as state or federal government—and then redistributes revenues to cities and counties on the basis of population represents a way to overcome entrenched inequities in resources between local jurisdictions.

Programs to ensure regional fair-share housing provision likewise form a mechanism to improve equity. In California, the state Department of Housing and Community Development requires cities to update their General Plan Housing Elements every five years to accommodate "fair-share" amounts of housing in different income categories, as determined by formulas developed by regional agencies. Courts in states such as New Jersey and Connecticut have likewise required cities and towns to accept affordable housing. In practice, however, localities remain very resistant to doing this and often fail to comply with such mandates. Unfortunately, relatively few penalties exist to compel compliance. Another strategy is for higher-level governments to offer incentives for local fair-share housing compliance. For example, the state of California has at times offered unrestricted grant funds to cities that exceeded a certain percentage of their past affordable housing construction. Meanwhile, some local governments have adopted "inclusionary zoning" requiring all large development projects to include a certain number of affordable units, generally 10 to 20 percent. Such mechanisms have the potential for increasing the equitable distribution of housing options.

These are a few of the main ways that planning can help address the equity goals so vital to sustainable development. The big question is how equity objectives are to be advanced in the face of a society whose electorate and political and economic leadership tend to be uninterested in them. Developing equity initiatives may require planners and other professionals to adopt proactive or advocacy roles—speaking up for under-represented constituencies, reminding decision-makers of the interests of groups not represented at the table, working to help minority and lower-income groups become more familiar with decision-making processes and able to articulate their viewpoints, and so on. Much advocacy on behalf of equity has also traditionally been done by local organizing groups such as ACORN, by labor unions, and by local community development corporations (nonprofit organizations run by local residents and dedicated to neighborhood improvement). Calls for "equity planning" have been sounded within the planning profession for decades, most notably by Davidoff and Norm Krumholtz, former planning director of Cleveland. Many practicing planners are deeply sympathetic to equity concerns. Figuring out how to inject them into planning debates more systematically and successfully will be an essential element of the sustainability planning agenda.

Many of the more visionary planning pioneers during the last century sought in their own ways to reconcile goals of environment, economy, and equity. The influential British garden city theorist Ebenezer Howard, for example, is well known for his vision of a balance between city and countryside, but he also sought to integrate equity concerns into his garden cities by having collective land ownership and social organization. He went so far as to work out the economics of how residents would jointly purchase land and build housing. This dimension of his work was overlooked by many of his followers.[20] Mumford likewise paid attention to all three themes, as have more recent authors such as Kevin Lynch and Jane Jacobs.

But too often in practice these objectives have become separated. Economic development specialists have assumed that any form of new business development would help the community—without taking environmental impacts into account or considering the nature and wage levels of new jobs. Local environmentalists have often bought into "slow growth" movements without realizing that without accompanying efforts to promote affordable housing these would exclude lower-income residents and generate inequities. And some equity advocates have brushed aside environmental considerations in the search for new development for their communities. The task ahead for sustainability planning is to figure out creative strategies for reconciling these perspectives, simultaneously meeting all three sets of objectives in the context of particular places.

5

SUSTAINABILITY PLANNING
IN PRACTICE

Given the general characteristics of sustainability planning discussed previously—a long-term perspective, a holistic outlook, active involvement in problem-solving, recognition of limits to growth, and acknowledgment of the importance of place—what are the hallmarks of sustainability planning on a day-to-day level? Needless to say, a great many different approaches are possible. Each of us has our preferred style and operates within a unique context. But a few sustainability planning strategies stand out. Without these applied skills, we run the risk of perpetuating what has come to be known as BAU (business-as-usual). With them, we can potentially bring about much more positive and sustainable futures.

Although the following principles may sound simple, they are not. In fact, the seemingly common-sense appearance of the sustainability concept overall has been one of its main drawbacks from the beginning. Almost everyone thinks they know what sustainability means, and on the surface it does appear that sustaining human civilization is a straightforward goal. However, the reality of helping do this every day is not at all easy. Being an effective sustainability practitioner means carefully honing approaches and skills over many years, learning from experience as you go, and constantly pushing yourself to better weave together professional skills, bodies of knowledge, strategic approaches, and methods of collaboration with others.[1]

UNDERSTANDING CONTEXT

The ecological worldview and holistic approach outlined in Chapter 2 emphasize the extent to which any situation is embedded within a dynamic, richly interwoven context. A planning issue, building, park, transit system, economic development initiative, or social policy does not stand alone, but fits into a complex and constantly changing world. When contemplating a particular action, the challenge is to understand as much as possible all the systems and interactions that this endeavor fits into so as to design the best possible intervention. The process of figuring out sustainable solutions depends heavily on such contextual understanding. Otherwise, actions may meet some goals but not others, may distract from more important steps that need to be taken, and may even be counterproductive in terms of overall progress toward sustainability.

Traditionally, professionals have focused on certain aspects of the contexts within which they work but not others. Architects have paid great attention to aesthetic and formal aspects of their buildings but far less to their environmental performance or the likely experience and health of building users and neighbors. Engineers have focused on the performance and capacity of infrastructure systems, but not on their character or environmental impacts. Economic policy-makers have analyzed the number and types of jobs created by different potential economic development strategies, as well as monetary flows through cities and urban regions. But often they have failed to pay attention to the quality of those jobs, the character of the resulting communities, the ways that their efforts were promoting an overly materialistic economy, or the extent to which economic development channels wealth and power into the hands of some groups as opposed to others. Highly constrained understandings of context, in other words, have often led to unsustainable development.

Although deeply understanding any given context is a challenging and ongoing process, this ability can be nurtured within education at all levels, and assisted through various formal methods and tools. On a conceptual level, the Three Es are a tool to help people integrate environmental, economic, and social equity dimensions of any given context. So are various diagrams such as Robert Thayer's "sunflower" graphic in Figure 5.1.

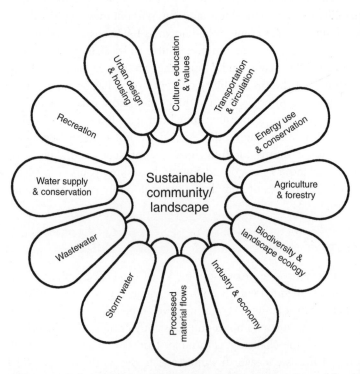

Figure 5.1 A "living systems" approach. In his community and landscape sustainability work, Rob Thayer weaves together many dimensions of a given context

Many past initiatives have sought to improve ways that professionals understand the contexts within which they're working. In urban design and landscape architecture, the overlay method of site analysis pioneered by Ian McHarg in the 1960s constituted one such method. McHarg and his colleagues attempted to study a wide range of environmental characteristics of a given site by laying clear Mylar sheets on top of a base map. These practitioners created different overlays for hydrology, soils, vegetation, fauna, slopes, geology, and other physical characteristics, and often determined what they thought was the ideal site for development by putting layers of constraints on top of one another and seeing what sites were left. Nowadays similar analysis is performed on computer using geographic information systems (GIS) software. The result can be a commendably thorough look at environmental characteristics of a given place. But this method has often missed out on social, economic, and aesthetic dimensions of analysis, and like any analytic tool may have led to a misplaced certainty that the context had been thoroughly studied. This method was also fundamentally incapable of asking questions such as whether any development on the site was desirable at all, given broader social, economic, or policy contexts.

Environmental review legislation, such as the National Environmental Policy Act in the United States and subsequent state environmental policy acts, has also required since the 1970s a relatively contextual evaluation of proposed projects. In addition to traditional environmental impacts these frameworks often require public agencies and in some places private developers to consider traffic, historical preservation needs, archeological resources, population, housing, recreation, and cultural resources. Agencies compile and analyze this information within environmental impact statements, reports, and assessments. In practice, however, such study often focuses on some dimensions of situations much more than others. Traditional ecological systems are usually well-studied, while reviews skirt politically sensitive questions related to social equity or economic strategy. Only after considerable effort have advocates been able to ensure that "new" topics such as greenhouse gas emissions are considered within these forms of mandatory environmental review.

Green design rating systems such as LEED in the US and BREEAM in the UK are yet further tools to promote a highly contextual analysis of any development project. These analytic frameworks require the developer to quantitatively evaluate many dimensions of the project, including site location, relationship to public transportation, ecosystems, water use, energy consumption, and indoor air quality. Green rating systems require a much more contextual analysis than McHarg's earlier generation of site analysis methods, but still leave many elements out. For example, it is inherently difficult to evaluate the experience of future users of any building or neighborhood, and the degree to which design will improve social connections and community. Some community-promoting elements can be required, for example a walkable streetscape, but the overall synergy of such components is difficult to determine. Early versions of LEED were also criticized for short-changing global warming and public participation topics. Still, green rating tools have expanded consideration of context in many ways, and their creators can potentially learn from experience and revise the systems periodically to better include other dimensions of development contexts.

The overall situation, then, is that much development decision-making has become more contextual over the past 50 years due to the rise of environmental evaluation tools and ecological mind-sets generally, but that it still has a long way to go. Any individual tool will be limited. What is most important is that sustainability professionals develop and continuously refine their own skills of understanding context rather than rely too much on any external method. It is crucial that we help each other with this process and be open to continually incorporating new understandings, new insights, and new ways of satisfying sustainability goals across contexts. Such learning processes are collective and dialogic rather than individual and solitary. It is particularly important to help clients, communities, and political decision-makers expand their own understanding of contexts, since they will be making many development decisions, and to create an institutional framework that supports such contextual understanding.

ESTABLISHING PRIORITIES

A related professional skill is to be able to focus on the sustainability priorities within any given situation. Time, energy, and resources are always limited, and we must constantly make choices about what to focus on. These strategic choices depend on a good understanding of contexts and coordination with others to see how resources can be mobilized. Good choices can lead to inspiring, cutting-edge actions. Poor choices may mean that a golden opportunity for improving sustainability has been missed, and resources wasted.

Projects that inappropriately emphasize one or two dimensions of sustainability at the expense of others may be counterproductive, misleading the public about what is needed, or convincing audiences that small, relatively ineffectual steps are enough. Many builders, for example, have put solar panels on large suburban houses to advertise them as "green" or "sustainable." Buyers may have rationalized their purchase decisions on that basis, without considering that the large size of these structures raises energy consumption and that their exurban location will require large amounts of driving. In the same manner consumer products industries often tout minor improvements in packaging, recycled content, or energy efficiency as proof that their products are ecologically appropriate, without thoroughly examining the whole chain of materials and processes that has gone into producing, distributing, and marketing the product.

The ability to see quickly the sustainability opportunities and priorities within any given situation is an extremely valuable skill that can be emphasized within education, professional practice, and everyday life. To a certain extent standardized tools can help with this process of prioritization. Sustainability rating systems such as LEED assign different point totals to different potential credits, reflecting the priorities set by a consensus of experts. For example, the location of a project, which affects many different environmental and social performance dimensions, receives a much higher weight than a more limited credit such as urban heat island reduction. But such weightings are always debatable, may fail to reflect regional or contextual characteristics, and may change over time. For example, reducing GHG emissions is now seen as a far greater priority than when the LEED system was first developed in the mid-1990s. In the end what is important is that each of us develops a good understanding of how to set priorities within any situation we are involved in, and continues to improve that understanding throughout our lives.

Priority-setting involves a careful weighing of opportunities, constraints, the scale and time frame of impacts, the various communities involved, and other contextual factors.

Seeing opportunities is one of the most exciting parts of sustainability work, since these can lead to exciting projects, dynamic collaborations, and rapid expansion of our collective sense of what is possible. But understanding constraints is essential as well, if only to avoid foreseeable disappointments and to figure out creative ways around barriers. Thus sustainability planning becomes a challenge of "pragmatic vision" or "visionary pragmatism"—a careful balance of pushing the envelope to pursue opportunities but staying just within the realm of feasibility. Of course, conceptions of feasibility change over time, and can be altered through strong leadership, compelling presentation of alternatives, coalition-building, and creative salesmanship, so this balance is never black and white. Rather, it is a continual dance of pushing what is possible.

Understanding the scale and time frame of impacts as well as the likely communities affected is also essential to priority-setting. In some situations it is best to go for limited near-term benefits; in others it's best to hold out for bigger but longer-term and more risky achievements. In some situations a project that will benefit just a small site or community is worth pursuing; in others it's important to affect many other communities at a greater range of scales. Experience, skill, and dialogue with others are all important in making such decisions.

For example, on some development sites, creating affordable housing may be the top sustainability priority, since workers in that community cannot find appropriately priced homes near their workplaces, and as a result have to drive long distances to get to and from work. The resulting traffic, air pollution, greenhouse gases, and lack of settled community are sustainability problems that make this land use a priority. Within other situations ecological restoration may be the top priority. City staff or a nonprofit organization sees, for example, that an old industrial site on a city's waterfront could be turned into both a restored wetland and a park that will become a centerpiece of outdoor community in the downtown, and is able to excite and organize constituencies around this possibility. On still other occasions a particular constellation of local skills and resources may present an opportunity to develop a new set of green businesses, which can likewise become a development priority. In each case the sustainability professional will be able to call people's attention to potential sustainability priorities, helping achieve the best possible vision for the future.

In any given context some sustainability actions may be vastly more important than others. Climate change, for example, has become an urgent priority in recent years (the issue of course has long been urgent, but for decades was overlooked or relegated to back-burner status even by many environmentalists). Since the future of humanity and many other species may be at stake, the need to mitigate greenhouse gas emissions and adapt to a changing climate should be a high-priority consideration for any project or policy currently, no matter what the local context. Likewise, a case could be made that social equity is a high-priority concern in many situations these days, given many decades of rising inequality and concentration of power in a few hands.

Other priority-setting is often less clear. Healthy, locally oriented food systems have become a goal for many activists in recent years, and who could argue against programs to support urban agriculture and local organic farming? However, the exact balance with other urgent priorities is open to question. It is by no means easy to develop efficient, reasonably economical food production on small parcels within cities. Though it may be worth trying anyway for symbolic or educational reasons, other urban needs for affordable housing, public gathering places, children's playgrounds, or outdoor recreation may be overlooked if too high a priority is set on this one goal. Figuring out how much to balance such priorities in any given time and place is a challenging process.

Different stakeholders within any debate will certainly have different goals, and an extensive process of dialogue and mutual understanding may be necessary to balance priorities. Sustainability professionals can help facilitate this process, while bringing into the picture other points of view that may not be represented, such as the needs of distant communities, future generations, or ecosystems. Developing common goals and priorities between groups is tough, and requires a deep knowledge of current needs and conditions, as well as different communities' points of view and opportunities for creatively meeting multiple objectives at once. But these understandings and skills can be improved over time, and are an important part of sustainability planning practice.

THINKING STRATEGICALLY

If the context of any situation is well understood and sustainability priorities are identified, the need becomes actually getting things done. This is far from easy. We live in large-scale societies with powerful forces wedded to continuing business-as-usual. Existing political systems often function badly in terms of addressing sustainability issues. People must be organized, organizations set up, resources found, laws changed, allies recruited, and the media cultivated. The obstacles often seem overwhelming.

Although quick, inspired action is important at times, the bigger need is usually to think strategically about putting the pieces in place for larger-scale change in the future. The point after all is to build a long-term campaign for a sustainable world. As with many political movements, techniques may include organizing drives, political coalition-building, the spreading of educational materials, campaigns of civil disobedience or protest, litigation, and/or lobbying efforts. During the twentieth century the labor movement, the women's suffrage movement, the civil rights movement, the environmental movement, and various peace movements employed such tools. But most importantly they developed strategy for themselves through which to build institutions, legislative change, and public awareness over time. More recently new, more rapid tactics have emerged, for example using social media to help alternative messages snowball quickly or go viral, or to quickly organize demonstrations and political theatre. But here too long-term strategy is essential if a political movement is to be long-lived and influential over many years.

Whatever the tools, a clear, strategic vision is essential. This can be codified in a written game plan or developed through the intuition and agreement of leaders. However, the more transparent and specific we can be about our long-term process of achieving sustainability goals, the better. Good questions to start with include: What are we trying to achieve? What are the pieces that must be put in place to get there? What is the message for the public and the media? What are potential ways around the inevitable roadblocks? What allies can be recruited? How can leverage be gained against powerful existing interests? Such questions are the foundation of strategic planning.

Strategic thinking can be applied at multiple scales and time frames at once. A given organization may have a long-term strategic plan, developed through a facilitated group process and agreed to by consensus, and be planning to take certain actions to meet its goals. Its near-term tactics—specific programs, actions, and daily organizing or advocacy—fit into that strategic framework. But it can show some flexibility if a different opportunity arises that also meets those goals. The point is not to be rigid or overly linear about one's course of action, but to get used to working backwards from long-term goals to build a framework for successful action.

That being said, progress toward sustainability in any given context is not always possible. Part of operating strategically is to figure out when it's not worth butting one's head against the wall, and to look for other times, places, issues, or methods in which to work. On an organizational level, this may mean rethinking mission, activities, and structure. On an individual level, a new assessment of obstacles may lead us to acquire new skills, knowledge, and connections that will be useful in different ways in the future, or to move to a different place where the context will be better. We do not have to know up front all the answers as to how our goals will be achieved. But we do need to hone intuitions and understandings of how various building blocks might fit together to promote progress in creative new directions.

WORKING COLLABORATIVELY

Although it is easy to think of ourselves as independent individuals, we are profoundly social creatures. So bringing about more sustainable societies requires working through organizations, networks, media, institutions of government, and other individuals—in short, collaborating with others in many different ways.

Sustainability planning depends on some specific types of collaboration skills. One of these is the ability to communicate across disciplines or professional boundaries. Interdisciplinary or transdisciplinary work is essential in order to achieve the multiple goals of any sustainability project, but is often difficult for those schooled in many traditional professions. Training in such fields emphasizes specialized terms, methods, and ways of thinking, and often implicitly or explicitly disparages other ways of looking at problems. Engineers and designers, for example, are often trained in radically different ways, and historically have often been antagonistic to one another. Designers typically view engineers as hopelessly bound by rigid standards and technocratic ways of doing things, while many engineers see designers as unaware of the practical aspects of materials and processes. Needless to say, each perspective has many valuable contributions to make to the other, and ability to understand both points of view and communicate across such professional dividing lines can help achieve the best long-term results.

Another type of collaboration concerns the development of strategic partnerships to achieve long-term goals. Such coalition-building and networking is particularly important within long-term efforts to reach goals across communities, building social capital as a resource to be drawn on over years of action. Collaboration of progressive organizations is particularly important in order to counterbalance the entrenched power of economic and political interests opposed to sustainability—oil companies, wealthy elites, conservative media, and the like. Making progress toward sustainability in the light of such entrenched interests depends on our ability to work together to neutralize them.

A final type of collaboration important to sustainability planning is the ability to develop working relationships with organizations and individuals that would otherwise be antagonistic. Wealthy societies tend to be full of well-financed and well-connected factions that are frequently at odds. This is also a characteristic of postmodern culture, in which societies fragment into a profusion of interest groups with different and often conflicting agendas. At a local level such trends can lead to Nimbyism, in which residents are hypervigilant about their own perceived rights being trampled on and often oppose any sort of progressive change. At higher levels of government these dynamics may lead to political dysfunction and paralysis, especially if different factions of society are focused on enhancing or preserving their individual rights without a corresponding emphasis on their collective responsibilities to one another. Getting beyond such impasses can require a wide range of leadership, facilitation, and collaboration skills. Of course, if politics becomes too oppositional then collaboration becomes impossible, and it becomes important to think about more radical changes to the system.

EVALUATING PROGRESS

A key part of sustainability planning is to constantly evaluate how much progress is being made, and to revise strategies if necessary to move faster. Not for nothing have indicators and benchmarks been a part of the sustainability movement since early on. Having strong, clear, easily understood ways to measure progress is essential. Such indicators serve a number of functions: to help guide the actions of participants; to remind everyone of ultimate sustainability goals; to give important feedback as to whether strategies are working or not; and to use as leverage against those who would undermine sustainability efforts.

The last point bears further explanation. Sustainability endeavors are often opposed by a wealth of short-term, more-or-less self-interested political forces, or simply by inertia within governmental or organizational systems. Calling attention to the need to reach an agreed-upon goal such as reducing greenhouse gas emissions or improving educational attainment can be a way to break political gridlock, reasserting the importance of long-term goals in the face of short-term interests. Particular indicators must be used carefully—emphasizing simple improvement in standardized test scores, for example, has been a counterproductive initiative to improve education in the US. But especially for basic measures of environmental health and social welfare, systematic ways to evaluate progress can be very useful in moving towards sustainability. Chapter 6 will further examine tools such as sustainability indicators, ecological footprints, and carbon calculators.

PUSHING THE ENVELOPE

To finish this discussion of day-to-day strategies, I want to come back again to the need to push the envelope, that is, to expand the boundaries of what is possible. Far too many individuals and organizations that are supposedly planning for the future have instead proceeded cautiously along the same paths as their predecessors have done for decades.

Moving to a sustainable society requires bold new thinking and visions. We need to excite and inspire people, and leverage our own creativity and energy into a broad and rapid movement. Going to work every day and doing exactly what we're told to isn't going to get us there. Existing responsibilities must be taken care of, but at every step we need to look for opportunities to do things differently.

Pushing the envelope means constantly pushing for better alternatives to existing ways of doing things, asking questions like "What if we did this?" "Wouldn't this other approach work better?" "Which would be better in the long run?"

Or we can simply ask "How do we get from here to there?" and begin a discussion with colleagues about how to launch an action, develop a demonstration project, or make an organization more sustainable. The success of such initiatives may depend on all the skills discussed earlier in this chapter. But the important thing was that right at the beginning we asked how things could be different. Getting the conversation going in that direction is the essential starting place.

6

TOOLS FOR SUSTAINABILITY PLANNING

Given that sustainability planning aims to bring about major changes in a range of areas, how can this be done? What methods might be particularly useful to planners, political leaders, or activists? Possible options include many traditional urban planning strategies—if handled in creative and proactive ways—as well as some new tools that have been developed particularly with sustainable development in mind, such as sustainability indicators, green development rating systems, and ecological footprint analysis. Educational and consensus-building processes, as well as old-fashioned political organizing and coalition-building, are also important.

The "toolbox" available to planners is increasingly large and varied. New mechanisms are invented almost daily. Each has strengths and weaknesses. Rather than becoming too entranced with particular strategies for their own sake, what is most important is to pick and choose appropriate methods for each situation, and to innovate or combine methods when appropriate.

The usefulness of tools depends in large part on what initial research questions and processes are guiding them, and on how time- and resource-efficient they might be. Some tools, such as complicated computer models of traffic generation, may fall relatively low on the scale of usefulness for sustainability purposes, or may be misused currently in ways that insert unquestioned assumptions into the decision-making process (such as that smooth-flowing traffic is always good) which obfuscate issues, or that distance the public from decision-making. Such methods must be rethought, and perhaps laid aside in favor of others. But such instances are probably relatively rare, and even then the tool itself may not be to blame so much as how it is used.

PLANNING PROCESSES

Plan-making

The most traditional planning tool, by definition, is the simple construction of a plan, that is, a well-supported collection of strategies for achieving desired results in the future.

Plans can be constructed at any level of government, though at local scales they tend to include more details of land use, transportation, economic development strategy, and the like, and at larger scales they tend to take the form of broader lists of goals, shared principles, and commitments to action. Much literature exists on the history and nature of planning. There are various models of this process, which usually include research and information-gathering tasks, analysis, generation of alternative future scenarios, and selection of a preferred scenario. Recently, public participation has become a priority as well. Traditional models include the famous "survey–analysis–plan" dictum of Patrick Geddes early in the twentieth century, and the later comprehensive rational planning model which is often seen to include four steps:

1. defining goals;
2. identifying obstacles to them;
3. identifying alternative solutions; and
4. comparing their merits.[1]

Specific types of plans particularly useful for sustainable development can include national "Green Plans" setting out environmental and development policy, regional plans to manage urban growth and/or develop transportation systems, municipal plans to regulate development of a particular area or the city as a whole, and neighborhood plans to establish specific zoning and design standards for a neighborhood. Development approval processes, through which governments approve specific projects supposedly in accordance with these plans, and creation of new programs and policies are ways to implement the principles that planning documents establish. Plan-making plus systematic implementation might therefore be seen as the essence of the urban planning field.

At the local level general plans (also called master plans or comprehensive plans) establish an overall vision and policy framework for a municipality. US states typically require each city or county to have such a plan and to update it regularly, and often specify key elements that must be covered, such as land use, circulation (transportation), housing, conservation, open space, noise, and safety. Cities may choose to add other elements as well, for example to cover urban design, economic development, environmental protection, and resource use. Though most of the plan consists of written goals, policies, and analysis of existing conditions, general plans usually contain a land use map for the city specifying areas for different land uses and densities. Such a map is in essence a physical development plan for the city and forms the basis for more detailed zoning regulations.

There has been a continuing debate in academic planning circles about the usefulness of general plans. Pushed strongly in the mid-twentieth century by figures such as Jack Kent at the University of California at Berkeley and Harry Chapin at the University of North Carolina as a vehicle to generate a consensus blueprint for a city, the general plan has been attacked by others as a formalistic exercise producing vague language with little effect on actual urban development. Critics point out that cities routinely amend their general plans to accommodate developers, and often fail to consult these documents when developing new programs. Originally oriented around specific visions of physical form in the early twentieth century, general plans frequently became more abstract and technocratic toward the middle of the century, with less power to galvanize public opinion or direct spatial development. However, toward the turn of the millennium such plans began to emphasize physical planning and urban design again, a hopeful sign for sustainability planning that depends heavily on a focus on physical place.

General plans vary widely in quality and effectiveness. Some consist of boilerplate rhetoric with lofty goals but little linkage to specific policies and programs. Such plans gather dust on shelves and are rarely referenced within planning debates. Others do contain good specifics but are simply not followed in practice. State legislation requiring consistency between plans and development practices can help in this regard. But potentially general plans can serve a powerful role in developing consensus around sustainable development directions, setting forth specific policies, and holding politicians accountable for achieving agreed goals.[2] As Michael Neuman points out in a classic article on the usefulness of the general plan, strong plans can help to portray collective hopes about the future of a city or town, allow necessary political conflict to emerge, build social, intellectual, and political capital within communities, and set agendas for powerful public agencies.[3]

Specific plans (also called area plans, neighborhood plans, sector plans, or precise plans) develop a planning vision for a particular area within a city, such as for a downtown, a transit station area, an older industrial district, or a neighborhood. Cities often develop these more focused plans with an intensive public process including workshops, meetings, and design charettes (workshops in which groups of participants develop potential designs for urban places). Specific plans may include a detailed land use vision at a parcel-by-parcel scale, particular economic development strategies, and recommended zoning changes and urban design guidelines to help bring about desired forms of development. Within a sustainability planning framework, specific plans provide a crucial mechanism through which cities or other public agencies can involve the public in developing a vision of how a particular place should develop. Although specific plans may cover broad areas within the city, increasingly cities are using much more narrowly focused versions to help guide revitalization and development of smaller areas, such as around transit stations or along key commercial corridors which can be retrofitted as more pedestrian-friendly, mixed-use areas (see Figures 6.1 and 6.2).

Functional plans develop a planning vision for a particular issue or topic area within a city, offering an opportunity for the city to explore sustainability planning directions in depth. Some of these policy documents may eventually be incorporated into general plans, or may amplify elements of existing general plans. Functional plans may be drawn up in a wide range of issue areas such as bicycle planning, transportation demand management, energy policy, urban design, recycling, economic development, or parks and green spaces. Many local jurisdictions worldwide have now prepared climate action plans, detailing steps to reduce greenhouse gas emissions and adapt to a changing climate; these documents are simply another form of functional plan created to meet a newly appreciated need. Sometimes state governments mandate that local jurisdictions develop particular types of functional plans. The state of California, for example, requires cities to update housing elements of general plans more frequently than other elements, and so these documents can be considered free-standing functional plans. States such as Oregon and Washington have required cities to develop growth management plans that establish UGBs; these may amend, replace, or augment portions of existing general plans.

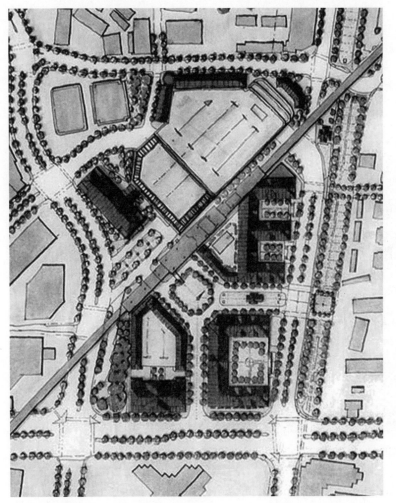

Figure 6.1 Plan for infill development at a transit station

Implementation

Far too often excellent plans are prepared but sit for years without being implemented or acknowledged within decision-making processes. Plan preparation is only part of the battle; the most important work is often to get the goals envisioned in the plan brought into reality. Whether or not a plan is implemented depends on many things: whether funding is available, whether staff time exists to work on the desired changes, whether different departments within city government or other agencies will cooperate with each other to get things done, whether elected leaders will pay attention or follow through, whether public interest can be sustained and opposition defused, and whether planners can keep the goals and commitments represented by the plan in front of everyone's eyes. Some of these things are beyond the control of local sustainability planners and advocates. But others are open to influence.

Figure 6.2 Perspective view of transit station development

Many steps taken during the plan-making process itself can help ensure that plans achieve their intended results. A plan that takes existing resources into account, that builds consensus between constituencies, that gets decision-makers and the public excited about future possibilities, and that includes specific budgeting or legally binding changes in city zoning code or laws stands a much better chance of being successful in the long run. Active efforts in such areas by planners, decision-makers, and members of the public will be necessary to ensure that sustainability-oriented plans achieve their desired result.

Visioning

Even before the creation of planning documents, the development of long-term, wide-ranging, and creative visions of alternative futures is important to help decision-makers and the general public understand that there are choices in how we develop our communities and society. Going beyond the general goals typically set forth in plans, such visioning can take place through the production of vision statements or reports, through graphic images that illustrate alternative futures, through the manifestos of particular groups, through films, through design charettes and public workshops, or through various other means.

Many of the most influential planning movements historically have been stimulated by particular visions of ideal urban environments. Howard's *Garden Cities of Tomorrow* (1898) presented a very carefully worked-out vision of alternative physical and social designs, together with diagrams that continue to be widely reproduced more than 100 years later. This vision influenced a whole Garden Cities movement in Britain, the Greenbelt New Towns planning in the US, and many individual planners and politicians. The modernist vision of architecture and community form, codified by the Charter of Athens in 1933, likewise developed an enormously influential set of principles and graphics promoting particular urban designs and policies.

In recent decades more environmental and humanistic visions have helped lay the groundwork for sustainable urban development. Ian McHarg's *Design with Nature* (1969) set out a vision of a more ecologically oriented landscape design process, Jane Jacob's *Death and Life of Great American Cities* (1961) reasserted a vision of traditional urban neighborhoods, and Dolores Hayden's *Redesigning the American Dream* (1984) developed a feminist vision of urban environments. The nonprofit New York Regional Plan Association has developed three visions (1927, 1965, and 1996) of development in the New York metropolitan area,[4] while Urban Ecology prepared a *Blueprint for a Sustainable Bay Area* in 1996 for the California region centered on San Francisco, Oakland, and San Jose. In the late 1990s Herbert Girardet and others developed an initial vision for a sustainable London, taken further in the 2000s and early 2010s as the city prepared for the 2012 Olympics. In the late 2000s Sustainable Melbourne and Sustainable Rotterdam groups developed web-based communications hubs to promote sustainability models, events, and visions related to their respective cities. At the most visionary end, ecocity theorist Richard Register in 1987 developed a book-length 100–200-year vision for the future reshaping of Berkeley, California, in *Ecocity Berkeley*, proposing that development be removed from many ecologically sensitive areas of the city through a program of ecocity zoning and placed instead in dense, walkable urban nodes.[5]

Vision documents can inspire others, inject new ideas into political discussions, educate the public, and spur planning reform. They can also help develop consensus on shared planning goals and how to implement them. Municipal general plans usually start with a vision statement for a particular city or town, although this is typically quite general and often does not mention objectives such as ecological sustainability or social equity. Much stronger, more visual, and more specific visions will probably be required to inspire people with new ideas and motivate them to work for change.

Of course, there is always the challenge of relating visions to reality. They must contain elements that respond to current problems and are broadly appealing, otherwise they may be ignored or dismissed as irrelevant. They must make clear the linkages between visionary goals and practical implementation, so that readers or viewers can see how such dreams might actually come about in practice. They should be written or presented in a way that is accessible to a broad audience and that does not turn people off with arcane language or bizarre concepts. It also helps if vision statements are timed to be maximally useful within particular decision-making processes or to achieve widespread distribution through the media.

Though they rarely consider themselves visionaries, planners can play an important role in making sure that such ideas are considered within debates. One prime place to do this is within environmental impact statements or reports (EIRs), dry documents to be sure, but processes that by US law must present alternatives to a proposed project. These alternatives are then reviewed and contrasted with the proposed project in terms of environmental impacts. EIR alternatives are often weak or trumped-up straw men, leading to much litigation. But strong, meaningful alternatives can help show decision-makers that different strategies exist for any particular project, and are a pragmatic way to insert alternative visions into the planning process.

Review of proposals and alternatives

Once courses of action are identified within planning processes, it is vital to review them and compare them with other alternatives. Environmental review is one way to do this. In some countries this process is known as strategic environmental appraisal; in the United States it usually takes the form of environmental impact statements (required by federal law), environmental impact reports (required by some state environmental policy acts), or environmental assessments (a lower-intensity form of analysis that can be required by either). The heart of such documents is generally the comparison of a project's impacts with various alternatives, including a "no project" alternative. A sizable consulting industry has sprung up to assist government agencies and private developers or corporations in preparing environmental reports. State or federal legislation requires that these documents take a broad definition of environmental impact, including many social impacts. In addition to analysis of land use, air quality, water quality, and hazardous materials they may contain sections on traffic impacts, archeological resources, noise, cultural resources, and employment.

There have been efforts to promote the use of social impact statements as well. The World Bank, for example, uses a poverty and social impact analysis process to appraise projects, including distributional impacts and "before and after" estimates of poverty in the relevant geographic area. Though the process is difficult, the International Association for Impact Assessment is attempting to develop guidelines on social impact assessment that establish standards and models for this sort of analysis.[6]

Cost–benefit analysis is a traditional way of assessing the impacts of any course of action in economic terms. This approach is widely used in corporate and governmental sectors. However, cost–benefit approaches typically minimize or leave out environmental and social impacts; these must be quantified in monetary terms to be meaningful to such models. Environmental economists have worked mightily to incorporate these essentially noneconomic dimensions into cost–benefit frameworks. But the process is unlikely to prove satisfactory in the long run.

Perhaps the most promising strategy is to develop a comprehensive sustainability appraisal of any proposed action that takes social, environmental, and economic impacts all into account. The British Department of the Environment, Transport, and the Regions took some steps in this regard under the Blair Administration.[7] Methodology for this is in its infancy, however, and such appraisal will place substantial demands on staff and decision-makers, requiring them to be able to forsee and understand the full range of impacts of a project. Among other things, such ability may require a broad and sophisticated education, and substantial experience.

Best practices

One way to make visions real for people—and to prove that they are feasible and can actually be carried out in practice—is to publicize existing "best practices" within particular areas of planning. Consequently there have been concerted efforts internationally to develop databases, websites, guidebooks, and exhibits containing examples of best practices of urban and suburban development, often mixed with sample guidelines or linked to annual awards programs.[8]

The United Nations Center for Human Settlements and other organizations first coordinated an international exhibit of best practices at the 1996 Istanbul conference. Displays from scores of countries showed a wide range of design and policy ideas connected with urban sustainability. At the visionary end, Australian planners showcased Adelaide's "Halifax Project," in which local nonprofit groups were spearheading redevelopment of an inner-city industrial block as a mixed-use, alternative energy-powered neighborhood. On a more pragmatic note, China exhibited models of very large social housing projects designed to provide livable, high-density residential environments. Other interesting projects included Vienna's plan for a regional greenbelt, Chattanooga's ecological cleanup and riverfront restoration programs, and initiatives for grassroots democracy in the city of Belo Horizonte, Brazil. Similar international exhibits have been staged at subsequent conferences. The government of Dubai has funded a prestigious international award program entitled the Dubai International Award for Best Practices in Improving the Living Environment. Awards are presented every two years and a database has been established with more than 4,000 best practices from 140 countries.[9]

In the United States, the federal government's Partnership for Sustainable Communities makes available case studies of sustainability action at its website www.sustainablecommunities.gov, while the US EPA has other materials at www.epa.gov/sustainability. These government resources of course are dependent on political backing; one of the first actions of President George W. Bush's Secretary of Housing and Urban Development, Mel Martinez, was to discontinue that agency's previous best practices program. But the value of such examples means that the number and variety of such resources is generally increasing. Many nonprofit organizations such as the Local Government Commission, the Congress for the New Urbanism, the Sustainable Communities Network, and the Resource Renewal Institute have also publicized best practices of community development.

SUSTAINABILITY INDICATORS

One widespread initiative within sustainability planning has been the development of indicators that track progress towards sustainable development. More than 25 large US cities have developed such frameworks, including Portland, Seattle, San Jose, San Francisco, Santa Monica, Austin, Chattanooga, Jacksonville, Tampa, Indianapolis, Milwaukee, Boston, and Cambridge.[10] The use of indicators within planning is by no means new, but sustainability indicators tend to be more inclusive than many previous sets of performance measures, and focus more directly on showing trends concerning crucial environmental and social problems.[11]

The best-known prototype for sustainability indicators is the Sustainable Seattle process begun in the early 1990s. This grassroots effort initially drew together several hundred leaders from a range of organizations within the community with the intent of developing consensus on 20 key indicators of regional sustainability. The group published an initial list in 1993, which later expanded to 40 indicators in several main categories.[12] Though the Sustainable Seattle Coalition had no institutional authority to plan for the region, these indicators helped influence the City of Seattle's 1996 General Plan and the work of regional agencies, some of which developed similar lists of indicators for their own use. The Sustainable Seattle process also became enormously influential internationally as a model of a community-based indicators effort. Interestingly, after updating and expanding its initial set of indicators in 1995 and 1998 the Sustainable Seattle organization revisited its indicator approach in 2000 and decided to change its strategy. Local officials had complained that indicators such as "wild salmon returning to spawn" involved factors out of their control, such as logging practices throughout the watershed, dam construction, and global weather patterns. Consequently in the 2000s Sustainable Seattle refocused its work on neighborhood-scale sustainability projects.[13]

Other influential sets of sustainability-related indicators have been developed by many local governments including Jacksonville, Florida, whose quality of life indicators were pioneered in 1986, Santa Monica, and the region of Hamilton-Wentworth in Ontario.[14] A leading citizen-based indicator set has been developed by more than 2000 residents of Calgary, Alberta, who have produced four "state-of-our-city" reports and agreed on 36 social, environmental, and economic indicators for that community. Over nearly two decades these indicators have shown some improvement in environmental quality, but much work remains to be done in areas of equity, sustainable economy, and resource use.[15]

Indicators potentially have great power to demonstrate problems, motivate action, educate the public, and show the positive effect of sustainability policies. They can be helpful in monitoring program effectiveness and in guiding revisions to policy over time. Some indicators such as levels of air or water pollutants have direct public health implications and are tied to state or national policy. For example, the federal Clean Air Act mandates regular measurement of air pollutants such as carbon monoxide, nitrogen oxides, ground-level ozone, and certain particulates. Other indicators are good reflections of the health of particular species or ecosystems and are routinely monitored by environmental agencies under the Endangered Species Act or other legislation.

Yet indicators are a tool that must be used judiciously. If they do not have political commitment and buy-in by the relevant institutions and leaders, their development may prove to be merely a symbolic exercise. Creating a set of indicators for a region can be difficult, expensive, and time-consuming. Some indicators are much easier to quantify and maintain than others. Public agencies already routinely collect data on some, such as air quality, water quality, housing affordability, and public transit usage, but not on others, such as inequities of wealth, income, or tax base within metropolitan areas. Some indicators such as quality of life or sense of community are qualitative in nature and may be difficult to measure (the usual approach is to conduct surveys through which people are asked for their subjective ratings). Some indicators are much better than others at capturing the public imagination and encapsulating the health of complex systems within a single image.

For example, Sustainable Seattle's criteria of "salmon returning to spawn" worked particularly well because it used a charismatic species that served as a symbol and actual keystone element of regional ecosystems.

Once indicator frameworks have been set up, they must be maintained over time and updated regularly. If the staff and resources are not available to do this, indicators may quickly fall into disuse and undermine the entire effort. Sets of indicators must be revised periodically to reflect changing priorities, for example the increasing importance of reducing greenhouse gas emissions. Perhaps most importantly, indicators must be linked to agencies or levels of government with the power to actually address the problems they measure. Preferably these groups should actually be involved in the creation of the indicators, so as to develop institutional commitment to them. If an indicator effort is not linked to implementation mechanisms and political power, it may serve little function other than short-term public education.

One last problem is that citizens may develop unrealistic expectations that sustainability indicators will show great improvement in response to changed policies. For some initiatives, such as air and water quality, such dramatic change may be possible over relatively short periods of time. But for others, especially those involving complex urban social or environmental systems, change may be much slower and may depend on a range of interwoven policies being put in place over time. For example, no specific action is likely to quickly reduce automobile use in metropolitan areas. What is required instead is that a constellation of land use changes, pricing policies, improved transportation alternatives, and other steps such as an improved balance between jobs and housing within each community be put in place over an extended period of time. Gradually then people will drive less. But those who look for immediate reductions in a measure such as "per capita vehicle miles traveled" as a short-term result of Smart Growth policies may well be disappointed. For these reasons indicator efforts must be approached with caution (see Box 6.1).

STANDARDS, BENCHMARKS, AND GREEN RATING SYSTEMS

A related set of tools to help bring about sustainable development consists of measures that establish standards for building or planning many elements of cities and towns. Rather than assessing problems or the effects of past policies, these benchmarks can help planners, architects, and many other professions improve their work in the future.

Historically, of course, building codes, road design standards, zoning frameworks, and other professional guidelines have often shaped urban development in directions that are unsustainable. For example, transportation engineering associations and state highway departments (later renamed transportation departments) established road design standards that institutionalized overly wide roads within suburban development. Even in residential neighborhoods 40- and 50-foot wide streets became the standard in the US, along with wide turning radii and other features appropriate for relatively high rates of speed.[16] Other agencies such as the federal Home Loan Administration also required particular types of street networks if new developments were to qualify for loans, in particular mandating subdivisions with a high percentage of cul-de-sacs (such neighborhoods were thought to be the modern ideal and likely to hold property values better than communities with old-fashioned gridded streets). The result of such standards was to enforce a particular

Box 6.1 SUSTAINABLE SEATTLE INDICATORS

Indicators of Sustainable Community 1998

SUSTAINABILITY TRENDS

Declining Sustainability Trend

Solid Waste Generated and Recycled
Local Farm Production
Vehicle Miles Traveled and Fuel Consumption
Renewable and Nonrenewable Energy Use
Distribution of Personal Income
Health Care Expenditures
Work Required for Basic Needs
Children Living in Poverty

Improving Sustainability Trend

Air Quality
Water Consumption
Pollution Prevention
Energy Use per Dollar of Income
Employment Concentration
Unemployment
Volunteer Involvement in Schools
Equity in Justice
Voter Participation
Public Participation in the Arts
Gardening Activity

Neutral Sustainability Trend

Wild Salmon
Soil Erosion
Population
Emergency Room Use for Non-ER Purposes
Housing Affordability
Ethnic Diversity of Teachers
Juvenile Crime
Low Birthweight Infants
Asthma Hospitalizations for Children
Library and Community Center Use
Perceived Quality of Life

Insufficient Data

Ecological Health
Pedestrian- and Bicycle-Friendly Streets
Open Space near Urban Villages
Impervious Surfaces
Community Reinvestment
High School Graduation
Adult Literacy
Arts Instruction
Youth Involvement in Community Service
Neighborliness

model of low-density, automobile-oriented suburban development that we now see as profoundly unsustainable.

In many cases these old standards must now be reviewed and changed. Zoning codes constitute one of the most central set of standards for urban planning, and will need special attention. Minimum lot sizes in the US, for example, are commonly set at 5,000 to

10,000 square feet by many cities, enforcing relatively low-density patterns of development. These might be reduced substantially, say to 2,000 square feet for townhouses, and
maximum lot sizes put in place instead to prevent wasteful use of land in urban areas.

In the past decade or two many countries have developed new standards to promote
sustainability. The US federal Energy Star guidelines, for example, set standards for energy-
efficient appliances. Consumers often receive rebates from utility companies for purchasing appliances that have Energy Star certification. Simply labeling appliances to show their
energy efficiency has been an important educational technique. To take another example,
the LEED (Leadership in Energy and Environmental Design) standards established in the
late 1990s represent an extremely useful way to promote green buildings and energy-
conscious architecture. Under these standards, buildings are rated on a variety of criteria
and receive basic, "silver," "gold," or "platinum" certification (see Chapter 14 for more
information on LEED).[17]

Despite the huge benefits that standards can have in terms of spreading sustainable design
practices into the mainstream of urban development and consumer choice, they have disadvantages as well. Standards can often be too rigid, and have difficulty keeping up with
changing technology and innovation. Initially they may prevent innovation, until codes are
changed to accommodate new techniques (for example, building codes in most communities prohibited straw bale construction until these codes were amended). Firmly established
standards may reduce creativity, in that design or development becomes a process of meeting established benchmarks rather than "pushing the envelope." Extensive formalized standards can also add cumbersome bureaucracy and paperwork if not developed carefully.

"Performance standards" represent one way to avoid rigid codes. Under these, a building or development must simply meet certain overall criteria, such as keeping energy use
below a certain level or maintaining a certain species diversity in an ecosystem. The exact
means are up to the developer or policy-maker, thus opening the door to creative new
approaches. Within building construction, future performance can often be modeled by
computer or analyzed by engineers. Within larger-scale urban development, care must be
taken that impacts are measured over time and aspects of the development adjusted to
achieve performance goals. A somewhat related mechanism is form-based codes, used by
Andres Duany, Elizabeth Plater-Zyberk, and other New Urbanists to specify the character of
new communities. These simple, graphic codes illustrate desirable building form, street
design, and neighborhood layout. Advocates believe that they provide a simpler, easier-to-
understand, and more intuitive approach to urban design than complicated written zoning
codes and street standards.[18]

ECOLOGICAL FOOTPRINT ANALYSIS

One of the most intriguing methods to quantify the environmental impact of human communities is the ecological footprint model. Initially developed by William Rees at the
University of British Columbia and Mathis Wackernagel at the Redefining Progress organization in Oakland, this technique seeks to turn various aspects of human resource consumption into equivalent amounts of land that would be required to produce such
resources. Each individual or community is therefore assigned a "footprint" in terms of
acres or hectares that represents their ecological impact on the planet.[19]

At a simple, intuitive level the ecological footprint model has a great deal of appeal as a way to dramatize the impacts of resource consumption or changes in materials use over time. Ordinary citizens can run online versions of footprint models to calculate the impacts of their own lifestyles,[20] and some analysts have attempted to calculate footprints for large urban regions or entire countries. Herbert Girardet, for example, calculates the footprint of the Greater London area at 19,700,000 hectares (about 48 million acres, or 76,060 square miles), an area almost as large as Britain itself. This city, in other words, would take the equivalent of a land area 125 times its size to meet its resource needs.[21] Other studies have found that the average American requires an ecological footprint of more than 12 hectares, while the average Briton requires 6 hectares and the average Indian just 1 hectare.[22]

However, the usefulness of such statistics as well as more elaborate "urban metabolism" projects is questionable. What does one do with such figures, beyond generally educating one another about excessive resource consumption? Moreover, such models involve many assumptions about how various forms of resource use or pollution translate into land area, and can become tremendously complex exercises. Also, many key elements of urban sustainability, especially involving equity, livability, and social well-being, are virtually impossible to incorporate into such a quantitative model. Ecological footprint analysis therefore seems a limited sustainability planning tool with applications more useful in public education than in specific policy-making.

CARBON CALCULATORS

A more specific and applied type of calculation is increasingly used to estimate greenhouse gas emissions from cities, industries, institutions, and individual households. At their most general, "carbon calculators" are simple, online tools to help individuals figure out their approximate contribution to global warming, and compare this to that of other people in the same or different locations. The US EPA, the British government (carboncalculator.direct.gov.uk), and the Cool Climate Network offer relatively good examples of such tools. These calculators involve many assumptions about typical lifestyles, diets, building construction and the like, and so are at best approximations of actual emissions.

On a larger scale, climate change policy in many places now requires local governments, businesses, and institutions to complete much more elaborate greenhouse gas (GHG) emissions inventories as a way of establishing and monitoring progress towards GHG reductions. Such inventories can take large amounts of staff time to prepare, and likewise involve a range of sometimes controversial assumptions. For example, should a local government consider itself responsible for the emissions from motor vehicles on a freeway passing through its territory? From an airport within its borders? (The usual answer is no for both.) How should the embodied emissions in consumer products and new buildings be accounted for? (No good answer for that one.) Emissions inventories, like household carbon calculators, are then at best an approximation of actual emissions, useful mainly to direct and motivate behavior, and to establish emission trend lines that can be updated year after year.

GIS AND MAPPING

One of the most rapid areas of growth within the urban planning field has come through increased use of "geographic information systems," or GIS. These computer-based applications provide a sophisticated ability to map and analyze many different spatial layers—land use, topology, roads, rail lines, census data, hydrology, soils, slopes, fragile habitats, endangered species, and so on—across urban or rural regions. Such mapping can be used for many types of analysis to support growth management planning, environmental protection efforts, environmental justice analysis, and many other sustainability-oriented endeavors.

GIS systems and computer modeling based on them have, for example, been used to support growth management planning in many urban regions around the globe. Planners and citizens have been able to study the implications of different metropolitan growth scenarios in terms of open space consumption, urban densities, population near transit, air quality, traffic generation, greenhouse gas emissions, and other variables. Extensive GIS systems are also being used by an enormous range of nonprofit organizations including the Nature Conservancy, to map ecoregions and prioritize lands to be protected from development, and worldmapper.org and gapminder.org, to make global information easily available on the web in a way that can help people understand the evolution of differences between countries. Academic scholars have used GIS to study topics ranging from less-educated job seekers' access to jobs in the Boston Metropolitan Area[23] to the emergence of suburban nodes of relatively dense housing around Seattle.[24]

GIS systems, though a useful tool for sustainability planning, are by no means a panacea. They can provide important information for planners and highly educational material for the public, but it can be time-consuming and expensive to set up such computer databases, and the data must be of high quality and well-maintained to be useful over time. Good design of GIS systems is very important so that they fit well with policy-making, contain the right information, and are understandable by the general public.[25] Some types of problem also do not lend themselves to quantitative spatial analysis, but require on-the-ground observation and work with local communities instead. The location of a particular building or activity, for example, should not just be decided on the basis of abstract analysis about proximity to other facilities and so forth, but on what the proposed site actually looks and feels like. GIS analysis should never take the place of common sense or hard-nosed organizing and action. With such a technology, the temptation is great to study problems at enormous length rather than taking action based on already-extensive data to address them.

ENVIRONMENTAL ASSESSMENT AND REPORTING

Environmental assessment, also called environmental impact assessment (EIA), is a science-based discipline that has emerged since the early 1980s in response to the passage of regulation regarding air quality, water quality, toxic chemicals, and other environmental threats. The focus is usually on assessing conditions and risks for particular facilities, industries, or watersheds. This assessment then becomes the basis for documents such as EIRs and for environmental management policies designed to reduce risk. Since its inception the field has evolved from measuring the effects of particular chemical pollutants on a single species to a more broad-based assessment of the impacts of multiple stressors on ecosystems.[26]

Environmental assessment relies on an increasingly standardized set of methods for assessing threats to ecological health. Agencies such as the US EPA's National Center for Environmental Assessment, the Canadian Environmental Assessment Agency, and Britain's Environment Agency have established guidelines for conducting ecological risk assessments.[27] This discipline is particularly important to certain professions such as real estate, which must know the status of certain properties, and to industries that fall under extensive federal or state environmental regulation.

A formal process of researching environmental and human impacts of new development projects is now legally required by many countries and states under varying titles that might be collectively labeled "environmental impact reporting." The US environmental impact statement (EIS) process is required under the National Environmental Policy Act of 1970 (NEPA), which was one of the first such pieces of legislation; it established the principle that the impacts of any federal action "significantly affecting the quality of the human environment" should be studied, and alternatives to this action considered that might have less impact. Most American states subsequently passed their own versions of this requirement in "mini-NEPA" legislative acts. The California Environmental Quality Act, for example, requires environmental impact reports to be prepared on major projects. State courts have extended this requirement to private sector as well as public sector projects having significant environmental impacts. Briefer environmental assessments or "findings of no significant impact" must be prepared for smaller projects.

Other countries have similar requirements. Canada requires environmental assessments on projects that are authorized, funded, or carried out by federal agencies. The European Union requires member states to adopt environmental impact assessment processes in line with the United Nations Convention on Environmental Impact Assessment, which entered into force in 1997.[28] As mentioned previously, such assessments can both provide information on environmental inputs and help evaluate alternatives to a proposed action.

One related method, used especially in Europe to assess how various paths of economic development might lead towards sustainable development, is development path analysis (DPA). Under this technique activities are categorized into one of six different potential development paths, depending on their impact on the environment. The paths range from business-as-usual to dramatically different types of economic activity. This method has been used particularly by the European Union to allocate structural funds within its Building Sustainable Prosperity program.

INSTITUTIONS AND POLICY MECHANISMS

The role of institutions

The institutions of planning—boards, commissions, processes, guidelines, regulations, and the like—are among the most important tools at our disposal. These entities can help shape the physical, social, and economic landscape around us in more sustainable directions, although they can also hinder change and become enormous obstacles if poorly designed or captured politically by particular interests. The process of carefully improving such institutions is essential to sustainable development.

Among the most important institutional tools are agencies of government, which may seem static and immutable but in the longer term can be and have been changed dramatically. Progressive-era innovations in the early twentieth century introduced notions such as open meetings, a professional civil service staffing local governments that was supposed to be immune from politics, and citizen commissions to advise city councils on policy and administer zoning. Regional agencies are another innovation that first arose in many metropolitan areas in the 1960s, but that still need considerable strengthening and refinement in most places. New metropolitan, watershed, or bioregional agencies may be particularly important to sustainable development, since many issues of urban growth and environmental protection are now seen to transcend the boundaries of past local government institutions. Election reforms are a somewhat different type of institutional change that is also essential, in particular to reduce the influence of big-money donors on government. These reforms are needed in order to ensure that equity and environmental interests have the same sway within government as economic constituencies.

Mechanisms through which institutions determine and set forth policy (that is, agreed principles, strategies, and standards) are open to change as well. The notion of an urban general plan was an innovation in the first half of the twentieth century, as were zoning, street standards, and the like. Some of these mechanisms may now be proving counterproductive (zoning for example is often seen as too rigid and enforcing an inappropriate separation of land uses), and may need to be revised. New mechanisms may also be needed. Urban design codes and guidelines have proliferated in recent years, for example, as planners and citizens have realized that the public sector needs in many cases to more proactively shape the physical form and character of communities.

Consistency provisions

One main problem with current planning, which governments have not yet thoroughly addressed, is the need for consistency between planning tools. If specific area plans and zoning do not follow city master plans or state growth management goals, for example, then these broader planning frameworks become meaningless. Also, if local code enforcement officials or city commissions do not uphold officially approved plans and codes, or grant numerous exceptions to them as frequently happens, then these planning tools become meaningless as well.

Consistency problems have arisen as various planning tools have been put into place in piecemeal fashion over the past century, often with somewhat half-hearted commitment by political leaders to the overall concept of planning. And certainly, the idea of a single, legally binding planning framework between different scales of government can be a frightening one. But on the other hand a lack of consistency between these levels can defeat the whole purpose of planning. So ways must be found to improve consistency and enforceability while also making the framework responsive to changing local conditions. Usually this is accomplished by setting broad goals and policy directions at higher levels of government, with more detailed policy and implementation occurring locally.

A number of US states have begun to adopt consistency requirements mandating different levels of plans to be consistent with one another, and giving the public the legal power to sue to enforce consistency. This provision essentially puts teeth into planning. States such as Florida and Oregon have been in the lead in this regard. City governments can also pass ordinances or adopt charters requiring consistency between different scales of planning. Such provisions are still weak in many locales, however.

Intergovernmental incentives and mandates

One of the most important toolsets related to sustainability planning has to do with "carrots and sticks" by which various levels of government can encourage each other to plan for sustainability.[29] Typically, higher levels of government establish general goals and provide incentives and mandates for lower levels to take action implementing these, but the process can sometimes work the other way as well, for example if particular cities or states adopt stringent environmental requirements that are then copied at the national level. In the current era in which citizens are often skeptical of government, it is unlikely that any particular level of government by itself will be strong enough to adopt and implement comprehensive sustainability policy. Strong regional agencies, for instance, have historically often been seen as the solution to metropolitan problems, but the fact is that with very few exceptions American urban regions have failed for more than 50 years to bring such institutions into being. So what is needed instead is for our existing, imperfect institutions to reinforce one another, working together to establish intergovernmental frameworks for sustainability planning.

States such as Oregon have relied extensively on an interlocking framework of incentives and mandates for decades, with generally positive results. That state first developed a set of statewide land use and environmental planning goals in the late 1960s and early 1970s, and since then has systematically worked with regional and local governments to encourage them to meet those goals. Such efforts have included direct grants to local government for planning and implementation that meet state goals, technical assistance through staff and data, and strong mandates and timelines requiring local governments to produce results.[30] The state of Maryland likewise has developed both incentives and mandates to encourage local governments to implement its Smart Growth program. The state will only fund infrastructure within "priority funding areas" designated by local government consistent with state criteria established under the Smart Growth and Neighborhood Conservation Act. These criteria emphasize planned urban growth in areas where infrastructure is already in place and preservation of the state's "rural legacy."

In contrast to the traditional model of top-down planning common to many European countries and some US cities in the past, such a decentralized framework of interrelationships between levels of governments seems more likely to succeed in the North American context. Moreover, much of the world is following the American model, with regional agencies losing power (those governing metropolitan London, Barcelona, and Copenhagen were dissolved in the 1980s), and an increasing emphasis on participatory, consensus-based, or market-based planning rather than top-down control. So like it or not, the new political model in many places is one of flexible governance in which a variety of public and private sector institutions work together to get things done. It is this environment within which most sustainability planning will take place.

EDUCATION, COMMUNICATION, AND CONSENSUS-BUILDING

Among the most important tools for long-term change are strategies of education, communication, and consensus-building. These include the whole range of activities that make up Friedmann's "social learning"[31]—action-oriented practice, education, group decision-making, and cognitive growth. But this set of strategies also includes more recently publicized processes of consensus-building, meeting facilitation, and networking which often are included under the "communicative action" label.

One common denominator behind such strategies is the recognition that, for political or social change to occur, people's beliefs, knowledge, values, and paradigms of thought must also change. This understanding has given rise to a wide range of educational activities by public agencies and NGOs, including extensive newsletters and publications, websites, planning workshops, design charettes, creation of demonstration facilities, and the like. Another foundation to this approach is the recognition that "social capital"— defined by Putnam as "features of social organization, such as trust, norms, and networks, that can improve the efficiency of society by facilitating coordinated actions"[32]—is necessary for institutions to work well. Building social capital has thus become an important goal of many sustainable development advocates.[33] Many of the same strategies apply to this objective as well. Particularly important are structured community processes through which participants can get to know one another over time, build trust, establish a common base of knowledge and data, and develop networks that can bear fruit in the long run. Examples include urban design charettes and the coordinated resource management programs (CRMPs) that have been undertaken in some western US states to bring all stakeholders in a watershed together to determine how best to protect resources in the long run.

As noted earlier, caution is in order when contemplating extensive consensus-building processes as a tool for sustainable development planning. Such processes are often lengthy, expensive, and time-consuming, and their outcomes depend on the willingness of all parties involved to participate in good faith and forgo unfair power advantages. "Participation" in general has been a mantra within planning circles for several decades now, and certainly is extremely important, but also has problematic aspects. Deferring too much to the desires of local communities may not be wise, if these groups are bound to a narrow viewpoint or limited self-interest (exemplified in many Nimby battles within local government). One particularly vocal minority may hijack public meetings, or assert an interest that runs counter to the interests of broader constituencies or under-represented groups. Planners must use discretion in structuring such processes and taking their input into account, and must seek opportunities to educate both themselves and others about the full range of groups and goals involved in any given planning debate.

ORGANIZING AND COALITION-BUILDING

Although advocates for sustainable communities can use many new and traditional planning mechanisms to bring about change, and may seek to restructure institutions and build social capital to create the context for sustainability planning, they must at some point also confront realities of political and economic power. They may then need to develop alternative sources of power in order to change the status quo. Perhaps this can be done by winning over current politicians and corporate leaders, but developing coalitions, advocacy groups, political parties, and public events calling for change may also be necessary.

Since organizations calling for progressive social change are often small and limited in resources, coalition-building has been a necessary skill in many places. Natural alliances are often formed between environmentalists, social justice groups, religious organizations, public health advocates, educators, affordable housing builders, and some labor unions. Themes such as "livable communities" can draw these and other groups together. In the Portland, Oregon, region, such organizations banded together to form a Coalition for Livable Communities. Beginning in the 1960s a broader, de facto alliance also came about in that region between environmentalists, family farmers, and progressive developers that resulted in statewide growth management legislation. In the San Francisco Bay Area, some 46 different local organizations banded together to form the Transportation and Land Use Coalition, now the TransForm, dedicated to reforming the region's transportation and growth policies. At a national level, the Surface Transportation Policy Project and the National Clean Air Coalition have been examples of coalitions that have succeeded in passing or altering significant pieces of legislation.

Progressive political parties are a somewhat different mechanism for organizing political power. Green parties in many countries have helped push a broad agenda (typically ten main points) that is very closely linked to sustainable development. Because of the strength of the US two-party system, the Greens have had a hard time making headway in the United States. But in countries where political institutions allow a greater role for minority parties they have been more successful. The German Greens stunned the German political establishment in 1979 by winning seats in the Bundestag with more than 5 percent of the vote, the level under the German system of proportional representation at which parties may be represented in the national government. Ever since then the Greens have exerted a significant influence on Germany's environmental and social policy. Green parties have also been active in France, Britain, and many other countries. In New Zealand, the Values Party that was formed in 1972 achieved more than 5 percent of the national vote in 1975 and pushed for sustainability-oriented policies. This constituency has evolved into New Zealand's current Green Party.

Coalition-building and political organizing is often less daunting at a neighborhood or local government scale, where just a few individuals can form an organization and a few organizations can represent a political movement. Sustainable community groups have arisen in a great many locales. Key challenges for these organizations are to figure out creative ways of gaining local press attention (a few media stories have a powerful multiplier effect for small political movements), gaining access to politicians (which at a local level can often be had just by asking), and maintaining a consistent presence over time. Attracting involvement by individuals or businesses with substantial resources is also an important task. Finally, a core task of sustainability organizing at any level is to establish a positive and proactive agenda, rather than simply opposing bad projects. A constructive agenda can have greater power to inspire people in the long run, and can accomplish much more than the draining process of fighting rear-guard actions against inappropriate projects initiated by existing forces.

Part Two

Issues central to sustainability planning

7

CLIMATE CHANGE PLANNING

A generation ago climate change planning did not exist in most places, and although scientists had known about the possibility of global warming for many years few policymakers were taking it seriously. Even today efforts to combat global warming are still in the early stages. But the threat is so great that this crisis more than any other is now driving much sustainability planning. Over the coming decades every jurisdiction on Earth—from entire countries down to the smallest towns and rural districts—will need to figure out how to nearly eliminate its greenhouse gas (GHG) emissions and how to adapt to a changing climate.

Heading off the most disastrous forms of climate change amounts to the largest planning challenge ever, and relates to virtually every other area of sustainability planning. Among other things, it will mean reorienting economies so as to no longer rely on fossil fuels, revising transportation systems to reduce needs for travel by motor vehicle, revising urban design so as to make walkable and bikable communities possible, redesigning food systems to make them more resilient in the face of a changing climate, and rethinking lifestyles so that they can be more equitably shared by the world's population. If we want to have a humane response to this crisis, for example avoiding famines in the developing world or environmental justice disasters like the US government's response to Hurricane Katrina, it will also be necessary to emphasize a wide range of preventative measures, in particular aimed at improving the status of the least well-off.

Many good resources exist on the basic science and history of the global warming problem, so we won't go into such background here.[1] Rather, we will focus on planning activities related to climate change, with the aim of highlighting key dimensions of this fast-moving field.

THE FRAMEWORK OF CLIMATE PLANNING

Although a unified global response to this crisis would be desirable, we have instead a worldwide hodgepodge of climate action planning from the international scale down to local levels.

All of these efforts are relatively new, having arisen since the early 1990s. Activists and planners have made some progress at most scales but their efforts are nowhere near sufficient to combat the problem. GHG emissions as of the early 2010s were rising faster than ever, and very little had been done to prepare for more intense storms, drought, flooding, higher seas, food scarcities, and many other potential impacts.

At the international level, the United Nations Earth Summit Conference in 1992 set in motion a long and ponderous series of international meetings oriented around establishing binding goals for GHG reductions. One product of this process was the 1997 Kyoto Protocol, an agreement setting modest emission reduction targets for 37 countries and the European Union by 2012. EU countries have generally done well at meeting these targets, although the economic slowdown that began in 2008 undoubtedly led to some of the decrease in emissions. In 2012, 19 of those 37 countries had met their targets, and the EU as a whole had reduced its emissions by 11 percent below 1990 levels. The UK had reduced its emissions by 24 percent, Germany by 23 percent, and Sweden by 12 percent; but Austria, Italy, Spain, Denmark, the Netherlands, Norway, and Switzerland had missed their targets.[2] Outside of Europe, most countries failed to meet their Kyoto goals. Canada's 1990–2009 emissions increased by 20 percent, the US's 7 percent, Japan's 3 percent, and Australia's 52 percent.[3] Under pressure from fossil fuel industries, the US and Canada withdrew from the treaty altogether before the target date was reached. Meanwhile, emissions rose rapidly from developing countries not included in the original treaty. Subsequent UN events in Copenhagen (2009), Durban (2011), and Rio de Janeiro (2012) failed to produce follow-on treaties. Some lesser agreements have been signed, for example a 2012 treaty to reduce emissions of black carbon, which when deposited on top of snow retains heat and so contributes to polar icecap melting. But international climate change planning mechanisms, if they can be called that, have been weak overall.

At a continental scale the European Union has taken one of the best-known international climate actions, an emissions trading system begun in 2005. This system covers some 11,000 power stations and industrial plants in 30 countries, accounting for about half of the continent's CO_2 emissions. It was expanded to cover airlines in 2012, despite vociferous objections from other countries such as the US and China. The Emissions Trading System (ETS) sets an overall cap on the allowable quantity of emissions, to be lowered over time, and then allocates emissions to these facilities so as to stay within the established limit. Companies that are successful in reducing their emissions below the allowances given to them can then sell those credits to others on an open market.[4] In the early years of the ETS, the EU gave away most emissions allowances, resulting in windfalls for a number of large corporations, a very low carbon price, and few if any emissions reductions. However, during its second and third phase officials took steps to tighten up the ETS, so there is hope that in the long run this will be an effective mechanism.

National planning for climate change began in the late 1980s, and is particularly important since nation states have the power to take a far more comprehensive approach to climate planning than lower-level governments. Sweden, Denmark, and the Netherlands were among the first to take action to reduce emissions, primarily through the promotion of renewable energy and tax measures to reduce fossil fuel consumption. Germany and Japan adopted policies early on as well. By virtue of a strong emphasis on energy conservation as well as hydroelectric and nuclear power, Sweden's electricity was virtually carbon-free by the end of the twentieth century. Sweden also instituted a carbon tax and made very extensive grants to local communities for GHG mitigation planning.[5] The British government passed a Climate Change Act in 2008 setting a target for 2050 GHG reductions of 80 percent compared with the 1990 baseline (perhaps the most common long-range target adopted by governments currently). However, due to leadership changes implementation is uncertain. China has clearly taken climate change seriously in some ways, launching enormous solar and wind power industries. However, the country continues to build coal-fired power plants at the same time, is rapidly expanding motor vehicle infrastructure, and in the late 2000s overtook the US as the world's largest GHG emitter.

Separately from national action, many state, regional, and local governments developed planning tools beginning in the 1990s to reduce GHG emissions and adapt to climate change. These actions initially focused almost exclusively on reducing emissions, with relatively little attention to adaptation. That balance has been partially redressed in recent years, though mitigation remains the priority. Typically these plans outline a range of actions such as requirements that utilities produce a certain percentage of their power from renewable sources, requirements that public agencies build green buildings and purchase fuel-efficient vehicles, changes to transportation systems to reduce driving and promote alternative modes of travel, programs to cap landfills so as to reduce methane emissions, efforts to increase recycling, and steps to promote compact development that will be less motor vehicle-oriented. Typically, though, there has been little money or legislative authority behind such strategies, and so the success of these plans is questionable.

MITIGATION: REDUCING EMISSIONS

In the language of climate change policy "mitigation" refers to any action to diminish GHG emissions, i.e. to mitigate or reduce the extent of the problem. Countless such actions are possible, from individual steps to turn off lights, travel less, and better insulate homes to municipal, state, national, and international policies and programs to reduce fossil fuel use, preserve forests (to keep stored carbon in trees), cap landfills (to reduce methane emissions), and phase out certain refrigerants (since those chemicals have powerful global warming effects). In practice, such actions must be bundled together at every level if we are truly to reduce emissions enough to head off global warming.

As with many other aspects of sustainability planning, the most important element of GHG mitigation is an overall strategy to reduce emissions systematically and comprehensively. Not surprisingly, "climate action plans" have thus become the planning tool of choice.[6] These have been promoted worldwide since the early 1990s by the nonprofit organization ICLEI (Local Governments for Sustainability, formerly the International Council on Local Environmental Initiatives), which advocates a straightforward approach organized around five milestones:

- measuring emissions through a greenhouse gas inventory;
- committing to GHG reductions in a target year compared to a base year;
- planning for reductions through a climate action plan;

- implementing that plan; and
- monitoring emissions reductions on an ongoing basis.[7]

To assist such efforts, ICLEI and other organizations, such as the Climate Registry, have developed specific protocols and models for measuring and analyzing emissions. More than 1,000 local governments worldwide have participated in ICLEI's Citizens for Climate Protection campaign, and hundreds more additional communities and states have developed other forms of climate action plans. For example, in California the statewide climate change planning framework adopted by the legislature plus actions by the state Attorney General's office in the 2000s meant that virtually every local government has had to develop climate analysis and planning documents.

Climate action plans tend to include anywhere from 20–60 main types of action for their communities, grouped into a number of categories. Typically these include mobility (actions to promote biking, walking, public transportation, and alternative fueled vehicles); energy (actions to encourage utilities to use more renewable energy, homeowners to improve the efficiency of their homes, and public and private sector developers to obtain green building certification for their structures); land use (promoting compact and transit-oriented development); consumption (improving recycling and reducing waste); water use (steps to conserve water and reuse gray water), and regional planning (coordinating with regional agencies and other local governments around Smart Growth, affordable housing, and other initiatives). Good plans will quantify the expected emissions reductions from each action, and put in place frameworks to monitor and evaluate the plan's performance.

Such climate planning frameworks are all well and good, but the problem so far has been that implementing actions are very loosely tied to emissions reduction goals.[8] It has been relatively easy for politicians to announce dramatic long-term targets, such as 80 percent reductions by 2050 or a return to 1990 levels by 2020, but these time frames are well after such leaders will have left office, and there is little incentive to launch near-term initiatives to actually get to such goals. Consequently the needed funding, regulation, and programs to actually reduce emissions are often lacking. This is particularly true for the largest potential emissions reductions, such as steps to dramatically reduce driving or shut down coal-fired power plants. Such efforts would be politically difficult, and so are often avoided.

Even with more energy-efficient technology, very deep reductions in GHG emissions will require lifestyle changes so that people consume less, drive less, buy far more energy-efficient houses, and change their diets away from meat. Few people currently want to embrace such new ways of living, and most societies still predicate their economies upon continued growth in material consumption. Transitioning to a more steady-state economy based on quality not quantity of goods and services would take vision, leadership, and public willingness to change. These elements are currently still lacking.

The situation is, then, one in which leaders and the media give much publicity to a limited range of GHG mitigation activities—solar panel installations, wind turbines, electric vehicles, and compact fluorescent or LED lighting—while avoiding discussion of more far-reaching approaches that could make far more difference in the long run. Outside of

Europe, movement has been slow towards broad policy mechanisms such as carbon taxes or cap-and-trade systems. Most political leaders have avoided any sort of discussion about the need to reduce consumption and change lifestyles, and in the US climate change deniers have dominated political discourse making any discussion of the topic next-to-impossible. Meanwhile, fossil fuel industries have promoted alternatives such as "clean coal" and carbon capture and sequestration (CCS) that are almost certainly unworkable and a distraction from the other steps that need to be taken. CCS for example is based on the belief that carbon dioxide could be captured from the emissions of fossil fuel power plants and injected deep into the earth where it would remain safely out of the atmosphere. However, despite billions of dollars of research this method has never been demonstrated for a large-scale power plant, and would almost certainly be fantastically expensive. In 2011 a major test project in the United Kingdom was shut down due to fossil fuel companies' concerns about commercial viability and their unwillingness to fund the research without continuing public subsidies. That same year an American utility shut down a test site in West Virginia, arguing that the technology hadn't been shown to be technologically feasible or commercially viable.

All is not lost, however. Some GHG mitigation successes have occurred at every level, and at least it is increasingly clear what needs to be done. Individual households or neighborhoods can move towards zero-net-emissions status by monitoring their own energy use, transportation choices, diet, and lifestyles. Architects and builders, both nonprofit and for-profit, can create zero-net-energy buildings. Industries can improve their energy efficiency. Cities and towns can develop climate action plans and then seek better ways to implement them, for example by tightening their building codes and ending suburban sprawl. States and countries can exert greater leadership, and incentivize local action through grants and technical assistance. Sooner or later all of these actions will reinforce each other and zero-net-emissions societies will come about. However, we will most likely have brought about a very different climate by that point (see Figure 7.1 and Box 7.1).

ADAPTATION: LIVING WITH CHANGE

A second, under-appreciated area of climate change planning focuses on the process of adapting to a different world. Adaptation efforts will only modestly blunt the impact of global warming—no set of preparations can really compensate for decades of drought, or intensified storms that dump a foot or more of rain, or sea levels that will eventually, centuries from now, rise more than 100 feet. But careful planning can help millions of people avoid various forms of personal harm from climate change-related events, and can also help save some species and ecosystems from global warming-connected damage.

The subject of adaptation was rarely mentioned within early climate action plans, but is increasingly seen as essential given that humanity will no longer be able to avoid global warming of at least 2° Celsius (3.6° Fahrenheit). Main adaptation strategies vary from place to place (as will the impacts of climate change), but typically focus on moving development away from possible flooding and rising sea levels, preparing for drought and increased wildfire risk, shading buildings and designing them so as to remain cool in hot weather, diversifying crops in agricultural areas, and guarding against potential

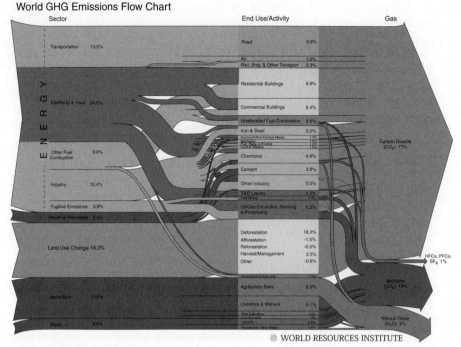

Figure 7.1 World flow chart of GHGs

Box 7.1 CALIFORNIA'S CLIMATE PLANNING FRAMEWORK

Since 2005 the state of California has emerged as one of the world's leaders in cli-mate action planning, and provides an example of the type of integrated action that might be taken elsewhere. Although the state had studied climate change in the 1990s and has long been a leader in energy efficiency and air pollution regulation, the current process began in 2005 when then-governor Arnold Schwarzenegger signed an Executive Order setting a target of reducing GHG emissions 80 percent by 2050. In 2006 the legislature passed AB 32, a bill committing the state to lower-ing GHGs to 1990 levels by 2020, a 25 percent reduction from emissions at that time. The bill established a sophisticated planning process to meet this goal, with requirements for emissions reporting, a "scoping plan" of feasible actions, a market-based "cap-and-trade" system, and consideration of environmental justice and economic impacts.

The state's well-oiled Air Resources Board (ARB) then swung into action, build-ing on 30 years of experience in regulating local air pollution. The ARB required all large companies and institutions to begin reporting their GHG emissions in 2008, and adopted 18 statewide strategies that became effective in 2010. Agency staff quantified the expected emissions reductions from each strategy. By far the most effective were GHG emissions standards for motor vehicles, improved energy efficiency requirements for buildings and appliances, a requirement that utilities produce 33 percent of their electricity from renewables by 2020, reformulated motor vehicle fuels, and regulation of refrigerants and other non-CO_2 GHGs. Equally important was the cap-and-trade program for reducing industrial emissions, which went into effect in 2013.

Other actions in the state complemented the work of the ARB. The state Attorney General's office threatened to sue any local government that didn't consider climate planning in its general plan or develop a separate climate action plan. The legislature passed a 2008 bill, SB 375, requiring metropolitan regions to create transportation and land use plans reducing emissions. Nonprofit organizations worked in many jurisdictions to raise citizen awareness of the need to reduce emissions. Green technology companies moved rapidly to develop ideas and raise venture capital. And the California Energy Commission funded extensive research into climate change strategies and impacts, resulting in a number of well-publicized reports.

To be sure, there have been obstacles to California's climate action planning, including an unsuccessful 2010 ballot proposition funded by out-of-state fossil fuel companies, the post-2008 economic recession, and a severe shortage of funds for new initiatives. However, state commitment to the planning framework has remained strong, and the synergies between California's different climate-related actions are likely to bear fruit in the years ahead.

food shortages in the developing world. Some adaptation planning attempts to take social justice into account by paying special attention to risks faced by disadvantaged communities, ensuring for example that low-income elderly residents have access to air conditioning during heat waves, and that levees are sufficient to protect poor communities from flooding. However, since climate impacts may vary widely from place to place, careful contextual study is the starting point for adaptation planning in any given location.

In many parts of the world, adaptation will mean increasing local self-reliance in terms of food. Global grain markets may be disrupted by drought, widespread flooding, changing temperatures, or severe storms. Even mild swings in the prices of essential commodities may put them beyond reach of millions of lower-income residents in developing countries. The globalization of agriculture during the past century has assumed relatively stable climatic conditions; however, multiple crop failures in different parts of the world could greatly stress food systems. Global agricultural systems also depend on price to distribute uneven harvests around the planet, and poor countries will inherently be at a disadvantage. So strong and diverse local agriculture can provide greater food security. Within any given country, diverse cropping systems are also likely to be more resilient than monocultures in which just one foodstuff is grown over thousands of acres. "Agroecologies" which maintain hedgerows and corridors of riparian (streambank) vegetation as well as crop diversity also offer adaptation advantages in terms of preserving wildlife, native pollinators, and habitat.

Climate adaptation planning along coastlines means a package of initiatives to protect against sea level rise, flooding caused by more severe storms, and saltwater intrusion into aquifers. The most basic action would be to prohibit new development on land within several meters of sea level or within inland floodplains. But this has been difficult to do historically. People like to live near beaches, on barrier islands, and in scenic locations near wetlands. Developers also profit from building in such places, and are frequently cozy with local governmental officials. However, even if coastal development protections cannot be enacted, other mechanisms can often be found to discourage building in coastal or inland areas likely to be flooded. One leading strategy is to deny flood insurance to such development. Also, governments can refrain from constructing infrastructure such as levees and breakwaters that may just encourage continued development within areas that will be difficult or impossible to safeguard in the long run.

Other coastal adaptation measures are likely to include relocating roads, sewage treatment plants, electrical transmission facilities, power plants, and other infrastructure away from areas likely to be flooded. Many airports in particular are currently located in places that will be swamped by sea level rise, since during the twentieth century dumping landfill into bays was often the easiest way to create the large amounts of land needed for runways. These facilities will need to be protected or relocated. The world witnessed the harm created by inappropriately located coastal infrastructure during the partial destruction of the Fukushima nuclear power plant by a tsunami in 2011. Relatively modest preventative measures in that case to shield the emergency generators from seawater could have saved tens of billions of dollars in cleanup expenses after the reactors melted down. As the warming climate produces higher seas and more severe storms in the years ahead, it will be imperative to avoid a repetition of such calamities.

Yet another main area of adaptation work has to do with existing electric and water infrastructure. Hotter temperatures are likely to mean increased electricity demand from air conditioning, and on summer afternoons demand may rise beyond existing supply. Increasing power plant capacity is the traditional strategy in response to rising demand, but more sustainable alternatives include reducing demand through energy conservation programs, and implementing aggressive "peak shaving" programs to provide incentives to large customers to turn off equipment during peak load times.

Water infrastructure has a different set of problems. Increased annual precipitation or more severe rainstorms may strain reservoirs and flood control systems, raising the prospect of flooding or dam failure. Rather than simply building larger infrastructure systems, the most sustainable response is likely to be a watershed-wide planning effort to reduce runoff, through good management of forests and agricultural land as well as effort to reduce impermeable surfaces within urban areas and retain runoff on site. Other locations may experience droughts. In those places new procedures to reuse wastewater and improve water use efficiency may be needed.

Avoiding environmental justice problems due to climate change will be one of the biggest challenges to adaptation efforts. Climate justice is a rapidly growing field of study with many implications that are discussed further in Chapter 15. Main strategies to head off injustices due to climate change overlap with social equity strategies in general, and include efforts to raise the income and quality of life of the world's poor, steps to decentralize economic power to local and regional levels, policies to keep the poor from having to live in the most disaster-prone geographic areas, and steps to ensure that emergency programs respond adequately to poor and minority groups after disaster strikes. Though all of these steps are important, if poor people simply have more adequate and secure incomes they will be able to cope with climate change in ways similar to better-off individuals. This basic step of poverty reduction is likely to be the most effective and flexible social equity response to global warming.

Additional areas of adaptation planning are likely to emerge in the years ahead. Dealing with the potential spread of infectious diseases or plant pests as species change their ranges is one such area of work. Preventing wildfires within landscapes that become hotter and drier is another. Global warming will bring many different challenges, not all them foreseeable, and much study and proactive planning will be necessary to minimize harm to societies and ecosystems.

Climate change adaptation and mitigation planning are of course connected, and in some cases the same action benefits both goals. Urban tree-planting, for instance, has an adaptive benefit by helping keep cities cool during hot weather, and a mitigating benefit since it will then also reduce energy consumption for air conditioning and sequester a modest amount of carbon in trunks and roots. Home energy efficiency also both reduces emissions and makes buildings more comfortable during extreme weather. Diverse, community-scale energy systems may increase reliability (an adaptation benefit) while reducing emissions. Diversified agriculture with lower fertilizer inputs can help make local food systems more resilient in the face of a changing climate while also helping reduce nitrogen oxide from nitrogen-based fertilizer. And so on.

However, at other times the two sets of actions will have very different priorities. Adaptation planning within a given community may focus on flood prevention, while mitigation planning requires a much broader range of steps to reduce driving, improve building energy efficiency, reduce utility and industrial emissions, and change lifestyles. Although adaptation planning is certainly important and has been overlooked often in the early years of climate action planning, mitigation must remain the overall priority if we are to avoid extremes of dangerous climate change for which there is no possible adaptation.

THE CHALLENGE OF CHANGING BEHAVIOR

The biggest question concerning climate change planning is how to change behavior so that people move towards zero-net-carbon lifestyles. Global warming is a priority for very few of the world's citizens. Most are preoccupied with daily survival, personal advancement, family and community life, and/or the perceived need to conform to social norms. Somehow climate change and sustainability in general must become priorities helping determine individual decisions, an ethical sea-change on an unprecedented scale.

One school of thought holds that human behavior only really changes in response to major crises. However, global warming may be happening too slowly to motivate such dramatic change. By the point when sufficiently large climate changes have set in for the public and corporations to re-evaluate their behavior, it will almost certainly be too late to avoid catastrophe, given the period of time in which greenhouse gases will stay in the atmosphere and the likelihood of positive feedback cycles through which additional carbon is released from other terrestrial sources.

Another school of thought argues that change can come about through determined social and political organizing. Many examples exist, including the civil rights movement, women's rights, gay rights, and peace movements in a number of countries. It is certainly possible that organizations such as 350.org can mobilize public opinion on the subject of climate change as well. But, given the enormous forces that benefit from highly consumptive, fossil-fuel-based economies, the task will be greater than in these other cases. Social and civil rights issues do not threaten the foundations of wealth within capitalist societies quite as directly. Even many other environmental causes, in particular preservation of species and wilderness, are less controversial than global warming. Climate change appears to be the most fundamental challenge to the sustainability of business-as-usual economic and political models, and so will be most vigorously denied and resisted.

There is, alas, no easy answer to the question of how we can individually and collectively change our lifestyles and behavior so as to avert disaster. Various sorts of social marketing campaigns analogous to the public health campaigns against smoking can be tried, but the change may have to be a generational one as young people grow up with awareness of a changing climate and a natural sense that something must be done about it. For the time being it seems that we need to proceed full speed ahead on all fronts—climate action planning through existing levels of government, organizing through NGOs and other social networks, construction of exciting models of zero-net-emissions living, ecological education of future generations so that perhaps they will live by different values, personal example so that we inspire others, and creation of new social networking strategies for change. We also, as I have argued in my book *Climate Change and Social Ecology*,[9] need to learn to see our world through a different lens offering a more ecological understanding of how human societies evolve and how through actions large and small we can shape that evolution in positive directions. Viewing our societies in such a paradigm, we can identify many different pressure points for action and understand how such actions can build synergistically on one another. In the long run, such a changed worldview may be the surest foundation for progressive social change—planning for societies that in turn can plan for sustainability.

8

ENERGY AND MATERIALS USE

Reducing human greenhouse gas emissions to virtually nothing, as necessary to head off the worst global warming scenarios, will require radical transformations in energy and materials use. These subjects were virtually unknown to the nascent environmental movement at the turn of the twentieth century, and only entered the public consciousness in a major way in the 1960s and 1970s. Now, in the era of climate change, they are at the center of sustainability planning. Many other reasons exist for reforming energy and materials use as well, including pollution of air and water, problems with toxic chemicals in the environment, and environmental justice impacts on disadvantaged communities.

These two major physical inputs to industrial society are highly interrelated. It takes energy to produce materials, for example to smelt bauxite into aluminum and then convert this metal into aluminum cans for soft drinks or components for consumer products. It also takes energy to dispose of materials, for example to haul waste to landfills or to recycle used material into new products. The reverse is true too, though to a lesser extent: it takes considerable amounts of materials (and energy) to produce energy. An oil refinery is a large consumer of both.[1] Compact fluorescent light bulbs contain mercury, a dangerous material. Nuclear power requires radioactive fuel. And so forth.

Most materials have a life cycle that is potentially near-circular; that is, they can be reused, recycled, or repurposed in various ways so that few of these resources are wasted after every human use. Energy, being subject to the second law of thermodynamics, cannot be reused or recycled. Entropy increases; the power contained within relatively concentrated forms of energy such as oil, gas, and electricity dissipates into less concentrated forms, such as heat. Energy can potentially be used for multiple purposes at once, as for example wood chips can be burned in cogeneration plants to heat homes and generate electricity at the same time. But it cannot be reused. So conservation is especially critical for energy, particularly if we are deriving it from fossil fuels. And reuse or recycling strategies come into play for materials, though the desirable mix of approaches will depend on the context.

SUSTAINABLE ENERGY USE

In terms of energy, two main areas of effort have dominated sustainability efforts since the 1970s: steps to reduce the amount of energy consumed, and steps to produce it through renewable energy technologies. In the energy field "conservation" is slightly different from "efficiency." The latter implies using less energy to perform the same tasks, through better motors, reduced friction, more appropriately sized gears and piping, improved design, and other strategies. The former implies also reducing the number of tasks that energy is needed to perform, for example by owning fewer vehicles, leaving fewer lights burning, heating houses to lower levels, and not making as many motor vehicle trips.

To promote both conservation and efficiency, utilities, states, local planners, and national governments have developed a wide array of "demand-side management" (DSM) programs to reduce energy use, including initiatives setting energy-use standards for appliances, promoting compact fluorescent lighting, helping residents insulate their homes, and providing energy audits to homes or businesses. National governments and some states have set fuel efficiency standards for motor vehicles, appliances, and buildings. City governments in cold climates have been particularly active in helping low-income residents weatherize their homes, since these residents frequently live in the most poorly insulated dwellings and may not be able to afford to do so otherwise. Such programs can be seen as promoting equity, by improving quality of life and saving money for the least well-off.

Industrial production processes have been another main area for energy conservation and efficiency. An extensive field of "environmental management" has emerged within business, concerned both with helping businesses meet environmental regulations and with making processes more energy- and resource-efficient. The ISO 14001 set of international standards represents a comprehensive approach to environmental management within industrial processes. These standards are employed on more than 700 sites in the UK alone, and their use is growing rapidly worldwide.[2]

Information plays a crucial role in enabling energy conservation. Real-time feedback to users can help them understand how much energy they are using for different purposes and encourage them to reduce such consumption. A classic example is the energy consumption dashboard screen provided on most hybrid vehicles. If the driver accelerates more gently or coasts up to a stoplight, they are rewarded by immediately seeing the miles-per-gallon bar shoot up to very high levels. Conversely, if they accelerate strongly they see their real-time mileage tumble. Such feedback can in effect train drivers to conserve fuel.

Similar real-time electricity consumption displays can be positioned inside the front door of homes or offices, or within individual rooms, to give occupants information on their energy consumption at any given moment. Submeters or usage monitors plugged into specific outlets can give even more direct feedback on the electricity consumed by particular devices. Feedback displays can provide plant operators information on energy consumed within industrial processes. Regulation and policy at various levels of government can encourage or require informational strategies of this sort (see Figure 8.1).

Meanwhile, a variety of forms of renewable energy hold promise for sustainability planning. Wind power is by far the largest alternative energy source in the world currently, and has grown enormously in recent years from a capacity of less than 5 Gigawatts in 1990 to 238 Gigawatts in 2011, or about 2.5 percent of global electricity use.[3] China, the US,

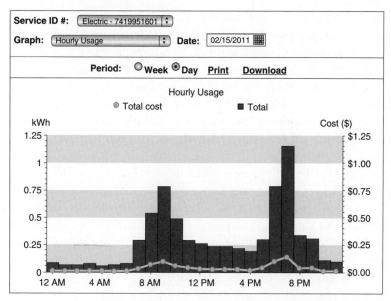

Figure 8.1 Informational strategies to reduce energy consumption. Letting consumers see exactly how they are using energy, as here with data from a Smart Meter, can help them reduce consumption

Germany, and Spain are the world's largest producers of wind energy. Areas with desirable wind speeds for wind farms are limited and often include offshore and coastal areas. Constructing wind farms in these locations at times comes into conflict with other local environmental goals as well as opposition from property owners. Such instances raise tough questions about priorities. A good case can be made that the need for renewable energy is so great that local environmental impacts are acceptable, if mitigated as quickly and extensively as possible. However, a case can also be made that large, centralized energy-producing facilities including wind farms are not the best way to go in the long run; that more decentralized, community-based energy systems are preferable. Probably both points of view hold some truth. Sustainability planning requires balancing such considerations within any given context and place.

Photovoltaics have grown enormously in extent during the early years of this century, accounting for 27.4 gigawatts of installed capacity as of 2012. Although panels used on rooftops or in solar farms are the most widely known form of photovoltaic installation, thin-film technology and concentrated solar power installations are important as well. The former approach applies thin photovoltaic layers to the surfaces of buildings or other structures. Thin-film photovoltaics are less efficient than rooftop panels, but far cheaper, and can potentially be applied in large quantities. Concentrated solar power installations focus the sun's energy on some type of thermal storage mechanism, often using water or molten salt. The heat is then used to run turbines, and these systems have the advantage of being able to store the solar energy for use at night.

Many questions exist regarding how utilities treat small-scale solar installations, in particular how they reimburse property-owners for excess electricity fed back into the utility grid. "Feed-in tariff" legislation in various countries and states requires such reimbursement, though utilities often argue that the small-scale electricity producer should help pay for its costs of maintaining the grid. Small changes in reimbursement rates can greatly raise the incentives for installation of distributed photovoltaic systems.

Small-scale solar water heating systems for building use became popular in the US in the 1970s, but have faded from view since that time. Such systems are often supplemented by an electric booster in temperate climates. They are very commonly used in China, Israel, Spain, and many other countries, and represent a relatively low-cost way to reduce the use of electricity or gas for domestic hot water.

Geothermal energy plants can be built in areas where hot rocks occur close to the surface (not to be confused with geothermal heat pumps which use the relative temperature of the earth 10 to 20 feet below ground to heat or cool buildings). Geothermal energy has relatively limited potential, since suitable areas are limited, but supplies about 11 gigawatts of electricity worldwide and about 5 percent of electricity in northern California. Tidal energy and river currents also offer potential energy sources, though major technical problems remain and applications are only in the pilot stage.

Lastly, a variety of biologically produced fuels can be said to be renewable. Wood pellets made from forest products industry wastes can be used for fuel in power plants. Methane from biodigesters that compost urban or farm wastes can also be used as a fuel. However, some biologically produced fuels have downsides from a sustainability point of view. In the mid-2000s many countries adopted incentives for ethanol production, usually to be blended with gasoline. By the early 2010s almost half of the US corn crop was being used for this purpose.[4] However, corn prices increased as a result, threatening turmoil in global food markets, and official enthusiasm quickly dissipated.

Germany's *Energiewende* (energy transformation) is an example of a holistic energy strategy combining conservation, renewables, and other types of policy.[5] Begun in 2000, with a legacy of Green Party support well before that, this transformation has involved substantial subsidies for wind, solar, and biomass power, making Germany among the world's leaders in these technologies. The country aims for renewable energy to account for 35 percent of electricity by 2020, compared with 20 percent in 2012. By the same year it plans to increase energy efficiency 20 percent over 2008. Also, after the Japanese Fukushima nuclear disaster in 2010, Germany decided to phase out its nuclear power plants by 2022, making conservation and rapid development of renewables even more urgent. Energy prices may rise with this transition, but that may not be a bad thing since higher prices will encourage conservation and probably more accurately reflect the true environmental and social costs of using energy.

A further question concerning renewable energy planning has to do with the scale of systems. Fossil fuel power plants have been primarily large scale, with generating capacities of hundreds of megawatts, and have been connected to grids covering entire continents under centralized utility control. Nuclear power plants are even larger scale, generally at least 1,000 megawatts in size. In contrast photovoltaic systems on residential roofs are usually less than 5 kilowatts. Individual wind turbines, small-scale hydroelectric systems, biodigesters, cogeneration plants, and other alternative energy facilities are also usually relatively small scale. Large utility companies are often not used to dealing with such small-scale generation, and have done relatively little to encourage it. These locally based systems don't easily fit into their model of a large, centrally controlled grid.

However, many renewable energy activists feel that decentralized, community-scale energy systems make the most sense in terms of sustainability. These systems are potentially more open to new technology, more subject to local control, more closely linked to individual choices around consumption, and more creative in their approach. Especially in rural areas and the developing world a great deal of experimentation is going on concerning microgrids and other distributed energy systems. Microgrids are localized clusters of energy use, generation, and storage that can be connected to larger-scale utility grids or disconnected for independent operation. Among other benefits, they help decrease the losses of electricity that occur over long-distance transmission, and increase reliability (especially in parts of the world where large-scale grids are not very reliable).

Another strategy has been for municipalities to own and operate their own utilities, providing greater public control (since utilities are then owned by the public rather than investors) and greater ability to meet local desires for renewable energy. US cities ranging from Seattle to Tallahassee currently own their own utilities, as do many rural towns and counties. This sort of "public power" can be seen as a way to detach energy from the capitalist market, making it easier to prioritize sustainability goals.

Such changes are difficult for large utility companies to deal with. They tend to oppose these small-scale or locally owned energy systems politically or by making it difficult for small-scale energy producers to link into their grid. Large-scale infrastructure does also have many advantages, for example transmission lines to take energy from wind farms or large solar installations to distant consumers. So new balances will need to be found preserving the advantages of large-scale grids while allowing a more diverse range of small-scale energy facilities and making management structures more responsive to their needs.

SUSTAINABLE MATERIALS USE

Whereas conservation is the number one sustainability strategy for energy, since energy cannot be reused, a variety of additional approaches apply to sustainable human use of materials ranging from metals to wood to water. The "Three Rs" of material use—reduce, reuse, recycle—form core strategies for sustainable handling of resources. However, a variety of other "Rs" can be pursued as well, including repurpose, recover, research, rethink, resist, and redistribute.

As with energy, "reduction" is the most important materials strategy, since this limits the volume of the materials stream from the beginning. Reduction occurs as we lower consumption, maintain and extend the lifetime of existing products, reuse what we already have, use less packaging, and diminish the waste associated with making products. Similar to energy conservation, reducing materials use lessens all subsequent problems with pollution, waste, and environmental justice. It can also have profound economic effects, helping move economies (if guided well by the public sector) away from the creation of unnecessary consumer products toward the production and distribution of things that people really need, such as decent housing, healthy food, and high-quality education.

"Reuse" encompasses all strategies in which products are used multiple times with only cleaning or refurbishment, such as when glass bottles are collected, washed at a beverage distributor, and reused for drinks, as was done routinely in most countries until the spread of plastic and aluminum containers in the 1970s. Energy and materials are saved through these multiple uses. Online "freecycle" networks have represented a major way to promote reuse in recent years; through these, people who are finished using a particular possession post a notice allowing others to come and collect it for their own use. Begun in the early 2000s, freecycle organizations now claim millions of members in 85 countries.[6] Such electronically facilitated sharing joins more traditional methods such as second-hand stores, flea markets, clothes swaps, and informal exchanges between family and friends.

Recycling involves a more extensive process of remanufacturing products from waste materials, as when paper is collected, shredded, mashed into pulp, and reprocessed into new paper goods. (With paper, recycling has limits as the length of fibers decreases with each remanufacturing, meaning that the recycled material can no longer be used for higher quality paper or must be mixed with virgin wood pulp.) Local governments began running recycling programs in the 1970s in North America and Europe in response to rising environmental consciousness and concerns about diminishing landfill space and pollution from waste incinerators. Aluminum has traditionally been one of the most successfully recycled materials; recycling this metal produces enormous energy savings over new production. Yet as of 2010 only about 50 percent of aluminum cans were recycled in the US. About 67 percent of steel cans were recycled and about 72 percent of newspapers, but only a third of glass bottles, 27–29 percent of plastic bottles, and 34 percent of the municipal waste stream overall.[7] Overall recycling rates are a bit better in the European Union (46 percent overall), Germany (69 percent), the Netherlands (69 percent), France (60 percent), and the UK (45 percent).[8] Countries such as Sweden burn a large amount of their waste in incinerators that produce energy, a strategy that has benefits in terms of energy but the downside of discouraging recycling. But in almost every society at least a third of all waste (and often more than half) goes into landfills. So there is much more to be done in terms of materials recovery.

Plastics recycling is highly controversial. Although plastics represent the fastest-growing part of the waste stream, recycling rates are persistently low, totaling only 8 percent in the US in 2010.[9] The plastics industry has long since stepped back from a 1991 commitment to reach a rate of 25 percent recycling, and only two of seven main types of plastic resins (PET and HDPE) are recycled to any degree. These resins are turned into products such as carpets, textiles, plastic lumber, and bottles. Partly as a result of the low rate of plastics recycling, many municipalities have adopted ordinances outlawing use of plastic bags in locations such as grocery stores. Other particularly important "Rs" include recovery (of energy or raw materials from waste disposal operations) and rethinking (of lifestyles, processes, and consumption patterns). Examples of the former include producing electricity from waste incinerators, collecting methane from landfills, or using waste heat from power plants to heat surrounding neighborhoods. In the developing world, large numbers of people make their living recovering raw materials or usable products from landfills and other sources of discarded goods. Recovery involves creating multiple benefits from what were formerly single-use goods and processes, a core ecological design principle. It requires rethinking the nature and use of products, as well as changing definitions of what we need and how we should live. Rethinking materials use also means moving from open-ended resource systems, in which resources are taken from the earth, consumed, and then released as waste, to closed-loop resource cycles, in which "waste" materials are reused or recycled, and overall material consumption is substantially reduced (see Figure 8.2).

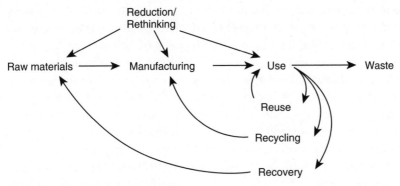

Figure 8.2 The urban waste stream

One of the biggest resource policy challenges for cities and local agencies is what balance to strike among the "Three Rs." Traditionally the greatest emphasis has been on recycling. These programs are the most public and visible type of sustainable resource use activity, something that ordinary citizens can easily participate in. However, much greater sustainability benefits can potentially be derived from reduction of the solid waste stream in the first place. Unfortunately these other alternatives require more fundamental changes in consumer attitudes, business procedures, and economic ideology. Yet such changes are not impossible—bottles and other materials were routinely reused in the United States until the spread of plastic and aluminum containers in the 1970s, and such reuse still occurs in much of the rest of the world. The Swiss, for example, require beverages to be distributed in certain standard-sized reusable bottles. In France, Guatemala, and many other countries one can see trucks carrying empty drink bottles back to the distributor. Throwaway containers are gaining ground in many places, but it is certainly not too late to reverse the trend.

Increasingly the focus of action by local, state, and regional governments is on "integrated waste management" in which all of these strategies are employed in varying combinations to achieve the greatest possible improvements in resource use. Municipalities typically run solid waste collection operations and are under the most direct pressure to reduce the waste stream because of rising landfill costs ("tipping fees"), a scarcity of landfill space, and/or state legislation requiring such a reduction. In 1989, for example, the state of California required cities to reduce their solid waste stream by 50 percent by the year 2000 (the actual reduction achieved was 47 percent). The state then raised redemption values twice during the 2000s, with the result that in 2011 consumers recycled 82 percent of beverage containers in the state.[10] In 2003 San Francisco planners set even more stringent goals of reducing the city's solid waste stream by 75 percent in 2010 and 100 percent in 2020, reaching a "zero waste" situation. In 2009 the Board of Supervisors voted to make separation of recyclables and compostables mandatory for all residents, and in 2010 Mayor Gavin Newsom announced that the city had actually reached a 77 percent waste reduction. The city has also increased monthly pickup fees for waste; undertaken curbside recycling for many materials including electronic components, hazardous materials, yard waste and food scraps; and initiated programs to reduce construction and demolition waste.[11]

In the UK, the Blair Administration announced in the early 2000s a goal of increasing the household waste diversion rate to 30 percent by 2010. Actual growth in recycling was strong in the mid- and late-2000s, leading to a recycling rate of well over 40 percent by 2011.[12] For their part, US states such as Vermont, Connecticut, Delaware, Massachusetts, Oregon, New York, Iowa, Maine, California, and Michigan passed bottle deposit legislation in the 1970s to promote recycling of beverage containers. Such deposit laws have been effective in increasing collection rates and reducing litter in participating states (and have spawned a cottage industry of "scavengers" who collect containers illicitly from household bins before municipal recycling trucks come around). However, deposit rates (typically $.05 per container) have not risen in many states since the 1970s, and the incentive has undoubtedly declined over time.

Local governments also frequently run programs to promote composting of kitchen scraps or to pick up and compost yard waste, since this organic material is a major element of the waste stream. Other programs collect used motor oil for recycling, promote recycling of construction waste, and even reuse surplus paint (Seattle has a program in which collected latex paint is blended and resold at a discount to local schools and hospitals as "Seattle beige"). Changing public sector procurement policies to require agencies to purchase reused/recycled products is a way to stimulate reuse and recycling. Eco-labeling programs also help establish norms for recycled content and improve consumer confidence that materials sold as recycled or reused are the real thing.

As part of their economic development activities, cities can promote eco-industrial parks in which industries use each other's byproducts as inputs. The most fully developed example of this remains the industrial ecosystem in Kalundborg, Denmark, in which a power plant, cement plant, oil refinery, pharmaceutical plant, wallboard factory, sulfuric acid producer, farmers, and residences use one another's waste steam, water, gas, fly ash, gypsum, and other materials.[13] With time and active coordination by local authorities, other such systems may emerge.

Liquid wastes pose a different set of recycling and reuse problems. Cities have great control over how wastes such as sewage and storm-water runoff are handled, even if these effluents are treated by a local or subregional utility and not the municipal government itself. Separating sewage and storm-water systems, so that heavy storms do not overwhelm sewage treatment facilities and flush contaminants into nearby waterways, is one basic step that many cities in North America are still working on. Stenciling storm drains (e.g. "No Dumping: Drains to Bay") to prevent people from dumping used motor oil and other materials into aquatic systems is another basic way to reduce improper waste disposal.

Some cities, such as Arcata on the northern California coast, have created wetlands for ecological sewage treatment, generally after primary treatment (settling of solids) and secondary treatment (aeration of the effluent) have been provided in a traditional facility.[14] A more radical method of ecological wastewater treatment has been undertaken by the Nova Scotia town of Bear River, which installed a Solar Aquatic "Living Machine" sewage treatment facility in 1995. This system handles sewage for 881 residents and consists of a series of large tanks inside a greenhouse to recycle liquids and sludge using plants such as water hyacinths, flower ginger, watercress, willows, mints, and grasses. The installation has proven a major tourist attraction in addition to performing its primary function.[15]

LIFESTYLE CHANGE

Although much progress has been made with energy conservation, renewable energy, and materials reuse and recycling, such efforts by themselves are not enough to set energy and materials use on a path toward sustainability. To a large extent growth in population and materialistic lifestyles has swamped the gains made by these means. Motor vehicles are more efficient and less polluting than they were in the mid-twentieth century, but there are eight times as many of them on the road globally as there were in 1960,[16] so the gains have been to a large extent cancelled out. Similar increases in use of appliances, personal electronic devices, and air travel threaten to overwhelm efficiencies in those areas. The prospect of extending resource-intensive lifestyles, currently enjoyed by perhaps a billion people worldwide, to seven to ten billion strains credulity. So in addition to improved energy and materials use, uncomfortable issues of population, equity, and lifestyles must be addressed, as will be discussed further in Chapter 17.

Which lifestyle changes are most important in terms of energy and materials? The answer to this question depends, as does so much else, on context. In cold climates in which the most energy is used for heating, smaller and more energy-efficient houses and apartment units may be the priority. In sunnier climates in which motor vehicles are the leading energy consumer, the need may be greatest to live and work locally so as to reduce driving (though building energy use should of course also be reduced). Needs to lower consumption of specific products will change over time as regulations, materials, and industry standards change. For example, in recent years there has been a great need to reduce the use of various refrigerants—especially chloroflurocarbons and hydrochloro-fluorocarbons—with a high global warming impact and/or destructive action on the stratospheric ozone layer. Efforts to find substitutes for such coolants are essential, but until these efforts fully succeed it is important to minimize their use through lifestyle changes, for example by purchasing fewer, smaller, and more efficient refrigerators and air conditioners. These appliances, even efficient models, are also among the leading users of electricity in households, so changing consumer decisions in this way is desirable simply from the point of view of energy conservation.

If the experience of sustainability activism so far has shown anything, it is that people are highly reluctant to change their lifestyles. It is also difficult for consumers to educate themselves concerning the relative environmental impacts of different choices, determine the most ecological purchase options, or afford greener models. So the public sector must step in to ensure that adequate information is available, provide strong economic incentives in favor of efficiency, set baseline requirements for efficiency, and even outlaw certain products. Such mandates are essential given the enormous incentives for industries to produce cheap, resource-intensive products. Governments can, for example, prohibit the sale of incandescent light bulbs, wasteful appliances, and gas-guzzling motor vehicles. Such action should not be seen as an unwarranted intrusion on individual choice, as conservatives would argue. Rather it is simply a way of compensating for the natural difficulty people have in changing their lifestyles and the inability of markets to encourage the change needed to protect the planet and make progress towards sustainability.

9

ENVIRONMENTAL PLANNING

One of the core tasks of sustainability planning is to help human communities coexist with the natural world. Although it includes the traditional goal of wilderness conservation, the scope of environmental planning has greatly changed over the years. Whereas 100 or even 50 years ago environmentalism sought primarily to protect and conserve existing wilderness, species, and ecosystems, now increasing effort goes toward restoring what has already been damaged or built upon, as well as educating humans about their relationship with the natural world. To put it another way, the eco-city of the future is one in which the gray urban world is radically greened and residents are reconnected with living systems. This process will improve not only environmental sustainability, but human health and quality of life within our communities.

Environmental planning thus focuses on these three main types of actions—protection, restoration, and education—through which we can reintegrate our living environments into ecological systems. Through such actions we can gain perspective on our relation to the planet, learn to coexist with its systems, and hopefully come to respect and care for the natural world. That sense of respect and caring, about both human and natural systems, is at the root of sustainability planning.

ENVIRONMENTAL PROTECTION

Although it has roots in North America and Europe, the environmental movement has had many successes worldwide in protecting land and ecosystems from development. Since the late nineteenth century these efforts have been of many sorts, and include both "preservationist" approaches to preserve wilderness in a more-or-less natural state, and "conservationist" approaches to manage landscapes so as to have both human use and a moderate degree of ecological function.[1] We will not go into the historical difference between these mind-sets here; suffice it to say that both are necessary in different contexts within sustainability planning and environmental protection.

During the Industrial Revolution of the mid-1800s, many early environmental plan-ners, designers, and activists sought to establish urban parks which were urgently needed within the rapidly expanding, dirty, unsanitary, and crowded industrial cities of that time. Often building on European ideals of pastoral landscapes with lawn and trees, these pio-neers created artificial green spaces within cities so that urban dwellers could seek recrea-tion and respite from urban life. New York's Central Park is the iconic example. This landscape of lawns, trees, walkways, ponds, and buildings bears little resemblance to the swamps, hills, and forests that formerly existed on the site, but offers enormous diversity of spaces for people and boundless charm. In many ways it is the heart of New York City, and its green spaces are the lungs. New York without Central Park (plus similar spaces such as Prospect Park in Brooklyn) would be a much tougher, grittier place. Quite probably fewer people would want to live there. In European cities similar urban open spaces such as Hyde Park in London and the Bois de Boulogne and Bois de Vincennes in Paris evolved instead from the hunting grounds of royalty and nobles. Occasionally such urban parks and landscapes also had a practical function such as flood control. Boston's "Emerald Necklace" including the Fenway is an early example.

Since the early industrial era preservationists have focused also on keeping scenic rural lands from development. Many methods have been employed, including the establishment of greenbelts around cities such as London, planning laws that tightly controlled locations of new urban development such as in France and Germany, and purchase of threatened lands by individuals or organizations interested in preserving them such as the Nature Conservancy and the National Trust. Often nongovernmental organizations have initiated the fight for rural land protection, and then once political agreement has been reached governments set up agencies to manage land so preserved.

One of the main planning strategies for environmental preservation has been to estab-lish national, state, provincial, regional, or local parks. Most countries now have multiple national parks—large, scenic areas with more-or-less intact natural ecosystems, usually operated so as to allow recreational use by citizens and tourists. Like urban parks these preserves are usually managed for extensive human use, with many constructed facilities. In the developing world, indigenous villages often continue to exist within parks, and some harvesting or hunting may occur. But by virtue of their official designation these areas are recognized as highly important natural assets, and protected to a large extent from development. State, provincial, and regional parks are similar in nature, though often encompassing less dramatic natural areas and with fewer resources.

Wilderness areas and wildlife refuges represent yet another preservationist strategy. Such areas are often quite different from parks. They are usually wilder, and operate more to pre-serve particular species and ecosystems than to serve human visitors. Agencies administer-ing wilderness lands typically prohibit motorized transport and the construction of roads and buildings. Sometimes human use is forbidden altogether, since this would disturb hab-itat and species. Long battles have been fought over the designation of wilderness land, since logging, mining, and ranching industries frequently wish access to resources. The presence of any roads whatsoever, even dirt tracks, has usually been enough to disqualify lands from wilderness designation. But still some 9 million acres (36,000 square kilometers) of land in the US has been set aside as wilderness, managed by four different federal agencies.

At a local level, municipal governments can systematically protect natural features in other ways. A creek preservation ordinance, for example, might prohibit development within 30 feet of any waterway, forbid culverting or channelization of waterways, establish a program for acquiring trail easements along creeks, and create guidelines for re-establishing native riparian vegetation. The cities of Berkeley and Oakland, California, have established such ordinances. A local wetland protection ordinance might go well beyond US Federal Section 404 regulation (which requires developers to obtain permits to fill wetlands) by actually prohibiting any fill, grading, or land clearance, and by regulating other structures such as docks, walkways, fences, and observation decks. An urban forestry ordinance might establish guidelines for street tree planting and park landscaping, and protect "heritage trees" above a certain diameter from being cut down. These and other steps can help local government preserve important ecological features within urban areas.

Environmental protection initiatives have also focused on air and water quality. These planning efforts have been moderately successful within many industrialized countries, primarily through legislation over the past 50 years that has greatly reduced pollution from industry, power plants, motor vehicles, and residences. In the 1950s London was known to much of the rest of Britain as "the Smoke" due to the sulfurous haze that hung over the city, produced by the widespread burning of coal for heat and industry. The air in Los Angeles was almost as bad, with pollution produced in that case by millions of motor vehicles with no emissions controls. Today both cities have much cleaner skies, though Southern California does still suffer from many bad air days. China is in a somewhat earlier phase of air quality regulation; the construction of hundreds of coal-fired electric power plants has resulted in a thick white haze of particulate emissions frequently hanging over many regions of the country. Hopefully the addition of pollution control equipment and the conversion of power sources to renewable energy will reduce this problem before long.

Increasing regulation since the 1960s has helped protect water quality in many parts of the world. One main focus has been on what is known as "point source" pollutants—pipes from industrial facilities, sewage treatment plants, power plants, and storm-water systems carrying foreign materials into streams, rivers, and bays. These sources are relatively easy to regulate because they can be specifically identified, controlled, and monitored. More difficult is non-point-source pollution—runoff from farm fields, roads, parking lots, and residential subdivisions that does not go through specific pipes or facilities. Such runoff carries fertilizers, pesticides, herbicides, bacteria, and motor oil into waterways. Phosphorus from farm fertilizers, for example, often leads to algae blooms in bodies of water. Preserving waterways from such non-point-source pollutants requires different sorts of regulation and a more general process of public education.

The field of environmental assessment analyzes industrial processes with an eye toward reducing pollution and waste. The ISO 14001 international standards, developed by the Swiss-based International Organization for Standardization and first adopted in 1996, have helped bring about environmentally appropriate industrial processes. In addition the US Environmental Protection Agency and similar institutions in other countries have worked actively with companies around pollution prevention, and some industries have undertaken such programs themselves internally simply as a matter of efficient management practice.

The "polluter pays principle" is a strategy to help control various types of pollution. This approach, potentially enforced by government, uses economic incentives to reduce environmentally risky or destructive behavior, assuming that potential polluters are aware of the cleanup costs up front. However, it can be extremely difficult to make polluters pay if the relevant industries are powerful and politically well connected. Extensive litigation has often been required, for example, to obtain cleanup assistance from polluters responsible for Superfund hazardous waste sites in the US. If much time has passed the companies may have disappeared or gone bankrupt, and the public sector has to pay for the cleanup. Government agencies such as the military are also frequently responsible for pollution, in which case the taxpayers must pay.

One basic debate in terms of environmental protection policy has to do with the relative merits of regulation compared with market approaches or voluntary action. Much mid-twentieth-century environmental planning focused on regulation, for example leading to the UK's Clean Air Act (1956) and the US's Clean Air Act (1963) and Clean Water Act (1965), both much strengthened later in the 1970s. However, by the 1980s neoconservative governments in Britain, the US, and elsewhere emphasized market approaches or voluntary action instead. Market approaches (in particular a cap-and-trade system) did, however, work fairly well in terms of reducing acid rain in the US during the 1990s, and are currently being used to try to limit greenhouse gas emissions in Europe and California. But they have not been able to address many other environmental protection problems, and raise questions about how "pollution permits" are distributed (if given away freely or cheaply these can be a windfall for corporations), and whether polluters should be allowed to pollute at all if strategies to reduce pollution are available.

Voluntary pollution reduction has had even less success. Yes, there is the occasional business owner, such as the late Ray Anderson, the visionary head of the international carpet company Interface, who undertakes initiatives on their own. But most do not. Especially if there is money to be made in producing goods in polluting ways, businesses are quite good at keeping such idealistic concerns at bay. Despite rhetoric from former US President George W. Bush and others, a purely voluntary environmental protection strategy runs a high chance of not working. In essence it is the same business-as-usual framework that has led to so many current sustainability problems.

Progress has been made worldwide in terms of environmental protection, but much more can be done. We are continually learning about new forms of environmental damage. For example, scientists only belatedly realized the risks to human health of small particulate matter, often labeled PM_{10} and $PM_{2.5}$ to refer to the size of particles in microns, and also referred to as aerosols. Initial air quality legislation in the 1960s and 1970s did not address these pollutants, which are generated from diesel engines, power plants burning fossil fuels, industry, and dust blowing off farm fields. It took a concerted effort in the 1990s and 2000s, more than 30 years after initial air quality legislation, to begin adding regulation concerning these pollutants.

ENVIRONMENTAL RESTORATION

While environmental protection efforts have been underway for more than a century, restoration planning is much more recent. The basic concept is simple: not just to prevent environmental damage from human activities, but to begin healing the damage that has been done in the past, in the process reintegrating nature into human lives and communities.

As with other sustainability issues, many scales of planning and design play a role in ecological restoration. Planning at the international scale has had some successes, for example producing agreements to protect and restore the Earth's ozone layer, to revive ocean fisheries, and to reforest degraded landscapes in the Amazon and elsewhere. NGOs often play a leading role in brokering these agreements. The International Union for the Conservation of Nature (IUCN), for example, has convened a global partnership for Forest Landscape Restoration, and aims to restore 150 million hectares (5,900 square miles) of degraded forest worldwide by 2020.[2]

At the national scale, restoration planning can reorient past habits of development and large-scale infrastructure. Instead of building large dams, roads, and power plants that destroy sizable ecosystems, national agencies can lead the way toward restoration. The US Army Corps of Engineers, for example, is an agency famous in the past for channelizing rivers, draining wetlands, building dams, and otherwise damaging ecosystems in the name of progress. However, the era of large-scale dam-building has now come to an end in the US, and the agency has adopted a set of "environmental operating principles" focusing on environmental sustainability. In 2002 it began a Sustainable Rivers Project with the Nature Conservancy to modify operations of dams on 10 rivers it manages in the US. Studies have shown that better management of water flow from dams can increase habitat in river corridors downstream and improve water quality, the amount of water available for human use, and production of electricity from dams.[3]

Costa Rica is well known for its forest restoration planning at the national level. Long plagued by deforestation, the country first established a national park system in the 1970s, and later began a concerted effort to encourage its citizens to use bamboo rather than wood as a building material. The national government has strongly promoted ecotourism, and gives cash grants to landowners willing to sign a contract for reforestation.[4] Beginning in the late 1990s it also used carbon offset funds through the Clean Development Mechanism established by the Kyoto Protocol to protect and restore forestland.[5] Many private and educational organizations have conducted restoration projects in the country, which now stands as a model of restoration planning at the national level.

State- or regional-scale restoration projects are increasingly widespread, in particular since these scales often correspond to watersheds or other natural areas of ecological significance. For example, in the state of Maine the Penobscot River Restoration Trust, a coalition of the Penobscot Indian Nation and several environmental groups, brokered a complex deal in the early 2010s with utility companies to remove several dams and add lifts to others so as to allow salmon and other fish access to 1,000 miles of river.[6] In true win-win fashion, the restoration groups allowed power companies to increase renewable electricity generation from other sites to compensate for the loss of power produced by the demolished dams. As is typical with complicated ecosystem restoration projects, the deal took more than 10 years to come to fruition, and involved many players, including one Indian tribe, six local, regional, and national environmental NGOs, several private companies and landowners, four state agencies, and four federal agencies.[7]

Increasingly, cities are taking a more proactive and systematic approach to restoring green spaces. Local actions by government or NGOs often include revegetating previously degraded sites or ecosystems, cleaning up "brownfield" industrial sites, unearthing culverted creeks and restoring stream function, recreating wetlands, reconnecting isolated patches of habitat with new wildlife corridors, and replacing asphalt with permeable paving that allows aquifers to recharge from rainfall. The end goal is to connect every bit of the city with the natural landscape in some way, in the process making communities far more attractive and livable.

The restoration process is not easy, of course. It requires a good knowledge of landscape ecology within the place in question—understanding interactions between soils, climate, hydrology, species, plant communities, and potential human use. It also requires excellent abilities to build political support and broker deals among many parties. Even with the best knowledge and skills, complete restoration may not be possible. Wetlands, for example, have proven very difficult to recreate with the same level of function as naturally occurring systems.[8] It is best to preserve existing wetlands in their natural form, rather than think that they can be recreated with similar function on the same or different sites.

An evolution in aesthetics may be required as well for restoration projects. Many traditional, widely popular park designs emphasizing lawns and scattered trees have little ecological value. Conversely, landscapes with high ecological value may strike members of the public as unkempt or messy. Native vegetation may go dormant at certain times of the year when visitors are expecting green and flowering plants, and may not be as showy or exotic as in conventional gardens. So a balance between human and ecological function may be needed at any given site, and environmental education becomes important to help public aesthetic preferences evolve and to ensure that members of the public understand processes of ecological restoration.

One important step in restoration planning is to develop integrated park or open space master plans that take a systems approach toward improving ecological function. These comprehensive documents will ideally inventory existing parks, open spaces, watershed elements, and habitat while developing a vision for connecting and enhancing these. Once-isolated parks and bits of habitat on private land can be reconnected through new greenways or green corridors in street rights-of-way. Such plans can create "green infrastructure" systems which manage storm-water runoff through green spaces, green roofs, permeable paving, swales, and retention basins rather than large-scale "gray" infrastructure focusing on piping systems carrying water away below ground. New York City adopted an ambitious Green Infrastructure Plan in 2010 that aims to reduce runoff by 3.8 billion gallons per year through such means.[9] This plan, in turn, is one piece of that city's sustainability oriented PlaNYC planning framework launched in 2007.

Local governments can take the lead in cleaning up contaminated or previously used sites, and in restoring key features of local ecosystems. Private industry is often reluctant to undertake such efforts, and in the absence of strong growth management policy (more on that in Chapter 10) may find it cheaper and easier to relocate to greenfield sites elsewhere. Local community groups and neighborhood associations usually do not have the resources to pursue cleanup activities themselves, although they can be an important source of volunteer labor for stewardship of restored ecosystem features. National initiatives such as the US Superfund program may assist with some of the most contaminated sites but do not cover others. So the leadership role often falls to local planners, elected leaders, or NGOs to initiate the reuse of previously urbanized land and manage the process.

Figure 9.1 The engineering approach. An industrial approach to development has resulted in destruction of natural landscape elements, as illustrated by this channelized stream in Las Vegas

Restoration activities are still preliminary in many places. Technical understanding is still developing of how to restore functional habitat, handle storm water on site, and mix recreational and ecological goals within landscape design. Political backing for such initiatives is often fragile or lacking. Public acceptance varies from community to community. But the field is growing, and the basic concept of restoring as well as protecting natural areas has become a centerpiece of sustainability planning (see Figures 9.1 and 9.2).

ENVIRONMENTAL EDUCATION

It would of course be best if humans lived harmoniously with the natural world from the start, rather than having to laboriously protect and restore elements of the environment after they are damaged. This is the challenge of environmental education—helping us learn to understand, respect, and care for the environments around us, and structure our lives, professions, and habitats accordingly. Although still a relatively minor part of official school

Figure 9.2 Ecological restoration. Sustainable design instead respects and works with the natural landscape, emphasizing ecological restoration. In this photo volunteers have unearthed a culverted creek in Berkeley and are restoring a naturalistic, meandering stream channel

curricula, not nearly as much emphasized as reading, math, and science, environmental education (and civic education, we might add) is crucial to a sustainable society.

Environmental education has roots going back to at least the nineteenth century, and has grown spectacularly since 1970, when the first Earth Day helped kick off such efforts on a more concerted basis. Yet it is still far from mainstream. A North American Association for Environmental Education was founded in 1971, and now helps coordinate 45 state, provincial, and local affiliates. An Intergovernmental Conference on Environmental Education, organized in 1977 by two UN agencies in Tbilisi, Georgia, and attended by 265 delegates from 65 countries, developed an influential manifesto on the topic. The Tbilisi Declaration states that

A basic aim of environmental education is to succeed in making individuals and communities understand the complex nature of the natural and the built environments resulting from the interaction of their biological, physical, social, economic, and cultural aspects, and acquire the knowledge, values, attitudes, and practical skills to participate in a responsible and effective way in anticipating and solving environmental problems, and in the management of the quality of the environment.[10]

The Declaration emphasizes that environmental education should be connected with real life and environmental problems faced by actual communities. Environmental education should be lifelong, the delegates agreed, should involve development of critical thinking and problem-solving skills, should be linked to decision-making, and should "help learners discover the symptoms and real causes of environmental problems."

To meet such objectives, several main dimensions of environmental education are important to plan for. These include programs within K-12 and higher education curricula, experiential learning programs outside the classroom for people of all ages, and the creation of a green urban environment that is profoundly educational for its residents within daily life.

For several decades now, dedicated teachers have pursued environmental education within existing schools and curricula. Although often treated as an elective or add-on subject, teachers have linked environmental topics to science classes, social studies, geography, literature, art, and many other subjects. Within a math class, for example, students might use math to calculate the biodiversity of different environments; within history they might explore different cultures' views of the environment.[11] Some classes also use project-based learning, in which students work on environmental projects in the community outside of the classroom. They might, for example, observe and monitor a local wetland, or participate in ecological restoration activities at a local park.

Experiential learning programs, in which people of all ages are immersed in educational experiences outside the classroom, can be a very effective form of environmental education. There are a great many types of these programs, ranging from those such as Outward Bound and the National Outdoor Leadership School (NOLS) in the US that emphasize development of self-reliance and leadership within wilderness settings, to those more focused on ecological science per se. Service learning programs are particularly popular, in which students or volunteer groups, for example, travel to Costa Rica to help replant forests or to Honduras to build clean water systems for mountain villages. Such volunteer activities are a core part of many local ecological restoration projects as well. Some programs focus more on engagement with social issues, such as poverty, housing, or public health, than on environment. But both sets of issues are important, and they fit together in multiple ways.

Finally, a great deal can be done to make our built environments educational. Greening school buildings and grounds is one place to start—a substantial movement for playground greening and edible schoolyards has been underway for some time.[12] One pioneering project was the Edible Schoolyard at Martin Luther King Jr. Middle School in Berkeley, California, spearheaded by famed chef Alice Waters and involving hundreds of students in developing a one-acre organic garden, a kitchen classroom, and community outreach programs. However, dozens of similar programs have sprung up worldwide, helping students participate in all aspects of growing, harvesting, and preparing food.[13]

In other places, schoolyards are landscaped with native or drought-tolerant plants, school buildings collect rainwater for later use in irrigation, and schools receive energy efficiency and renewable energy upgrades. All such initiatives can be done in conjunction with environmental curricula for students.

Beyond school buildings and grounds, there is no reason why every bit of cities and towns shouldn't be educative in a way that promotes understanding of sustainability. People should be able to see flows of water through the city, with rainfall running from buildings and streets through swales to infiltrate into the ground or into holding tanks for irrigation. People should be clued into flows of energy, seeing it generated locally through photovoltaics, wind turbines, or biodigesters, and receiving real-time information about its use within buildings and vehicles. People should be able to observe the restoration of natural habitat near where they live and work, seeing hawks circling above city buildings and foxes prowling stream corridors. In terms of social issues, urban residents should be able to learn about the history, culture, and economic character of cities simply by living within them. Architecture, urban design, and public art can help create this educative environment. If an architect is designing a building for industry, perhaps large windows can let the passers-by see the workers and machines. If a landscape architect is designing a public space, they can "make nature visible" (in the words of Sim Van der Ryn and Stuart Cowan) by careful use of water flows, vegetation, and energy systems. If an urban designer is laying out a new neighborhood, he or she can design multiple walking paths throughout the landscape so that people can both get exercise and observe the daily activities of the community and local ecosystem.

Above all, an educational environment is a humane environment, helping people understand themselves, the dynamics of the society around them, and interactions with nature. Such understanding can potentially lead toward a more informed public and a more enlightened electorate. But on a daily basis such an environment is also likely to provide a far higher quality of life for residents, perhaps relieving some of the stress of current lifestyles and improving both physical and mental health.

10

LAND USE AND URBAN GROWTH

Changing land use patterns are a pressing challenge for sustainability planning. The reasons are multiple: to preserve important natural habitat, wilderness, and agricultural land; to reduce energy and materials consumption as well as GHG emissions; to rechannel investment into existing urban areas so that those areas thrive and social equity improves; and to create more balanced, diverse, and livable communities for ourselves. Urban regions in both industrial and developing countries obviously cannot expand forever in the extremely rapid fashion that they have for the past century and a half. Their growth causes many problems related to motor vehicle use, pollution, GHG emissions, congestion, quality of life, and social equity. But managing urban growth is not just a question of stopping suburban sprawl. Equally important is the need to improve land use and urban design within existing urban areas. The two strategies go hand-in-hand. Only by revitalizing existing urban areas and accommodating more people within them can pressures for outward expansion be reduced and growth management policies succeed.

Sustainable land use planning must start with the question, "What is the most appropriate form of development for any particular site?" The answer may well be "none." It may be best not to build on a given parcel at all if it is ecologically important or located far from existing communities. The most appropriate use may be instead intensive agriculture or restoration of the site's degraded ecosystems to something approximating their pre-human state. Or if the site is strategically located within an urban area it may be appropriate to build more intensely than was originally intended. The most appropriate building or type of use may also be different than originally considered. If a parcel is located within an area already rich with jobs, it may be best to build housing to create a better local balance, despite existing zoning for commercial uses (the compatability of building types must also be considered). And vice versa.

Some land uses may be generally inappropriate in most locations. Big-box retail development, for example, which has been pursued extensively in many industrial countries beginning in the late 1970s, may not be desirable anywhere within a more sustainable society. This form of development uses land inefficiently, tends to generate high levels of motor vehicle use, drives smaller, locally owned stores out of business, and creates an over-scaled, pedestrian-unfriendly environment. Sprawling office parks may not be a particularly desirable model either, for similar reasons. Rural sprawl—houses on 1–10 acre lots—should probably be restricted as well since it gobbles up farmland and wilderness while requiring inhabitants to drive long distances.

Sustainable land use is likely to emphasize instead compact, balanced, mixed-use communities rather than these single-use monocultures. Changing land use in this way may require new concepts of property ownership, moving away from the notion that people should be able to do anything they want with a piece of land which is "theirs," towards land use that balances individual, collective, and ecological interests. Such a transition represents a profound ethical change, to say the least. But it is an essential one if sustainable land use is to come about.

THE COMPACT CITY MODEL

The "compact city" has been a goal of many planners for decades, especially in places such as North America and Australia where sprawl is rampant and communities have a low population density. Although a few theorists at the libertarian end of the political spectrum argue that sprawl is not bad, that it is desirable in terms of enhancing individual choice and mobility,[1] such arguments usually leave out the immense social and ecological costs of this form of development and the fact that sprawling cities often do not give residents any choice at all to live in good, affordable housing in green, safe, well-located, and walkable urban neighborhoods.

Compact cities represent a radically different model from most twentieth-century urbanization in North America (and, to a lesser extent, Europe). If pursued rigorously this approach would call for virtually all new residents and businesses to be housed within the existing urban envelope through infill development, and would involve a range of other urban design strategies to improve the livability of such environments. The goal would not just be compactness or density per se, but improved land use mix, public spaces, street design, transportation patterns, housing options, and the like. Decent schools and safe neighborhoods would be essential elements as well. Some compact communities might include high-rise buildings and some not. But all would focus on infilling and improving existing urban places, rather than allowing development to sprawl across the landscape.

Infill development includes building on vacant lots within the urban area, redeveloping underutilized lands where, say, small or deteriorating buildings exist, and rehabilitating or expanding existing buildings. The opposite of infill is often known as "greenfield" development in that it frequently takes place on agricultural fields at the urban edge. There are no official statistics on infill development in the US (and the term is defined differently by different analyses), but most likely about 30 percent of new development is accommodated through infill in even the most infill-oriented US metropolitan areas.[2] One researcher has estimated the percentage of "population infill" during the 2000–2007 period even lower, ranging from a high of 20 percent in Portland Oregon and parts of southern California to a low of −7 percent in Detroit (which is losing population).[3] The United Kingdom, with its more limited land area and more urban tradition, set a goal of 60 percent infill under the Blair Administration and achieved a 57 percent level in some years.[4] Friends of the Earth in the UK has advocated increasing the infill target to 75 percent.[5] Some British authors complain that government-designed infill projects lack green space and integration with the natural landscape; Peter Hall has warned that compact city efforts may amount to "town cramming."[6] However, improving design, de-emphasizing motor vehicle infrastructure, and creating more parks, gardens, and restored ecosystem features within urban areas should be able to at least partially address these concerns.

In many Asian cities infill is problematic since cities there have traditionally been extremely dense. The density is of several kinds: often dilapidated but pedestrian-friendly low-to-mid-rise traditional urban neighborhoods; new high-rise apartment towers in more motor-vehicle-oriented environments; and informal settlements built illegally on a variety of types of public and private land. The issues are different in each case, but there is a common need to add green space and amenities while decongesting streets from the rising tide of motorized traffic. At the same time, these cities also increasingly see lower-density North American-style subdivisions on their outskirts, including upscale enclaves and a growing amount of rural sprawl. Discouraging those built landscape types is an important sustainability planning goal in such metropolitan regions, even though some urban expansion may be necessary to house the growing population.

Infill provides many benefits for the environment. For example, a 2007 US EPA study found that shifting just 8 percent of Denver's jobs and households over time toward 10 regional centers could reduce traffic congestion by more than 6 percent and air pollution emissions by 4 percent, while lowering motor vehicle trips by half a million per day.[7] Another review by leading planning scholars that same year found that compact development has the potential to reduce vehicle miles traveled (VMT, the leading measure of driving) by 20–40 percent relative to sprawl, and greenhouse gas emissions by 7–10 percent.[8] If combined with better transportation options, revised economic incentives, and more locally oriented lifestyles (which compact development makes possible), benefits could be even greater.

Compact development can also save local governments money that would otherwise have to be spent to build and service roads, sewers, water supply infrastructure, schools, and other services. Although there has been much debate about the exact costs of sprawl, they are substantial. In a review of more than 475 other studies of sprawl, Rutgers professor Robert Burchell and others found significant savings in compact development. The three studies bearing most directly on this topic found savings of 25 percent in roads, 20 percent in utilities, and 5 percent in schools compared with sprawl development.[9] Another study in the *Journal of the American Planning Association* found that residents of compact developments are in effect subsidizing those in more sprawling developments by $17.24 a month in infrastructure costs, or $2,567 over a 30-year period.[10] Development that is compact, contiguous, and well connected will also be easier and cheaper for public transit to service. In these and many other ways, sustainable urban form can prove cheaper and more cost-effective than the alternatives.

URBAN FORM VALUES

The need, however, is not just to focus on compactness, but on a whole set of urban form criteria that can help create sustainable places. These "good city form" values, to use the terminology of urban designer and MIT professor Kevin Lynch, can guide land use and growth management planning.[11]

In past centuries urban form has gone through a distinctive evolution in North America and much of the rest of the world. Medieval cities and many pre-industrial settlements are characterized by organic street patterns of tightly connected, winding streets that fit the needs of those traveling by foot or horse. Gridded street patterns with origins in Greek and Roman planning reappeared during the Renaissance in Europe and were repeated through-out the New World by the colonizing powers, being well adapted to the quick establish-ment of new cities for military or speculative land development purposes. Somewhat looser, rectangular-block grids appeared in late nineteenth-century North America along the new streetcar lines, and are often referred to as "streetcar suburbs."[12] Developers wish-ing to create tony upper-class neighborhoods also experimented with "garden suburbs" characterized by lower density and curving roadways during this time, beginning with Frederick Law Olmsted's community of Riverside near Chicago in 1869. In the twentieth century, especially after the Second World War, the spread of the motor vehicle encouraged increasingly disconnected subdivisions with cul-de-sacs and loop roads, influenced by previous garden suburb design.

As the twentieth century progressed urban regions became vastly larger and more dis-persed, and new types of urban form appeared, such as the office park, the shopping center, and the regional mall. Since these types of urban form accreted onto one another in each metropolitan area, the result in many cases was a pattern that might be labeled "Vienna surrounded by Phoenix," as one observer has characterized Toronto[13]—relatively dense, tightly connected grids of streets created in the nineteenth and early twentieth cen-turies surrounded by a much more loosely connected sprawl of suburbs built in the mid-to late twentieth (see Figures 10.1 to 10.3).

Figure 10.1 Well-connected and poorly connected street fabrics. These nine-square-mile areas compare an urban fabric created between 1880 and 1920 (eastside Portland, Oregon) with one created in the late twentieth century (suburban Washington County). Street patterns in the latter are far more disconnected. Many of the large open areas contain office parks

Sacramento

Modesto

Fresno

Stockton

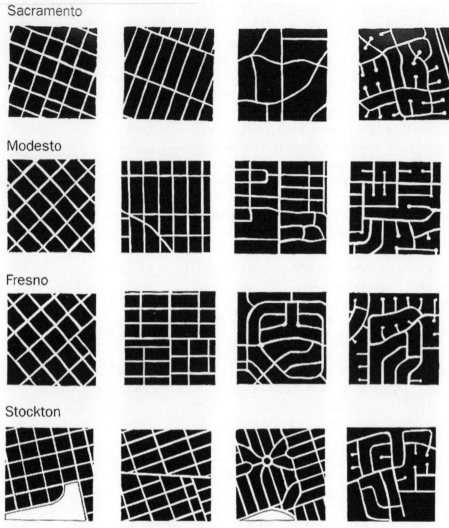

Figure 10.2 Evolving urban form. Most western US cities have followed a similar pattern of physical evolution toward more disconnected urban fabrics. The one-square-mile areas here represent neighborhoods created in the mid-1800s, the late 1800s, the early 1900s, and the mid- to late 1900s

Looking ahead, certain urban form values are likely to be particularly essential in the future for sustainability goals such as preserving open space, reducing automobile use, enhancing equity, and improving community vitality. The work of Kevin Lynch, especially his landmark *A Theory of Good City Form*, is an important precedent to such analysis, in that he was among the first to systematically analyze the values and characteristics of different types of urban form. But Lynch did much of his writing before the influence of the modern environmental movement had been fully felt, and his work needs to be updated in light of current sustainability concerns. Expanding on his efforts, five urban

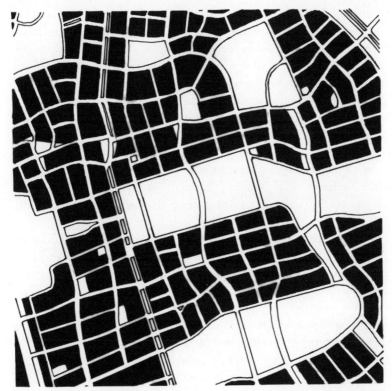

Figure 10.3 Connected street patterns. Cornell, a New Urbanist development in the Toronto suburb of Markham, shows a return to fine-grained street patterns, along with a generous amount of open space. Compare this urban fabric with those in Figure 10.2. This development suffers, however, from its suburban fringe location far from jobs and shopping

form values now seem particularly important to land use planning for more sustainable cities and towns (see Figure 10.4):

- *Compact urban form* limits suburban sprawl and makes more efficient use of land than in conventional suburbia. The challenge as previously mentioned is two-part: to preserve open space and to design a more efficient, compact, and livable urban form

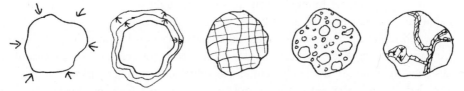

Figure 10.4 Sustainable urban form values. Sustainable urban form is likely to be compact, contiguous, connected, diverse, and ecological (integrated with the natural landscape)

inside growth limit lines. Regions such as Portland, Oregon, which have sought to manage growth through an urban growth boundary (UGB) have learned this lesson the hard way. Much sprawl has occurred inside the Portland UGB because it was set too far out initially and the characteristics of new development were not a major focus for the first 20 years. Somewhat belatedly the region was forced to rethink how to design more compact development within its existing urban area.

- *Contiguous urban form* implies that new expansion takes place next to existing urban areas. If new development projects are not contiguous, then inefficient, disjointed land use patterns are likely to result as the spaces between projects fill in haphazardly, and street connections between subdivisions are likely to be poor. The opposite of contiguous development is often referred to as "leapfrog" growth, in that development jumps from place to place across the landscape to wherever developers can find cheap available land.

- *Connected urban form* features good street, path, and visual connections within the region, and is also relatively "legible" and easy for people to find their way around. Without these connections, a disjointed landscape is created in which walking, bicycling, using public transit, and even driving are difficult and involve circuitous routes. Arguably it then also becomes more difficult for residents in disconnected subdivisions to gain a sense of participation in the broader society. The nineteenth-century, square-block grids at the core of many older cities provide an extremely high degree of connectivity, promoting travel through the city by a variety of transportation modes. Not surprisingly, winding suburban street patterns feature very low connectivity.[14]

- *Diverse urban form* contains a mixture of land uses, building and housing types, architectural styles, and prices or rents. Preferably this mix is what Lynch termed "fine-grained," i.e. with elements mixed together at a small scale. If development is not diverse in these ways, then the result is a homogeneous built form, monotonous urban landscapes, segregation of income groups, and increased driving, congestion, and air pollution. Nineteenth-century neighborhoods with diverse building types and land uses are today among the most vibrant, attractive, and popular districts in many North American cities. Twentieth-century, single-use zoning was a major force preventing diversity of urban form and enforcing more coarse-grained land use mixes within cities. In addition, the large scale of recent homebuilding and office park construction often prevents the creation of a diverse urban fabric, in that each builder is often unwilling to create more than a single type of land use.

- *Ecological urban form* integrates features of the natural landscape into the form of the city in a way that protects and restores local ecosystems while providing recreational amenities for residents. In most urban areas little thought was given to this urban form value until the last third of the twentieth century. Developers simply bulldozed hills, culverted streams, and generally treated the landscape as a slate to be wiped clean for human use. Even garden suburb developers treated ecosystem elements primarily as aesthetic amenities for human benefit, not as valuable entities in their own right. In the past few decades, however, planners and citizen activists have begun seeking ways to protect or enhance ecosystem elements during the urbanization process. Regional and local planning agencies have designated park and greenway networks, placed some wetlands and stream corridors off-limits to development, and changed zoning codes to require park or open space dedication for most projects of any size. These agencies increasingly seek to identify key areas of ecological concern well in advance of development and integrate them into local or regional planning frameworks. Where simply zoning land off-limits to development is not an option, officials may choose to negotiate with developers, environmentalists, and other constituencies to develop "habitat conservation plans," under which some land is protected while other sites are developed.[15] This controversial approach has the advantage of leveraging protection for some areas without enormous expenditure of public funds, but the disadvantage of allowing much other development to go forth. In other cases local governments, park or open space districts, or NGOs such as the Nature Conservancy may purchase fragile habitat or conservation easements on the land to prevent development. Such emphasis on incorporating environmental concerns into the development of urban form, however, is still in its early stages and is far from universal.

These urban form criteria vary independently. Development can be contiguous to existing urban areas without being connected to them. It can be compact without being contiguous. And it can be all three of these things without having the diversity, human scale, or environmental elements that help make places attractive, livable, and sustainable. Nineteenth-century urban fabrics tend to be fairly compact, connected, and diverse, though not always contiguous or ecological. Twentieth-century fabrics tend to score low on all criteria. Recent New Urbanist-inspired planning tends to rate more highly on each, though still facing many obstacles because of the large scale at which large development companies like to work, the need to accommodate the motor vehicle, and the lack of institutions helping to implement these design values on a regional scale.

Municipalities in most countries have been weak in terms of taking a proactive approach toward shaping land use and urban form—adopting a profusion of zoning and subdivision codes to be sure, but then letting private developers take the lead in determining land use mixtures, block and lot layout, street patterns, street design, and open space configuration. The traditional land use and zoning maps found in municipal plans are far too general, and do not address these details. The more progressive jurisdictions may work extensively with developers on design questions internal to projects, for example requiring sidewalks and attractive landscaping, but usually don't focus on the bigger picture of how each subdivision will relate to existing development around it, and how its form will be sustainable. The result is a landscape of fragmented land uses and inwardly focused neighborhoods that do not relate well to one another or the city as a whole. Overall land use densities and mixes are often inappropriate for sustainability objectives.

TOOLS FOR MANAGING GROWTH

Since the mid-twentieth century many urban regions worldwide have engaged in "growth management" planning, that is, developing strategies to slow or stop outward growth and organize land use to better fit with transportation systems, ecological and recreational needs, and existing development. Growth management can be accomplished through a wide variety of planning mechanisms applied at different scales. Basic strategies start with community visioning exercises, development of land use plans, and regulations for zoning and subdivision control. Other more specific tools include urban growth boundaries, urban service limits, agricultural zoning, acquisition of conservation easements, transfer of development rights, and purchase of open space for parkland.

The basic mechanism through which cities in most developed countries regulate allowable types and densities of development, following land use guidelines established in planning documents, is zoning. Every parcel of land is given a zoning designation determining allowable uses, forms, and requirements for development. Developers can then seek exceptions to these designations by appealing to local boards and city councils. In countries such as the US, zoning is the primary form of land use control and is almost entirely developed by local governments. British planning, in contrast, places a smaller emphasis on zoning and instead relies on the discretion of local authorities in interpreting the public interest when deciding whether to permit development. National authorities in the UK may also review and decide upon local development applications for projects deemed to be of national interest.[16]

Zoning occupies a central place within the set of local planning tools that can promote sustainability. However, this mechanism has been problematic since New York City adopted the first comprehensive citywide ordinance in 1916. The case against zoning is strong. In North America and Australia it has often helped to institutionalize sprawl by mandating low residential densities and designating large areas within cities for strip development, regional malls, and office parks. Parking standards contained within zoning codes typically mandate that new development must provide several parking spaces per residential unit, encouraging a suburban style of development, raising housing costs, lowering densities, and promoting motor vehicle use. Zoning typically prohibits apartments, duplexes, townhouses, and many other forms of higher-density housing from being built in single-family home neighborhoods, reducing the range and affordability of housing types and further reducing urban densities. Most troubling of all, many communities historically used zoning as a device to keep poor people and minorities out of affluent areas,[17] usually by establishing requirements for large lots and setbacks that will ensure that only expensive housing is built ("exclusionary zoning"). For these and other reasons, rethinking zoning must be a key element of sustainability planning.

Extensive zoning codes are a relatively recent planning tool. Throughout the nineteenth century and the early decades of the twentieth, land development in most parts of the world was far less regulated than at present. Local governments did not restrict the allowed uses for given parcels of land or regulate the height, volume, density, or setbacks of buildings, although local cultural traditions and available building materials may have restricted such elements in other ways. But the scale and pace of development were generally much more limited then, before the motor vehicle, the elevator, and new construction methods enabled much more rapid and large-scale forms of building.

During the Progressive Era the idea arose that government should regulate land use planning in order to protect the health, safety, and property values of homeowners. Citizens often felt a need to protect residential neighborhoods from smokestack industries or rapid encroachment of apartment buildings. The rise of scientific management approaches to city planning also contributed to the attractiveness of zoning—it provided a set of formalized regulatory tasks to rationalize the emergence of a new profession (planning) that would use scientific analysis to manage urban development.

Pioneered in German cities such as Frankfurt in the 1890s, zoning laws first appeared in the US in the late 1910s and 1920s. These early versions were relatively simple, dividing cities into a few basic categories of allowed land uses. In the decades after the US Supreme Court upheld their constitutionality in the landmark 1926 decision *Euclid v. Ambler*, zoning schemes became increasingly detailed. Planners added requirements to cover building setbacks from lot-lines, maximum lot coverage ratios, minimum parking requirements, and many other aspects of site and building design. Today virtually every city and town has an extensive zoning code in place, with a few notable exceptions such as Houston. Zoning even extends to rural areas; counties typically zone agricultural land and open space outside cities for minimum parcel sizes of 5 to 100 acres or more. Zoning is the main tool for implementing the land use policy contained in a general plan, and state law generally requires local zoning designations to be consistent with this general plan land use vision.

Unfortunately, zoning has proven to be primarily a negative planning power, in that its main function is to prevent certain types of unwanted development from happening. In the United States this restrictive tool was generally not balanced by positive and proactive planning to ensure that desired forms of development do happen. Such an active public sector role in planning, land ownership, and development has happened more frequently in European countries, but is a challenge everywhere.[18]

Given its unfortunate impacts in the past, should zoning be eliminated or reduced in scope within sustainability planning? The answer is probably no. For one thing, eliminating zoning, although enticing as an idea, would be greeted by storms of political opposition in most communities, and may not be desirable given the need to respect existing neighborhoods and homeowners. For another thing, zoning can be useful for sustainability purposes, for example to protect historic neighborhoods and ecological features. The need is not so much to eliminate zoning altogether, or for that matter to add a host of new zoning requirements, but to use the zoning tools more wisely. In some cases that will mean loosening up the overly stringent and bureaucratic requirements that have been set up over the past century, for example by allowing a greater range of land uses or housing types in a given area. In other cases it will mean adopting new standards to nudge an often recalcitrant building industry towards more sustainable development, for example by adding minimum densities for residential development and requirements for water-conserving landscaping and setbacks from creeks. Maximum floorplate sizes for commercial buildings might also be set, to prohibit "big box" retail stores in many areas.

New Urbanist architects such as Andres Duany and Elizabeth Plater-Zyberk have developed a somewhat different approach of establishing a detailed urban design code for each new development that takes the place of zoning. Such "form-based" codes provide easy-to-understand diagrams showing what street, lot, and building designs are permissible. Form-based codes may be easier to understand and implement—and may better respond to needs for livability and to mesh with architectural traditions—than traditional lengthy, bureaucratic zoning codes.[19]

Since the 1990s many communities have been revising zoning codes to implement principles of the New Urbanism and Smart Growth, and further changes are likely in the years ahead to promote sustainability. Major areas of focus include:

- producing a greater mix of land uses by reducing large areas of single-use zoning and allowing or requiring shops, workplaces, and community facilities within residential districts, typically at transit stops or neighborhood centers;
- increasing residential densities and the diversity of housing types by allowing second units, duplexes, townhouses, and small apartment buildings to be mixed with single-family homes;
- establishing minimum densities and building heights in many locations (especially downtowns or infill sites) rather than focusing on maximums; and
- adding provisions to require development to be set back from creeks and shorelines, and to preserve areas of important wildlife habitat.

Box 10.1 provides a summary of some key modifications to local zoning codes that may help create more sustainable communities.

The development approval process represents the "front lines" of much urban planning. How and where development occurs—the subdivision of land and construction of buildings—is a central concern of sustainability planning. Development decisions made now will determine the form and character of communities for centuries to come, as well as influencing how cities or towns relate to features of the natural environment, how they use energy and natural resources, and where and how different groups of people can live (especially those of different income groups).

City governments issue a variety of permits allowing property owners to legally develop their land and have a significant degree of control over what gets built. At minimum the development approval process in the US typically includes a review of building plans by city staff at the zoning or "current planning" counter and issuance of permits for site grading, construction, and/or building use. But if a project requires any variances from zoning ordinances it will probably also be reviewed by a zoning adjustments board, which can deny, approve, or require modifications to the application. Large development projects may be scrutinized by a design review committee in some cities. If they require amendments to a city's general plan these projects may be routed through the municipal planning commission, whose job is to oversee such plans. Decisions of the zoning board and planning commission can be appealed to the city council or county supervisors, resulting in yet another layer of review. So developers typically face a lengthy, multi-stage process to acquire permits for building.

Although the development approvals process often simply checks to ensure that projects meet the city's zoning and building codes, there is frequently leeway in the process for planners and public commissions to shape the character and impact of development. If a project requires zoning variances for certain things, planners and commissioners can use these approvals as leverage to negotiate with the developer about other changes to the project. They may end up allowing the zoning variances if an overall package of changes is made. There may be negotiation as well around environmental mitigations and various fees

Box 10.1 SAMPLE ZONING CHANGES TO PROMOTE SUSTAINABLE DEVELOPMENT

	Typical current practice	Smart Growth alternative
Minimum lot sizes	6,000 sq. ft or more	1,500–4,000 sq. ft if any
Maximum lot sizes	Rarely regulated	5,000 sq ft or less for single-family homes in many infill locations
Dwelling units allowed per lot	Most urban land zoned for single-family detached housing	Allow second units on existing lots; allow multiple units on vacant lots in single-family districts if building design conforms to neighborhood context
Allowable densities, downtown areas	Many suburban cities specify maximum residential densities of 20–40 dwelling units per acre even in high-density zoning districts	Eliminate maximum densities; rely on height, bulk, and/or design restrictions instead. Institute minimum densities of 20–30 dwelling units/acre.
Allowable densities, residential areas	Many suburban cities have maximum residential densities of as little as 1–4 units per acre in low-density zoning districts	Establish minimum residential densities of 8–10 units per acre for new single-family development and 20 units per acre for multifamily development; allow residential infill at this level
Height restrictions, downtown areas	Often 2–3 stories even in town centers; no minimum	At least 3–5 stories in downtowns and neighborhood centers; a 2–3 story minimum
Height restrictions, residential areas	2½ stories or 30 feet	At least 3½ stories or 40 feet
Lot coverage	Often less than 50 percent of the site	No maximum if parks and other public open spaces are nearby; encourage use of rooftops for open space
Floor area ratio	Often 0.50–0.80 maximum in downtown locations	At least 1.0–2.0 maximum, 0.5 minimum in downtowns, or height limits instead.
Front setbacks	Often 20–40 feet minimum except in downtown areas; no maximum	No minimum necessary in many areas; consider adding maximum
Side setbacks	Often 5–15 feet	Allow zero-lot-line construction with appropriate design
Setbacks from creeks	Usually none	At least 30 ft from the centerline of the creek

continued

	Typical current practice	Smart Growth alternative
Lot widths	Some cities require minimum widths of at least 50 feet for single-family housing, 70 feet for duplexes	No minimum necessary
Mixture of land uses	Only homes, stores, or workplaces allowed across large areas of cities	Allow a finer mix of land uses to reduce driving and enhance community vitality; allow housing and shops to be added to office parks, offices and shops to housing districts
Mixed-use buildings	Not permitted most places	Allow mixed-use buildings within neighborhood centers and along arterial strips; provide incentives for these
Secondary units	Prohibited or subject to conditional use permits	Allowed as of right in single-family residential districts
Parking for downtown or transit-oriented locations	1–2 spaces per unit minimum	1 space per unit maximum; car-free housing allowed in certain transit-oriented locations; car-sharing encouraged in large projects
Parking for residential neighborhood locations	2 off-street spaces per unit minimum	1 off-street space per unit minimum; 1 additional on-street space required for larger unit sizes; consider parking maximums
Parking charges	None mandated	Mandate a monthly fee per space for rental and condominium units
Parking for retail	3–4 spaces per 1,000 square feet minimum	1 space per 1,000 square feet minimum in downtown, transit-oriented, or neighborhood center locations; businesses allowed to contribute in-lieu fee instead of providing parking on site; 2–3 spaces per 1,000 square feet in other locations
Parking for offices	3 spaces per 1,000 square feet minimum	No minimum in downtown, transit-oriented, or neighborhood center locations; 1–2 spaces per 1,000 square feet in other locations; employers required to charge for parking and provide incentives for alternate travel modes; local hiring policies encouraged

that the city charges for transportation, affordable housing, infrastructure, or other potential project impacts.

Development projects that require subdivision of land—the creation of separately owned parcels—trigger an additional approvals process governed by state law and typically overseen by planning commissions. This "subdivision mapping process" often involves the creation of new streets and blocks, and is vital in helping to establish a compact urban form, a connecting street pattern, a diversity of lot sizes and housing forms, adequate park space, and protection for creeks and other ecosystem elements. Through this procedure city officials establish the density of new development by approving certain lot sizes and building types (single-family homes, duplexes, townhouses, apartment buildings, or other building forms).

Subdivision review can also influence how much land is set aside for parks, greenways, schools, and other public facilities, and how well the new development relates to existing urbanized areas. Setting a pattern of street and pedestrian connections to existing neighborhoods helps make a neighborhood pedestrian- and transit-friendly and creates a unified urban fabric. Typically during the past 50 years new subdivisions have been inwardly focused with little relation to other urban areas around them, helping to create a fragmented, unwalkable landscape. Much the reverse must happen if we are to create more sustainable communities—each subdivision must be planned with the larger city and region in mind.

Much new urban fringe development in recent years has taken place within districts zoned for "planned unit development," a situation in which there are few pre-set zoning requirements but planners and the developer must agree on an overall plan for the new neighborhood. Such a plan includes subdivision of land, creation of new streets, selection of lot sizes and building types, and addition of parks, schools, services, or other community facilities. In this case planners have much more flexibility to ask for changes that will improve community sustainability.

Environmental review is a crucial part of the development approvals process. In California it is required of all projects under the California Environmental Quality Act (CEQA); in other states it may only be required under the National Environmental Protection Act (NEPA) for projects that receive federal funding. City staff may issue a "negative declaration" (finding no significant environmental impact), require developers to implement a modest package of mitigations for environment impacts, or force them to prepare an environmental assessment or environmental impact report that will guide more extensive mitigations and compare the proposed project with alternatives. These documents are often subject to litigation by outside parties who contend that impacts were not properly analyzed or alternatives not properly considered. NEPA and CEQA are often the only legal handles environmental groups possess to challenge development projects. However, these environmental review laws only require that potential impacts of the project and alternatives be studied—they cannot stop projects on the basis of their impacts.

Redevelopment is a separate land use planning mechanism through which cities coordinate revitalization of areas designated as "blighted." A redevelopment agency within the city government then takes responsibility for carrying out extensive improvements within the targeted area. These improvements may include assembly of multiple parcels of land into a smaller number of easily developable sites, and provision of infrastructure such as roads, sewers, water mains, parks, schools, and other community facilities. In some states and countries, cities have been allowed to use the mechanism of "tax increment financing" to pay for new infrastructure or purchase of property in redevelopment areas.[20] This allows them to issue bonds to fund initial improvements such as new streets, sewers, parks, and urban design changes. The bonds are then paid off from the increment of increased property taxes that will presumably come from the designated area as a result of new development. This financing mechanism has proven a powerful way for cities to leverage improvements in otherwise run-down neighborhoods, building urban sustainability by revitalizing downtowns, promoting transit-oriented development, and developing more mixed-use urban neighborhoods.

Historically, redevelopment—especially the mid-twentieth-century form known as "urban renewal"—was often misused to bulldoze working-class neighborhoods containing racial and ethnic minorities and replace them with generic modernist development for middle- and upper-class residents. Racist motives have been present at times; during the 1950s and 1960s "urban renewal" was known colloquially as "Negro removal" because so many African American neighborhoods were targeted. As Jane Jacobs pointed out in 1961 in The Death and Life of Great American Cities, planners failed to understand the benefits and potential of older urban neighborhoods, which often contained dense and vibrant social networks as well as historic buildings and a pedestrian-oriented, human-scaled building and block pattern. In these great urban planning tragedies of the twentieth century, residents displaced from these neighborhoods were often forced into far more difficult living conditions without their social support networks. Ironically, the fine-grained street, lot, and building fabric of bulldozed older urban neighborhoods is now an ideal that New Urbanist planners strive for.

Despite this checkered history, redevelopment powers can in principle be used by cities to bring about more sustainable forms of development. Redevelopment mechanisms may be particularly useful in coordinating transit-oriented development around new rail lines, in cleaning up and rebuilding older industrial areas with brownfield sites, and in catalyzing investment in abandoned downtown districts. US local governments also have the power of eminent domain to take private property (with compensation) for public purposes. Although this tool must be used with caution there is no reason why it couldn't be employed more actively to create bike and pedestrian trails, to ensure a supply of affordable housing, or to serve other sustainability purposes. British planner Paul Winter argues that in the UK local authorities need to be more aggressive in using their powers for compulsory acquisition to ensure that sufficient housing is built,[21] and perhaps to save open space and bring about better-designed development as well.

The city of San Jose, California, represents a good case study of both negative and positive forms of redevelopment. The city's downtown was extensively damaged by mid-twentieth-century urban renewal as neighborhoods were razed, parts of the fine-grained street grid turned into "superblocks," and vast parking lots and empty lots created in the belief that development would soon fill them. These sites languished in the 1950s, 1960s, and 1970s while suburban sprawl elsewhere in the Santa Clara Valley (which became known as Silicon Valley) sucked businesses and residents from the central city. With an influx of funds from Silicon Valley development and a new understanding of the need to create human-scale, pedestrian-oriented urban spaces, the city's Redevelopment Agency began actively rebuilding the downtown in the late 1980s and 1990s, constructing a host of new buildings, parks, greenways, public plazas, and streetscape improvements in conjunction with a new light rail system. In effect the agency was repairing the damage of earlier decades, seeking to recreate the vibrant downtown that had existed 50 years before.

Beyond these basic regulatory and planning processes, there are many growth management tools to protect open space within and around cities, each with its pros and cons. In the United States agricultural zoning is probably the most common method. Areas around a community are simply zoned for agricultural use only, usually with a minimum parcel size that may be from 5 acres to 160 or more. The disadvantage of this method is that zoning can be easily changed by a vote of the city council or county commission, usually in response to developer pressure. Landowners frequently attempt to subdivide large farms into 5 or 10-acre "hobby farms" that amount to very low-density sprawl. Once agricultural land is subdivided in this way, traditional agricultural uses are often no longer viable and political pressure may mount for further subdivision into low-density suburban development.

Urban growth boundaries (UGBs) have been adopted by many cities in Oregon, Washington (both under state mandate), and northern California, as well as a few other jurisdictions in Tennessee, Pennsylvania, and Colorado. They are a more stable, long-term way to protect open space at the urban edge, usually being established for a period of 20 years with periodic reviews to allow modest expansions of the urban area if population pressure warrants. Often a vote of the electorate or a two-thirds majority of a city council or county board of supervisors is required to change the boundary. The UGB prevents subdivision of land outside the boundary and/or extension of infrastructure that would make development possible. Because of their relative permanence UGBs are a favored strategy of environmentalists. Opponents argue that UGBs lead to housing price increases within the boundary as the supply of greenfield land diminishes, and that development may simply leapfrog to the next county. A potential solution to the former problem is to establish strong municipal policies encouraging space-efficient infill development and the provision of affordable housing; a potential solution for the latter concern is to broaden the area covered by growth management protections or to move to a statewide framework for protection (see Figure 10.5). Outside of the US, greenbelts established by national or regional governments in areas such as London or Seoul do much the same thing as UGBs: keep a ring of relatively undeveloped land around the city. Once established, greenbelts must be zealously guarded, since governments tend to place roads, airports, or other large facilities within them.

Municipal purchase of open space land for local parks or wildlife preserves is another option that has the advantage of being permanent but the disadvantage of requiring large amounts of public funds. Private groups such as the Nature Conservancy and the Trust for Public Land often make such purchases or provide loan funds to allow local organizations to do so. Municipal acquisition of parkland has also been immensely important in most cities, but is often a difficult and slow process, as funds may need to be raised through a bond measure or other procedure. Not infrequently by the time a bond measure has passed and the funds are available, the price of the land in question has risen so much that the original intended purchase is no longer affordable.

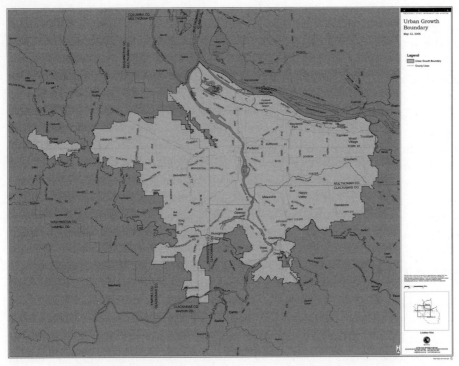

Figure 10.5 Urban growth boundary (UGB). Portland, Oregon's UGB, first adopted in 1979, prevents subdivision of land outside the designated urban area.
Source: Portland Metro

Purchase of conservation easements is a less costly way to ensure that land remains undeveloped. A community group or city purchases the development rights to the land from the owner (a conservation easement on the land), and a deed restriction is then recorded, preventing development. This measure has been widely pursued in recent years, usually with nongovernmental land trust organizations holding the easements. However, there are some concerns about whether such mechanisms will be enforced in perpetuity, and if violated these easements would depend on an environmental organization being willing to litigate the violation.

Transfer of development rights (TDR) is a relatively rare method in which a local government essentially establishes a market in development rights, allowing rural landowners to sell their rights to urban developers, who then use them to obtain higher densities than would be allowed by zoning for their projects in urban "receiving" areas elsewhere. This method has been in place in Montgomery County, Maryland, since the 1970s, and has been relatively successful at channeling growth into transit station locations. However, TDR frameworks are based on the assumption that urban zoning is less than optimal, raising the question of why the municipality in question does not permit denser development in the receiving area in the first place.

Taxing land at a higher rate than improvements (buildings) is yet another mechanism local government can use to encourage compact urban form and infill development.

This step would increase pressure on landowners to make more productive use of vacant lots, parking lots, or leftover areas between existing developments, and could do a great deal to counter the current wasteful use of land within suburbia. Nineteenth-century reformer Henry George advocated taxing only land, the "single tax" as he called it, as a way to more fairly recover wealth created by the entire community (as reflected in rising property values) and to reduce the tendency of speculators to hold vacant land for years hoping for price increases.[22]

Finally, land banking has been used extensively in Europe historically as a growth management tool. The mechanism is not used so much in the United States, although some municipalities employ the strategy to assist in redevelopment activities. Essentially the municipal government buys up land at the urban fringe, holds it, and sells or leases some of it for development in times and places that it thinks make sense. (Some of this land could be preserved as open space as well, or allocated to urban agriculture.) The local government must capitalize such a fund initially, but in theory it could then be self-sustaining or could even provide the city with a return on investment. The city of Stockholm, for example, at one point owned more than 60 percent of the land in its metropolitan area and managed it through land banking mechanisms.[23] This method has the further advantage that the public recaptures the value of its investment in infrastructure such as rail or road systems, since it rather than speculators profits from the corresponding increase in land values.

THE CHALLENGE OF INFILL

Economic, practical, and political considerations often work against infill development for developers. Urban land is generally more expensive than land outside the existing urban area, and the task of designing and gaining approval for an infill project is more difficult. Housing must be built at somewhat greater densities to be economically viable, and ways must be found to make those densities equally attractive, for example by providing a good range of outdoor open spaces. Most North American communities oppose high-rise development, but densities of 12 to 40 dwelling units per acre—double to quadruple the suburban densities of many places—can be achieved through efficiently designed two- to three-story development. Much higher densities can be achieved through five-story development, a mid-rise form that is more politically acceptable than high rise in many places. The dream of settling on previously unbuilt land lies deep within the psyche of many cultures, and the ideal of owning a country estate has great appeal within many cultures. So compact development requires a rethinking of the ideal of country living, and development of a desire instead for compact, walkable communities that contain many urban amenities not present in suburban or rural locations (see Figure 10.6).

Infill development can be facilitated by a number of strategies. Local governments can map vacant or underutilized land, and assemble databases concerning these sites to inform potential developers. Authorities can prepare area plans that develop detailed visions for infill neighborhoods, and can revise zoning codes and develop design guidelines so as to facilitate the best possible forms of infill. They can lessen the burden of environmental review by doing some of it themselves in the context of such plans. They can improve roads, sewers, water systems, and pedestrian infrastructure in areas to be redeveloped.

Figure 10.6 Infill development. Infill development in Portland's Pearl District makes use of a number of different building types as well as rehabilitated historic structures

And they can seek to streamline development approval processes for projects in line with stated municipal goals for such sites.

Greening infill development as well as existing urban neighborhoods is a sustainability priority. Even though such construction is dense, it should be highly green and livable. Keeping traffic to a minimum is one starting point (units might be substantially cheaper, for example, for residents who agree to forgo a parking place). Adding a variety of public, private, and semi-public outdoors spaces is another area for initiative. Community gardens, interactive landscapes along streams and shorelines, green streets, and other such strategies can help create a livable and attractive environment.

Better land use planning is essential to help us move towards sustainability. By itself, however, it will not be enough to solve problems caused by urban growth, as long as the number of residents in many areas continues to rise, not to mention the number of cars per resident, the size of houses, and the use of resources. Ultimately issues of population and consumption will need to be addressed as well. Doing so will involve rethinking fundamental social values regarding growth and the nature of progress. But better land use planning is very much a part of the overall picture, and can help "ground" sustainable societies in specific places and landscapes.

11

URBAN DESIGN

Many aspects of sustainability planning are a question of design, that is, the creative arrangement of elements within a given context. This process draws upon linear, rational thought, but also upon the brain's more creative sides, including intuitive understandings about how to balance physical and less tangible aspects of a situation in relation to one another. As ecological architect Bill McDonough has argued, design permeates human systems and good design can often solve sustainability problems far more elegantly than an emphasis on engineering or technology.[1]

Although policies, programs, industrial systems, and entire societies can be designed in various senses of the word, design applies most directly to the physical world around us, a process often known as urban design. Urban design includes not just the arrangement of land uses, public spaces, streets, neighborhoods, and homes, but the configuration of less traditional spatial elements such as greenway systems, regional growth patterns, transportation networks, water and sewer systems, and flows of energy and materials. Designing such systems requires thinking about how they relate to all other elements of a place, combining physical planning (concerned with land use, infrastructure, buildings, and the design of public places) with public policy frameworks (including tax systems and economic incentives) that can support such changes. Thus sustainable urban design is a creative weaving-together of elements in a particular place to improve long-term human and ecological well-being.

URBAN DESIGN VALUES

At its heart sustainable design is based on human and ecological values, rather than a value set dominated by economic efficiency and profit. This means creating urban places much like the ones Jane Jacobs described in 1961 in *The Death and Life of Great American Cities*: communities that are walkable, human scale, diverse, and oriented around a fine-grained and vibrant mix of housing, shops, and public facilities. These places do not have to be the highly urban neighborhoods that Jacobs described. They can also be neighborhoods in smaller towns, new districts at a variety of densities, or rural groupings of buildings.

To fulfill the environmental goals of sustainability, the physical design of such places should reflect local climates, ecosystems, materials, and flows of energy, water, and resources. These environmental elements can better integrate human communities with the natural landscape, reduce automobile dependency, use resources more efficiently, and engender a sense of place identity. But to fulfill social and economic sustainability goals, urban design may need to emphasize other goals as well. These include enhancing diversity of housing forms, creating appropriate residential densities (so as to be able to support local businesses as well as reduce driving and GHG emissions), developing a diverse range of public and semi-private spaces, creating spaces for a diverse range of green, locally oriented businesses, and incorporating educational strategies to help people learn from the context around them. In these ways urban design and land use strategies can work hand-in-hand to create the sustainable city.

Sustainability-oriented urban design might focus on restoring streams and greenways through cities. It might create neighborhood and village centers, and rebuild urban downtowns in cases where those have stagnated. It might make streets safe and pleasant for pedestrians and cyclists through a variety of street design techniques. It might design compact communities that mix various types and prices of housing. Perhaps most difficult of all, it might transform the desolate commercial strips, malls, office parks, and industrial sites that currently plague our cities and towns into places that nurture human community as well as ecological health.

Kevin Lynch, mentioned in the previous chapter, analyzed a number of key urban design values that he believed appropriate for contemporary cities:

> So what is good city form? Now we can say the magic words. It is vital (sustenant, safe, and consonant); it is sensible (identifiable, structured, congruent, transparent, legible, unfolding, and significant); it is well fitted (a close match of form and behavior which is stable, manipulable, and resilient); it is accessible (diverse, equitable, and locally manageable); and it is well controlled (congruent, certain, responsible, and intermittently loose). And all of these are achieved with justice and internal efficiency.[2]

These concepts may sound abstract and academic, but all have real and tangible implications. "Legible" urban form, for example, refers to city design that can be easily understood by the average resident, that can be "read" without undue difficulty. Grids and other relatively regular, well-connected street patterns tend to reinforce this characteristic, as does the existence of central squares and landmarks. Meandering, fragmented suburban street patterns are often not particularly legible. "Well-fitted" urban form is one in which buildings, block sizes, public spaces, and other physical elements of the city match human needs and activities, typically through creation of a relatively fine-grained, human-scale urban environment. Enormous modernist buildings and "superblocks" have a difficult time meeting this criterion; contemporary malls and housing tracts have difficulty as well.

"Accessible" urban form refers to buildings and spaces that can be entered, used, and even modified by people—in which space is not rigidly controlled, as in gated communities or many office parks, but is open to public use and can be adapted to the needs of different users over time. (Similarly, in his book *How Buildings Learn* Stewart Brand argues that the most interesting and livable buildings in the long run are ones that can be adapted and modified over the years by a wide range of inhabitants.)[3] Lynch's city form values are certainly not the last word in terms of a normative urban form philosophy. Most of his work, for one thing, took place before the full impact of the environmental movement had reached the design professions, and environmental dimensions of sustainability are largely left out. Yet Lynch played an enormous role in inspiring designers to rethink urban and regional form in ways that lay the groundwork for more sustainable development. More than anything, he pointed out the importance of discussing ideal urban form values, studying evidence about the actual effectiveness of urban design strategies, and figuring out ways to achieve urban design values in new development and planning (see Figures 11.1 and 11.2).

Another relevant framework of design principles, prepared at about the same time as Lynch's, was the "pattern language" developed by architect Christopher Alexander and colleagues in the architecture department at the University of California at Berkeley.[4] These writers identified a range of design characteristics that they saw as universal in good urban areas around the world. In terms of regional design, particularly important themes included "agricultural valleys," "city country fingers," "scattered work[places],"a "mosaic of subcultures," "density rings," and a "web of public transportation." In the view of Alexander and his coauthors, urban development is best pursued through an organic, incremental process in which each builder follows these time-tested patterns and considers the entire context surrounding his project before beginning development.[5] This thoughtful, Zen-like

Figure 11.1 A city's "living room." On the site of a former three-story parking garage in Portland, Oregon, urban designers created a multi-purpose space (Pioneer Courthouse Square) that has been called the city's "living room"

Figure 11.2 Public art. Art has long been a way to add vitality and humor to cities. The "Cows on Parade" concept pioneered by Zurich has been emulated by Chicago and many other cities (often using local symbols). The City of Chicago commissioned artists to decorate 320 fiberglass cows that were placed around the city and later auctioned off for charity. The city's Department of Cultural Affairs estimates that the exhibit drew more than a million visitors over a single summer, with a $200 million economic impact

approach is quite at odds with contemporary development, in which extremely rapid and large-scale building is the norm. Yet it does point towards an alternative urban design philosophy, at the regional scale as well as that of the neighborhood and individual buildings, that can potentially be more responsive to cultural, ecological, and historical contexts. Historically, building-by-building development has often produced far more interesting places than large-scale, master-planned construction of entire neighborhoods at once. Perhaps ways can be found to return to this slower-paced style of community creation, or to mix an environmentally sensitive large-scale planning framework with more incremental and creative addition of elements within it.

ECOLOGICAL DESIGN

What specific urban design strategies can be considered sustainable or ecological? Different designers have had different answers to this question, and have produced a wide array of built forms, including such bold visions as Paulo Soleri's Arcosanti project in the Arizona desert, the Earthships outside of Taos, New Mexico, and the spiritually oriented community of Auroville in southern India. On a smaller scale, many architects and landscape architects have designed individual buildings and landscapes that are green in one sense or another. Many score well on green design rating scorecards such as LEED or Sustainable

Sites. But others simply employ creative strategies of weaving flows of water, energy, or vegetation into human places. The exact form is not as important as the function. Does the design in question integrate well with the climate, the hydrology, the ecology, and the materials of the site? Does it serve the inhabitants or users well? Does it help create a balanced local community with a light footprint on the earth and a high degree of human welfare? Does it draw upon local knowledge and empower residents by involving them in the design and creation? These are the types of question that should guide choices of form, materials, and processes within sustainable urban design (see Figures 11.3 and 11.4).

The overall effect of a given design, following our original definition of sustainable development, should be to improve long-term human and ecological welfare. As much as possible designers can seek to create places that improve hydrological function (for example by keeping storm-water runoff on site), increase habitat (by recreating plant communities that might have occurred on the site prior to development), reduce urban heat island effects (by minimizing paved surfaces and maximizing vegetation), allow for human interaction (by creating welcoming public places and paths, and minimizing space devoted to motor vehicles), and serve the needs of diverse user groups (by ensuring that a wide range of people will have access to the site and that the design meets their likely needs). Green design rating systems provide additional, more detailed design considerations. The exact mix of design priorities will always vary depending on context. But a conscious effort to articulate and achieve such design goals is essential to sustainable urban design.

Figure 11.3 Ecological neighborhood design. Village Homes in Davis CA represents one of the world's most ecological suburban neighborhood designs. Houses are oriented south to promote passive solar design. The design includes three levels of bike and pedestrian circulation, narrow streets, and extensive community gardens

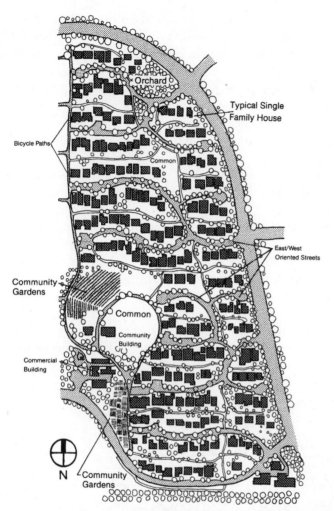

Figure 11.4 Greenways. The Village Homes neighborhood is structured around a greenway system featuring edible landscapes and swales which handle drainage on site. Residents are not allowed to erect fences on the greenway side of their homes

In *Regenerative Design for Sustainable Development*, John Tillman Lyle developed a useful comparison between what he called "industrial design" and "regenerative design."[6] Regenerative design is enmeshed in ecological and social systems; industrial design is often separate from them, treating the problem in isolation. Regenerative design seeks to meet multiple goals at once; industrial design is focused on one objective at a time. Regenerative systems allow nature to do the work; industrial systems try to do it through technology and engineering. Regenerative design seeks to reintegrate activities; industrial design separates them. Regenerative design uses information and feedback to guide appropriate-scaled systems; industrial design relies on raw power often at excessively large scales. And so forth. Such contrasts can be simplistic, but they also help dramatize the very real differences between paradigms. Much modernist design proved highly counterproductive in that it created buildings, neighborhoods, infrastructure, and industrial systems that ignored their contexts and produced a great range of social and environmental problems. A different approach is needed that emphasizes context, integration, feedback, involvement of users, and appropriate scale and technology.

Authenticity is a frequent concern within design. In many green development projects the designer has tried to emulate a natural landscape, but with key components missing. Landscape architects, for example, frequently create water features designed to look like streams. Typically the waterway has two or three sizes of cobbles trucked in from other locations and artfully arranged to lead towards a stagnant pool framed by a few trees and shrubs. The result may be pleasant and serve the local residents well as an outdoor green space—a very important goal to be sure—but has little ecological value. The waterway is not connected to local hydrology or to runoff and gray water flows from the surrounding buildings. Plants have not been chosen for habitat value or grouped in plant communities native to the site. This artificial green space is not part of a larger-scale green infrastructure that might create wildlife corridors throughout the city or at the very least a functional network of habitat islands for birds and native pollinators. Equally unfortunately, the design provides little way for human users to learn about natural systems or to sense their own interconnection with them.

Sustainable urban design must do better. The designer should think through the full range of human and ecological function in the place, and strike the best possible balance within the context of the site. Often this will mean pushing the envelope of what initially seems to be possible, for example by convincing the site owner to connect water flows from the buildings to the landscape, by working with the local government to green the streets surrounding the site, or by greening rooftops and walls in addition to the ground plane. Whether or not designs are an authentic recreation of the original ecology of landscapes, they can at least have much of the same function, and can help connect urban dwellers to the natural world.

Humanistic and ecological design must be made financially feasible, and this will require a long-term evolution of the marketplace and public attitudes toward caring for the common realm. Many urban design improvements are possible without adding greatly to costs, for example by connecting street networks within subdivisions and placing new commercial buildings along the street to improve the pedestrian environment. Some ecological design steps can even save money. Orienting buildings south and incorporating passive solar design features (described in Chapter 14) reduces energy bills. Handling storm-water runoff on site through swales and permeable paving reduces costs of storm-water infrastructure. Narrowing streets saves money required for paving and maximizes the amount of buildable land within new neighborhoods.

However, other sustainable urban design measures will inevitably require public investment or reductions in developer profit. Requiring greenway systems within development decreases land available for sale. Creating attractive plazas, streets, bikeway systems, playgrounds, and community gardens may take substantial public investment. Changes in development markets and attitudes toward the role of public investment will be needed for these things to happen.

Changing economic incentives can help. True-cost pricing of energy and motor vehicle use might be one such step. Differential fees to penalize sprawl development might be another. Institutional changes are important as well, for example changing zoning codes and subdivision regulations to require ecologically appropriate landscape design, pedestrian-friendly street design, and the provision of well-designed public spaces within new neighborhoods. Slowly, step by step, such changes can alter the reality of what is economically or pragmatically feasible.

In a world that is dominated by large-scale, conventional forms of development, the need is for urban design initiatives that can establish alternative forms of place-making. On a small scale, new design projects can add color, art, plants, and ecosystem function back into public spaces. Urban design of this sort isn't a question of creating a few showcase projects; it is a task of creating an "everyday urbanism"[7] that is often low-key and unpretentious but still strongly supportive of a more sustainable city.

On a larger scale, urban design projects can create new districts, neighborhoods, or street corridors that work well for people and ecosystems. Every new increment of development, as Alexander said, can help heal the whole. It can help show how the physical world around us can be different—more sociable and human-scaled, more rooted in the unique culture and history of a place, more functional in terms of promoting walking, bicycling, and public transportation, and more ecological in terms of restoring habitat value, protecting hydrological systems, cleaning air and water, and reducing large-scale impacts such as climate change. Not least, every urban design intervention no matter how small can inspire people to live more sustainably, more creatively, and more cooperatively with others.

12

TRANSPORTATION

Transportation systems have been a powerful force in determining the form and character of development throughout history. New cities and towns have grown up around transportation routes; changing technologies have led to major shifts in community designs and densities. This role has been especially powerful in the past 150 years, when railways, streetcars, and then the automobile helped decentralize cities and revolutionize everyday life within them.

Some level of transportation infrastructure is certainly necessary. But many human environments around the world are now completely dominated by motor vehicles—in terms of the volume of traffic, lifestyles, the motor vehicle-oriented design of streets and surrounding land development, the conversion of roads from multi-use public spaces to vehicle thoroughfares, and the generation of noise, pollution, and greenhouse gases. Rethinking transportation needs and systems thus becomes a central sustainability requirement.

Addressing unsustainable transportation systems does not necessarily mean getting rid of cars and trucks altogether. But it does mean using them far less. The challenge is to reverse the continual increase in vehicle miles traveled (VMT) in both absolute and per capita terms. This can be done through action in four main areas:

- providing good alternative modes of travel, in particular the ability to get around by walking, bicycling, and public transit;
- changing land use and urban design policies so as to support these alternative transportation modes and reduce the number and length of trips that people need to take every day;
- reforming transportation pricing to incorporate the full social and environmental costs of driving into the prices of fuel, road use, parking, motor vehicles, and vehicle registration; and
- rethinking the need for rapid mobility in the first place, changing our behavior and lifestyles to travel less, combine trips where possible, and reduce the overall demand for rapid mobility.

These four types of initiatives can reinforce one another, synergistically producing greater benefits than any one strategy alone.[1] Together they can result in radically different transportation patterns that are far more sustainable environmentally and lead to far more livable communities.

MODES OF TRAVEL

For most of the twentieth century, transportation planning at different levels of government in North America focused on providing private motor vehicles with high-speed, uncongested roads and free parking. Authorities widened existing streets, constructed new freeways and arterials, designated streets one-way to speed automobiles through downtowns, and added enormous amounts of parking in surface lots or parking structures. These efforts combined with the automobile industry's massive promotion of motor vehicles led to a meteoric rise in vehicle ownership between the early 1900s and the 1980s. Although private vehicle ownership reached the saturation point in many developed countries during the late twentieth century, developing countries such as China and India are still in the early stages of this motorization, with vehicle ownership and travel rising rapidly. And although vehicle ownership may have plateaued in many industrialized countries, per capita miles traveled has continued to rise except during periods of economic recession. Only during the past couple of decades have governments begun conscious efforts to reduce VMT, based on a desire to reduce road congestion, pollution, and fossil fuel use, as well as a recognition that communities designed for motor vehicles do not work well for human quality of life or the environment.

Sustainability planning inverts the traditional hierarchy of transportation priorities. Instead of focusing on private motor vehicles first and foremost, and considering public transit, bicycling, and walking only as afterthoughts, sustainability planning starts with the lowest-tech, human-powered forms of travel and only after possibilities with those modes have been exhausted moves up the ladder to others. Human-powered forms of transport are preferred whenever possible, especially for short-range trips, which will become a greater percentage of overall travel as the land use changes mentioned earlier bring destinations closer together.

Planning for pedestrians requires a comprehensive re-evaluation of land use and urban design. First and perhaps most important is creating a street and path network within which pedestrians can travel easily from one point to another. In many forms of suburban neighborhood design there is no direct walking route through neighborhoods. When developing zoning and design guidelines for new developments, cities can require a certain number of street connections per mile (Portland, Oregon, requires five), maximum block sizes, or the addition of pedestrian and bicycle trails within new developments. Within existing urban areas, cities can sometimes improve the connectivity of the street network for pedestrians by acquiring a key parcel of land or right-of-way and creating a new road connection or path.

Sidewalks, which are deficient or completely missing in many newer communities and some older ones, are essential for pedestrians if there is any significant volume of vehicle traffic on streets. Often developers of new neighborhoods construct only a four-foot sidewalk if they include one at all; this is too narrow for two people to walk abreast.

Six-foot-wide sidewalks are much preferable. If they are separated from the street by a planting strip (typically three to six feet), then the pedestrian is further insulated from traffic on the street, and the grade of the sidewalk doesn't change as driveways cut through it (driveways often dip sharply to meet the street in their last few feet, which can create discontinuities in a sidewalk). Sidewalk curb cuts at intersections to allow wheelchair use are desirable as well, and required in the US under the Americans with Disabilities Act of 1991.

Intersections are another key area for pedestrian design. The person walking must be able to get across easily and safely. As intersections have become monstrously huge in many urban and suburban areas—with multiple lanes of traffic on each street, complicated turning movements, and very long cycles for signals—the simple task of crossing a street has become increasingly complicated and stressful. One obvious solution is to keep streets small and distribute traffic across a grid of streets rather than channeling it all onto a few enormous roads. But other things can be done as well. Pedestrian-activated crossing signals, count-down and/or audible signals, pedestrian islands half-way across each street, and bulb-outs at corners to decrease the distance pedestrians need to travel and improve their sight lines are all strategies that can help.[2]

The pedestrian-friendliness of a street depends on many other factors as well. Having buildings lining the streets in commercial areas helps create a more intimate and human-scale environment than the current situation in many suburbs where parking lots front the street, creating a vast landscape of pavement through which pedestrians must wander. Cities can also require buildings to have a human scale along the street, and include windows, entrances, small plazas, attractive landscaping, closely spaced street trees, pedestrian-scale lighting, and occasional benches to add to the pedestrian environment. Prohibiting blank building walls along the street and encouraging street-front retail and sidewalk seating for restaurants and cafes also helps. Between the sidewalk and the street, planter strips and parking can help buffer pedestrians from fast-moving vehicles. Short blocks, human-scale buildings, and a relatively fine-grained mix of land uses also make for a more interesting and functional pedestrian environment.

A variety of traffic-calming measures on the streets themselves can improve the environment for pedestrians, cyclists, and neighbors who live nearby. An international traffic-calming movement has in fact been underway for several decades.[3] Early examples included Copenhagen's pioneering pedestrian street, the Stroget, created in the 1960s, and the Dutch *woonerf* designs in which road space is shared in residential neighborhoods between very slow speed motor vehicles and residents. Most European cities then created downtown pedestrian districts in the 1970s and 1980s. Some Latin American and Asian cities, including Curitiba (Brazil) and Bogata (Colombia), followed suit for at least a few downtown streets. An early wave of 1970s pedestrian streets in the US did not work particularly well, since surrounding land uses in declining central cities didn't support intensive pedestrian use. But many North American communities then adopted lower-key traffic-calming measures such as speed humps and traffic circles in residential neighborhoods, as well as pedestrian-friendly curb bulb-outs in some downtown areas (see Figure 12.1).

Bicycle planning is a rapidly growing field in most cities. The issues here are often similar to those for pedestrian planning—a street environment that is pedestrian-friendly will usually be bike-friendly as well. But other steps often need to be taken in addition. A variety of measures may be needed to provide bicycle space on streets, to slow traffic, to

Figure 12.1 Alternative transportation modes: walking. Like most European cities, Verona has
        pedestrianized large portions of its downtown

improve visibility, and to provide off-street bike paths and parking. Three main categories
of bike travel corridors can be created: on-street bike routes which simply mark preferred
bike streets and perhaps take some measures to slow traffic and improve safety at intersections; on-street bike lanes that designate a bicycle travel space at the edge of the road; and
off-street bike paths that are separate from motor vehicle traffic. These three types are
useful in different contexts. Grade-separated bike paths may be an ideal in many cases, but
there is not always room to create them within existing cities and towns. The only solution
then is to figure out better ways of accommodating bikes on streets through bike lanes or
routes. Also, on streets with very low traffic volume or speed, it is not necessary to have
separate bike lanes, and the ideal becomes to have the street shared safely by multiple travel
modes (see Figure 12.2).

Around the world we have seen a large resurgence in planning for cyclists, although
unfortunately some countries such as China that traditionally have had high rates of bicycle use are now discouraging this in the name of progress. Rio de Janeiro, for example, has
developed more than 84 kilometers of bicycle paths, including a trail from a large middle-
class residential area in the south of the city through downtown. Inspired by Dutch bicycle
planning, a municipal Working Group for Cycle Systems began in the 1990s to determine
priority routes and gather public support. Despite initial resistance from motorists and
shopkeepers, the city succeeded in creating 140 kilometers of bicycle paths by the late
2000s including a 3-meter-wide cycle track between the Ipanema and Copacabana neighborhoods, instituted a bike rental program in 2008, and developed programs to promote
bike use in its lower-income *favelas* in 2012.[4]

Figure 12.2 Alternative transportation modes: biking. Amsterdam has created many bike paths separated from vehicle traffic

Among public transportation modes, "bus rapid transit" (BRT) systems are increasingly popular. Under this approach, high-tech buses travel on routes designed to speed up service by offering the buses separate lanes, queue-jumper lanes at lights, signal-pre-emption abilities (the driver is able to trip the light by remote means), and quick-loading bus stops designed like light rail boarding platforms (see Figure 12.3). The buses themselves may be low-floor models to speed boarding, and may use alternative propulsion methods such as compressed natural gas or fuel cells. The idea is to provide a service comparable to light rail systems at a fraction of the cost. One extremely successful bus rapid transit program has been implemented in Los Angeles, where passenger travel times have decreased up to 29 percent on BRT lines. Several years after implementing the technology transit agencies found that ridership had grown 42 percent in LA, 66 percent in Oakland, and 84 percent in Boston.[5]

Light rail systems are also popular in many cities, and generally offer faster service and benefits in terms of encouraging more intensive land use development along transit routes (property owners have greater certainty that rail transit lines will be permanent). These systems vary from streetcars on roadways making very frequent stops, to more suburban-serving trams with their own rights-of-way, longer routes, and more widely spaced stops (see Figure 12.4). Heavy-rail systems, with heavier-duty trains, higher speeds, and infrequent stops are generally used to meet longer distance commuter needs. Frequently cities have placed these commuter rail lines in freeway medians to save money compared with acquiring separate rights-of-way through built-up areas. However, this strategy has disadvantages in that the environment for transit-users is then loud and unpleasant, and transit-oriented development around stations is more difficult due to the presence of the freeway.

Figure 12.3 Alternative transportation modes: bus. Turkey has one of the world's best inter-city bus systems, which includes this hub in Ankara

Figure 12.4 Alternative transportation modes: rail. Unlike most other North American cities, Toronto never tore out its streetcar system, and now has some of the best public transit on the continent

A wide variety of other transit modes holds promise as well.[6] One of the most useful in low-density suburban areas may be on-demand service, where a transit user requests a pickup by phone or online, and a van or small bus detours to pick him or her up on its next pass through. Many large employers have organized vanpools and carpools to workplaces. Efforts are underway to make mode combination trips more feasible, for example by providing bike racks on buses and shuttle services to commuter rail stations. In developing-world cities, a variety of privately run transit options, known as paratransit, often provide a better level of service than official agencies. Typically, vans or microbuses ply main corridors picking up passengers wherever they appear (see Figure 12.5). The *peseros* of Mexico City and the *dolmus* in Turkey are examples of such service. Taxis are an important travel option in many cities, particularly those that are relatively dense and have destinations close together, to keep fares down. Finally, car-sharing networks have proliferated in Europe and North America over the past decade, allowing members the access to private motor vehicles without the burden of owning one. Pods of car-share vehicles are stationed throughout the city; individuals wishing to use them simply reserve them online and receive a code to unlock them. Similar bike-sharing systems are also spreading in major cities (see Box 12.1 and Figures 12.1 to 12.5).

Box 12.1 CHANGING APPROACHES TO TRANSPORTATION PLANNING[7]

Traditional approach	Sustainability-oriented approach
Engineering perspective	More holistic perspective
"Traffic oriented"	"People oriented"
Focus on large-scale movements, often ignoring local trips (within zones)	Concern for local movements, small-scale accessibility
Automobile as the priority	Pedestrians, bicycles, and transit as priorities
The street as traffic artery	The street as public space with multiple uses
Economic criteria for decision-making	Environmental and social criteria as well
Increase road capacity to handle projected demand	Transportation demand management (TDM) programs to reduce demand
Consider road-user costs and benefits	Consider other costs and benefits as well
Focus on facilitating traffic flow	Calming/slowing traffic where necessary
Segregate pedestrians and vehicles	Integrate pedestrian and vehicular space where appropriate through careful design (e.g. boulevards, *woonerven*)

Figure 12.5 Alternative transportation modes: informal transit. Small, privately operated vans are a primary source of transportation in the developing world

Some modes of transportation (in addition to the private motor vehicle) are problematic in the era of sustainable development. High speed rail, for example, works very well for long-distance travel in countries with high urban densities around station areas. Europe, Japan, and China have developed impressive systems with trains typically running around 180 miles per hour. However, a proposed high-speed rail system for California is much more questionable. Only downtown San Francisco and potentially Los Angeles have the urban densities to support the service, and, even though a system might promote station-area development, it is likely to be very slow in much of the state. Moreover, the proposed system is fantastically expensive (estimates ranged around $100 billion in the early 2010s). Major equity questions exist concerning whether it is worth spending that kind of money on behalf of a relatively few affluent long-distance riders when the state has consistently starved more local forms of public transit serving much larger numbers of people from less well-off communities of color. Lastly, a major rationale for the system is to reduce air commuting between San Francisco and Los Angeles, and we can ask whether anyone should be routinely shuttling back and forth between such cities anyway. Such hyper-mobility is unlikely to be sustainable by any mode, and more local ways of living should quite probably be encouraged instead.

Air travel is the most destructive form of transportation in terms of carbon emissions. Exact emissions depend on factors such as the length of the journey, the number of stops, and whether planes are filled to capacity. However, they average around one pound of CO_2-equivalent per passenger for every four miles traveled—a rate comparable to drive-alone motor vehicles, and many times higher than for train or bus transport. A round-trip trans-continental trip may therefore produce the same volume of emissions as driving your car for half a year. Although some carriers such as Virgin Airways have experimented running jet engines on biofuels, those can entail environmental impacts as well, and there seems to be no good substitute for the jet engine overall in terms of air travel. As part of an overall strategy to reduce their ecological footprint, sustainable societies of the future are likely to be much more local, with air travel being an occasional luxury rather than a routine part of life.

LAND USE

As discussed previously, land use changes—the second main requirement for reduced automobile use—help reduce the number and length of trips that people need to take. Local governments can play a major role by ensuring that development is relatively compact, mixed use, contiguous to existing urban areas, and served by a well-connected street system. Achieving a jobs–housing balance within communities is a related goal. Or rather it is important to balance neighborhood land use so that all functions that residents will need in the course of daily life are easily accessible by foot, bike, or public transit. The price of housing must also match the wages of local workers; otherwise massive cross-commuting occurs between widely spaced communities and the job destinations employing those residents.

Land use planning strategies to decrease automobile use typically cluster development around transit facilities, a strategy known as transit-oriented development (TOD). A similar pattern emerged in the late nineteenth century before the advent of the automobile when streetcar operators developed new rail lines and subdivisions together.[8] Those streetcar suburbs and other neighborhoods are today some of the most attractive and sought-after living environments in North America, and include neighborhoods such as Dupont Circle, Adams Morgan, and Cleveland Park in Washington, DC, Jamaica Plain and Brookline in the Boston Area, North Oakland and Berkeley in the San Francisco Bay Area, and the Annex, the Beaches, and Kensington in Toronto. New Urbanist writers such as Calthorpe are proposing a new emphasis on TOD as a way to combat climate change (by reducing transportation-related GHG emissions).[9] It is important not to be simplistic about such assertions—TOD by itself will make only a modest difference in driving and emissions. Also, many TOD projects have not been well designed, creating relatively sterile complexes of large-scale buildings and corporate offices that do not function well as a neighborhood. But if well designed and combined with alternative travel mode choices (including bike and pedestrian options), economic incentives to reduce driving, and lifestyle changes, such land use visions have a much better chance of promoting sustainability.

PRICING

Economic strategies to reduce motor vehicle use include raising the cost of driving (by measures such as gas taxes, registration fees, parking charges, road use tolls, or fees on total mileage driven) and reducing the cost of alternative modes of transportation (by free or subsidized public transit, company rebates to employees who do not use parking spaces, and other means). Strategies such as high gas taxes face fierce political opposition in the

United States but are the norm in many other countries, where gas typically costs at least 50 percent more. Indeed, the fierce reaction of Americans to any increase in the price of gasoline, no matter how minor, shows just how far we need to go to change public attitudes towards transportation. Driving is seen as a right by many people; the personal vehicle is a cultural icon, status symbol, and indicator of personal freedom. Meanwhile, the negative effects of excessive vehicle use are commonly ignored. These attitudes are deeply entrenched.

Yet the days of unfettered adherence to the motor vehicle culture are over, as shown by growing movements for walkable and bikable cities and opposition to road expansion. Local and regional governments are paying much attention to strategies to reduce driving. Although the world discovered the importance of energy conservation in the 1970s, transportation planning somewhat belatedly embraced demand management beginning in the 1990s. This approach seeks to reduce the demand for mobility rather than increasing the supply of roads, motor vehicles, trains, buses, and aircraft. Comprehensive "transportation demand management" (TDM) strategies may include not just the pricing, land use, and alternative mode strategies mentioned earlier, but also techniques such as providing better information to drivers about available garage parking (to avoid drivers circling round and round in urban areas looking for a parking space), promoting carpools and vanpools, and providing employees who don't drive to work a bonus equal to the cost of the "free" parking that other workers take advantage of. Higher parking charges and "eco-pass" programs (under which employee transit use is free or subsidized) are also effective tools to reduce automobile use and have been adopted by a number of municipalities. None of these strategies alone can solve the problem of automobile dependency, but together they can help address this archetypal challenge of sustainability planning over the long term (see Box 12.1).

RETHINKING MOBILITY

Any form of motorized transportation is going to have environmental and social impacts. Most sustainable of all would be high capacity public transportation run entirely on renewably generated electricity. Small, private vehicles similarly fueled might be part of the mix as well. But still the vehicles themselves would have impacts through their materials and embodied energy, and through their operations and infrastructure requirements. Even if all vehicles could be miraculously converted to electricity, a city chock-full of electric cars, with drivers backed up in traffic jams and whizzing too fast down residential streets, would not be a particularly pleasant or safe place to be.

The lifestyles that high degrees of mobility lead to tend to be highly consuming, stressful, and rootless. They undermine attachment to place and local community. Although there are advantages to being able to travel around the world and experience different cultures, and to visit far-flung friends and recreational sites, in other ways such hypermobility dehumanizes us. It undercuts the bonds of place that have been part of human existence from the dawn of the species. It may also lead to stress and ungrounded ways of life, as portrayed in the 2009 film *Up in the Air* starring George Clooney. Just as the worldwide spread of fast food led to a "slow food" movement, originating naturally enough in Italy, so radical mobility may lead to a new desire to move slowly, traveling by foot or bicycle instead of motorized vehicle, and traveling less overall in the course of daily life. "Light green" approaches to sustainability planning may emphasize alternative modes of high-speed travel, but deep green perspectives question mobility as a value in the first place.

As with everything else, sustainable transportation planning requires a sense of balance. There are times in life, particularly when one is young, when it is important to travel in order to broaden one's horizons. And some long-distance trips may be called for at any time of life, for example to see friends and family. But lifestyles that involve commuting dozens or hundreds of miles to work or second homes on a regular basis are profoundly unsustainable. Steps to support more local lifestyles include ensuring that people can find affordable housing in decent communities near where they work; promoting clusters of small-scale neighborhood shops and restaurants; creating local parks, greenways, community gardens, and recreational facilities; and encouraging small to mid-sized businesses throughout urban areas. Conversely, very large-scale facilities such as regional malls and massive office parks can be discouraged, since these encourage more travel from farther distances. In the long run such movement toward slower-paced, more local ways of living may be far more important than new public transit systems or economic incentives for not driving in reducing the pollution and resource consumption impacts associated with transportation.

13

HOUSING, FOOD, AND HEALTH

Although some of the world's people now live very well many others do not, and even within affluent countries a sizable fraction of the population lives in low-quality housing, lacks access to healthy food, and suffers from poor physical and/or mental health. Ensuring high-quality environmentally friendly living conditions is key to a sustainable future. These conditions would be better than today's for most people and more equally shared, with far lower environmental impact. How might we plan for such improved quality of life?

Eco-communities would contain many elements. Among these are decent, affordable, and well-located housing as well as amenities such as shops, restaurants, public spaces, and parks. Good quality schools and public services are essential. A safe, clean, and unpolluted environment is important for public health, as is access to healthy food and opportunities for exercise. The overall environment would be calmer, greener, and healthier than in many of today's cities and towns. Indeed, citizens around the world are increasingly aware of the many unhealthy and stressful aspects of contemporary life, and many are looking for alternatives. Hence movements for slow food, urban agriculture, bicycle and pedestrian travel, healthy living, and voluntary simplicity.

HOUSING

The question of how human populations should best be housed is age-old, and has led to a great many innovations over the years. The tenement house, the high-rise apartment tower, and suburbia itself were all innovative steps to improve housing quality, albeit each with their own problems. Even within affluent countries much of the population is still housed badly—in shoddily built structures with little sense of character or quality, in units that are too expensive relative to income, in places far from jobs or schools, or in neighborhoods with few amenities and little sense of community. A sizable number of individuals and families in many countries cannot afford housing at all and are homeless. Others make do with crowded or substandard conditions, even in the midst of affluence—often with multiple families sharing just a few rooms. Even those of us with adequate housing often live in bleak, homogeneous neighborhoods with excessive dependency on motor vehicles and little connection to culture or the natural landscape.

Decent housing can be seen as a basic human right. In fact it was declared as such by the first United Nations Conference on Human Settlements held in Vancouver in 1976, which concluded that

> Adequate shelter and services are a basic human right which places an obligation on governments to ensure their attainment by all people, beginning with direct assistance to the least advantaged through guided programmes of self-help and community action."[1]

There were efforts to insert a similar principle into the official statement of the second UN Human Settlements Conference held in 1996 in Istanbul, but they were strenuously opposed by the United States, which argued that adequate housing should be left to the market rather than guaranteed by the state.

Historically many countries including the US have spent large amounts of money to provide subsidized housing for less affluent citizens. These units are known as "social housing" in much of the world (and "public housing" or "subsidized housing" in the US). Britain's county councils built an enormous amount of social housing during the first half of the twentieth century, and in Sweden, the Netherlands, Canada, and many other countries governments played a similar role. Sometimes such housing turned out well, though at other times these efforts suffered from deficient design or funding, producing relatively sterile modernist apartment blocks without the services and amenities needed for healthy living. In the US particularly, federal agencies adopted provisions ensuring that federally funded housing would be built with bare-bones designs and the cheapest possible materials. Very little attention was given to landscape design, among other things, which is crucially important to the livability of higher-density developments. The apparent intent was to stigmatize lower-income groups by ensuring that only low-quality housing was available to them. Not surprisingly, many of these housing projects failed and had to be torn down.

Opponents often argue that governments cannot afford to ensure decent housing for all residents. However, in affluent modern societies there are sufficient resources and regulatory mechanisms to ensure better homes and living conditions for everyone. It is a question of priorities. Military budgets, tax breaks for the wealthy, and subsidies for corporations consume much public funding. Meanwhile, for-profit builders have fought against many forms of development regulation. But, even with limited budgets and patchy regulation, good architecture and urban design can improve the livability and sustainability of housing, in particular by using land efficiently, varying the form and character of individual units, employing high-quality non-toxic materials, providing light, air, and private or semi-private outdoor spaces, improving the environment of street and public spaces, and giving residents opportunities to personalize their surroundings. "Housing as if people mattered," to borrow a phrase from Clare Cooper Marcus and Wendy Sarkissian,[2] simply applies thoughtful, common-sense design and planning practices to the creation of everyday living environments.

Various levels of government can play a role in creating the context for better housing. National or state agencies can supply funding and set general guidelines for structural quality, livable design, energy efficiency, and public access to housing. They can also collaborate with NGOs to improve housing quality, as Australian government agencies have worked with Livable Housing Australia to develop livable housing guidelines and a livable housing rating system with silver, gold, and platinum levels.[3] Local governments can work with builders and communities to ensure that such housing gets sited well and actually built, and can adopt design guidelines as well as building codes that set appropriate standards. Municipalities can also help ensure that neighborhoods are safe and contain decent schools. All levels of the public sector can provide incentives and funding to encourage NGOs to construct housing affordable to lower-income residents, or can set requirements that private sector builders include this within their market-rate projects, a strategy known as inclusionary zoning. Through all of these steps, communities can move towards housing that better meets human and ecological needs (see Figures 13.1 and 13.2.)

Leaving the provision of housing entirely to the market has clearly led to many problems. For-profit builders tend not to provide sufficient quantities or qualities of affordable housing for lower-income residents, and their projects may not meet other criteria of sustainable development such as reducing energy consumption, creating green buildings, providing public space, and linking appropriately to jobs and community facilities. Private sector housing is also often not in the right place—far from jobs, services, and public transportation. So a strong public sector role is required in regulating housing development and/or supplementing the efforts of the market, and this task generally falls to local government.

Figure 13.1 Well-designed affordable housing. Affordable housing, such as this development in San Jose, California, can be made attractive and human-scale

Figure 13.2 Compactly designed single-family housing. Single-family detached housing is possible on small lots within compact neighborhoods such as this one in Mountain View, California

Providing sufficient quantities of housing is a major challenge in many parts of the world. Otherwise housing prices escalate rapidly, desirable areas become enclaves of wealth, and the poor are displaced through processes of gentrification. They must then pay a large portion of their income for housing, double up in crowded conditions, or travel long distances to jobs from distant communities with more affordable housing. With development becoming more difficult in many areas because of decreasing supplies of centrally located vacant land, increasing public resistance to new projects, and tougher environmental review requirements, many urban areas are simply not producing enough housing to meet population needs. This situation highlights the need to stabilize population, but it also highlights the need to increase housing production in the short term. Appropriate housing densities are also needed in each neighborhood, as discussed further in Chapter 24.

Local governments can address this need through a variety of strategies. They can directly construct housing themselves, a strategy that has fallen into disfavor in the US and many other countries after decades of cheaply built, poorly designed public housing. They can provide loans, grants, and other assistance to nonprofit builders. They can require for-profit developers to include a certain percentage (usually 10 or 20 percent) of affordable units within market-rate developments through inclusionary zoning. And they can seek to ease the process of housing development by speeding up permitting processes, preparing plans that indicate sufficient locations for new development, making sure that zoning codes do not prevent local development of appropriate intensity, and acting as intermediaries with concerned neighborhood groups.

These activities all require heightened attention to good urban design and planning, especially if housing is to take place in infill locations. Finding ways to placate "not in my backyard" (Nimby) opposition from neighbors and to ensure that new infill development provides benefits for existing communities is especially important. Also crucial is the need to change the mix of organizations doing housing development. In recent decades much homebuilding has been taken over by large-scale, mass-production builders (in North America, companies such as KB Home, Pulte, D.R. Horton, Lennar, Trammel Crow, and Shea Homes). In many places these companies have displaced the small and medium-sized builders who once built the majority of new housing units, and who potentially have greater flexibility in responding to local contexts and environmental needs. Active local government support of small and nonprofit builders, especially those willing to do green construction and infill development, may be needed to counter the centralization trend within the housing industry. Many small builders are nonprofit corporations founded over the past several decades; helping these NGOs build even larger amounts of affordable housing is especially important.

FOOD SYSTEMS

In the past couple of decades awareness has grown of just how centralized, mechanized, polluting, and unhealthy our industrial-scale food systems can be. For example, as of the mid-2000s just four companies slaughtered 84 percent of cows and 64 percent of hogs in the US, while just two companies accounted for nearly half of all chickens.[4] Four companies also accounted for about half of all grocery sales, and 72 percent of grocery sales within the 100 largest urban areas.[5] Farm size has expanded to the point where very high degrees of mechanization are required. Meanwhile, small-scale family farms have dwindled in number and the average age of farmers is well above 50 in many places, leading to questions about who will farm in the future.

Such centralization means that companies have had enormous marketing power to promote standardized products, in particular processed foods containing large amounts of high-fructose corn syrup, sugar, salt, and other unhealthy ingredients. The massive advertising of soft drinks and fast food also leads to overconsumption of such products. Writers such as Michael Pollan and films such as Food, Inc., Supersize Me, and Forks Over Knives have pointed out the many downsides of such diets.

At the same time that millions of people in industrialized countries are eating too much of the wrong foods, millions more in both wealthy and impoverished countries don't have enough to eat. "Food security" is a pressing issue both within sub-Saharan Africa and North American inner cities and rural hamlets. In some parts of Africa and elsewhere in the world warfare, lack of access to land, soil depletion, and corrupt food distribution systems keep people hungry. Emphasis on export-oriented agriculture rather than farming for domestic consumption also contributes to hunger. Meanwhile, within inner cities of wealthy countries as well as in poor rural and suburban areas there may be no local grocery stores selling healthy food, or families may not have money with which to purchase it. As many as 16 million children annually go hungry in American society at one time or another,[6] often depending on subsidized school lunches and breakfasts for much of their diet during the week, and then scraping by on very little during the weekends.

Planning for healthier and more sustainable food systems means many things. In terms of agricultural policy, for example, it is important in the US to end today's subsidies to large-scale producers, while encouraging small-scale, organic production whenever possible. Some of those subsidies are direct payments to farmers; others are indirect, such as the federal government's provision of low-cost water to agribusiness in the West. It is also important to reduce consolidation in the food industry, allowing a range of small and mid-sized distributors to give farmers more options for moving their crops to market. Tighter regulation of agriculture and food processing operations such as slaughterhouses is essential to ensure that workers and consumers are protected. A ban on sale or advertising of unhealthy food products, especially to children, would be an excellent idea. Municipal governments can also develop farmers' markets, encourage grocers to locate within underserved areas, make city-owned land available for community gardens, and assist small-scale local organic farmers with local and regional distribution (see Figures 13.3 to 13.5).

Within agriculture itself, farmers can reduce use of nitrogen-based fertilizers which produce the greenhouse gas N_2O as well as polluting runoff into local waterways. These fertilizers are often over-applied, and substantial reductions can be made without hurting crop yields. For many crops, tillage (tractor passes over fields to prepare soils) can be significantly reduced as well, saving energy and reducing GHG emissions. Integrated pest management programs can replace application of potentially unhealthy pesticides. Organic production can be encouraged, in particular for fruits and vegetables, though it is not without environmental impacts of its own such as greenhouse gas emissions from application of manure.

HEALTHY ENVIRONMENTS

Apart from food systems, many other public health issues are of increasing concern within affluent societies. Although specific things can be done to address each of these problems individually, together they require rethinking physical environments and lifestyles in ways that mesh perfectly with other sustainability goals.

One longstanding concern has been exposure to pollution of various sorts, including toxic chemicals, air pollution, and radiation. Such threats have almost always been part of civilization—smoke from cooking fires within poorly ventilated homes has been a hazard within indigenous cultures since pre-history, and the Romans observed the deleterious effects of lead on the human body.[7] However, since the Industrial Revolution we have surrounded ourselves with smoke and chemicals at a much larger scale. This exposure

Figure 13.3 Community gardens. Many European cities have a long tradition of "allotment gardens," such as this one here in Basel, Switzerland, that offer urban residents large plots and even one-room cottages in the midst of the city

Figure 13.4 Cold-weather gardening. A family's greens growing in cold frames in Albuquerque, New Mexico

Figure 13.5 A farmers' market. Farmers' markets are a way for cities to support local agricultural producers, ensure residents access to healthy food, and build community

expanded after the Second World War with the rise of petrochemicals and nuclear weapons testing. The 1963 Nuclear Test Ban Treaty ended the above-ground testing which deposited radioactive strontium and iodine in the bodies of every living person. This treaty still stands as one of the premier cases in which the world's countries were able to take strong, unified action to safeguard the human environment. However, other risks continue from nuclear power, nuclear waste, and everyday uses of radiation such as for X-rays.

Air and water pollution as well as toxic chemicals have traditionally been addressed through a patchwork of national, state, and local regulation. For the most part these threats are local or regional, and can be reduced through requirements that pollution should not exceed certain levels and that hazardous materials should be disposed of properly. But allowable levels of pollutants are open to debate, industries lobby strongly against regulation, and enforcement is often spotty. Whereas drug companies must prove that a medication is safe before agencies allow them to market it, industries can create new chemicals and unleash them into the environment unless and until they are proven dangerous. So regulation often lags any particular threat by many years. A more comprehensive strategy to improving the health of our environment would take a precautionary approach, restricting new substances unless they were shown to be safe. It might also de-emphasize entire categories of products, for example plastics, which are likely to be associated with pollution or long-lived waste.

The threat posed by obesity and growing rates of diabetes came to international attention in the early 2000s. During the preceding decades obesity rates had skyrocketed in the US, and by 2010 more than one-third of the adult population was obese.[8] Other societies began experiencing similar problems, even the more affluent classes within developing countries such as China. Like most health problems, obesity has multiple causes including changing diet (in particular the rise of fast food, soft drinks, and prepared foods using high-fructose corn syrup), sedentary lifestyles, increasing time spent indoors with electronic media, and the lack of outdoor environments in which to walk or exercise. Addressing the issue is not simple, since each of these factors requires separate strategies. Some communities have enacted restrictions on the sale of soft drinks in schools or other public venues. New York City, for example, in 2012 adopted a ban on large sugary drinks in restaurants, street carts, and movie theaters. Clear labeling of calorie content in foods is another step that state and national governments can pursue. Another basic starting point at the local level is to design communities that are walkable, bikeable, and full of healthy living opportunities such as parks, trails, and sports facilities. Simple steps such as making stairways rather than elevators a central, easy-to-use feature of buildings would help promote greater physical activity. Making physical education a more central part of schooling would help as well, as would banning advertisements of unhealthy foods aimed at children.

A healthy environment is one that promotes mental as well as physical well-being. Stress, depression, anxiety, and eating disorders are unseen plagues in many societies. People often feel burdened with economic insecurity, excessive commuting, intense schedules, competitive pressures, negative self-images, and/or lack of community and support networks. It doesn't help that the material conditions of people's lives are often very different than the glamorous images seen on television, and that individuals often feel little control over their own futures. Such circumstances too frequently lead to frustration, anger, hatred, and suspicion of others—undermining social capital and democratic politics in addition to individual health.

Again, there is no single answer to such problems, but a range of planning strategies and programs can help. Potential initiatives include basic worker health and safety protection, livable wage ordinances, affordable housing programs, low-cost public childcare, counseling and health services, and environmental design improvements. But many of these threats to mental and physical health come from the fundamental values and priorities of societies, and can only be addressed by changing those. Slower-paced, less materialistic lifestyles could help greatly. A stronger social safety net could help to reduce insecurity. Workplace policies such as flexible work hours, decent vacation benefits, and a calm and supportive workplace environment would be excellent. Positive opportunities for education, personal advancement, meaningful work recreation, and connection with others could help reduce the loneliness and alienation often found within affluent societies.

A number of local governments or civic associations have adopted comprehensive "healthy city" programs that attempt to think holistically about creating healthy living environments. In addition to focusing on food and lifestyle, these initiatives may include topics such as emergency preparedness, civic health (for example democratic participation), and cultural preservation. Specific emphases will vary from place to place. The Chamber of Commerce in Emeryville, California, has a Healthy City Initiative that pays strong attention to earthquake preparedness, given that community's proximity to several geological faults.[9] Philadelphia adopted a Get Healthy Philly program in 2010 following that city's designation as one of the most overweight US cities.[10] The city of Udine, Italy, has a "No all solit" (say no to loneliness) program that has mobilized 1,000 volunteers to provide services for the elderly people who make up more than a quarter of the town's population.[11] And Belfast, Northern Ireland has a Shaping Healthy Neighborhoods for Children program that seeks to gather children's views on how to create a more kid-friendly city.[12]

On a broader scale, the World Health Organization (WHO) sponsors a global Healthy City program that emphasizes the need to address urban poverty, inequality in health, and the social, economic, and environmental determinants of health. This program thus takes the same sort of holistic or ecological approach to health as many current public health practitioners, seeking to address all of the factors that improve the livability and healthiness of human communities. This type of approach seems most likely to meet human health needs and desires for a high quality of life as well as dovetail with other sustainability initiatives.

14

GREEN ARCHITECTURE AND BUILDING

Buildings are one of the world's principal users of energy and materials, accounting for about 40 percent of all energy used in the United States.[1] Building construction consumes 25 percent of the world's wood harvest, and demolition of built structures accounts for 44 percent of landfill waste.[2] Together, building construction, heating, and cooling account for about half of carbon dioxide emissions in the United Kingdom;[3] the figure is probably similar in other countries.

Buildings also profoundly affect our relation to the natural world, either insulating us from it within highly artificial environments, or providing us with a more integrative experience that takes into account natural light, ventilation, climate cycles, and landscape as well as local culture and history. Buildings affect our social life as well—the ways that we arrange structures on sites and group them within urban areas can bring people together and promote civic life, or separate groups of people and undermine shared quality of life. For all these reasons it is important to rethink current architectural and building practices within a more sustainable society.

For more than 40 years now pioneering architects have developed green design principles and practices, though these strategies remain far from mainstream. Municipal building codes have also become substantially greener in recent decades, for example requiring greater energy efficiency and water conservation. But such efforts have still only scratched the surface. Even newly constructed buildings are still usually not very green, and the large existing stock of homes, offices, and institutional structures around us will need to be retrofitted or replaced. Many buildings are also not located in appropriate places, for example requiring their residents to drive long distances in everyday life, or putting them at risk from flooding and (in the long term) sea level rise. Those structures may need to be replaced as well. Given the rapid pace of building construction in many parts of the world and the fact that buildings last for many decades or centuries, there is a need to immediately and dramatically improve their sustainability performance.

ZERO-NET-ENERGY BUILDINGS

One of the most urgent needs is for buildings to become zero net energy (ZNE) as soon as possible. Given buildings' longevity, to reduce GHG emissions 80 percent by 2050 all new structures from now on need to be extremely energy efficient, and if possible to generate energy on site to offset their remaining energy use. ZNE structures are possible with existing technology. The BedZED housing project in London, completed in 2002, was perhaps the world's first example, and West Village in Davis, California, is the first US example of a ZNE neighborhood (see Figures 14.1 and 14.2).[4]

Efficiency is the starting point. Generally this goal is achieved by highly insulating the building shell, carefully sealing any opening to the outside (doors, windows, ventilation shafts, and utility conduits), and using very energy-efficient appliances and heating, ventilation, and air-conditioning (HVAC) systems. Using light-colored roofing materials can also reduce heat absorbed from the summer sun and keep the structure cool. Installing a green (vegetated) roof can likewise help insulate the building as well as retain storm-water runoff on site.

Along with energy efficiency, the set of strategies known as passive solar design can help reduce building energy use. Buildings must be oriented more-or-less south (or north in the southern hemisphere) to maximize solar exposure during cold months. Large windows on the south side allow low-angle winter sunshine to heat rooms, while sunshades, overhangs, or deciduous trees keep out high-angle summer sunlight that would otherwise

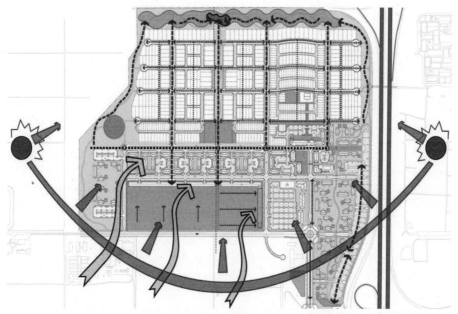

Figure 14.1 Passive solar orientation. This site plan shows how the zero-net-energy West Village neighborhood in Davis, CA is oriented to maximize use of the sun's energy for winter heating and prevailing breezes for summer cooling

Figure 14.2 Zero-net-energy buildings. These structures are twice as energy efficient as state code requires, and their energy use is offset by power from photovoltaics on the long south-facing roof slopes

overheat interior spaces. Designers typically also use thermal mass—thick walls or floors—to retain winter warmth or summer cool, keeping the building's temperature naturally more constant during the daily temperature cycle outdoors. Fans or carefully designed natural flows of air (using the tendency of hot air to rise) can then help distribute warmth throughout the building or channel excess heat outside.

Complementing passive solar design, geothermal heat pumps can use the inherently stable temperature of the ground 2–20 meters below the Earth's surface to heat or cool structures. Water circulates through a grid of pipes buried in the ground (the desirable depth will depend on soil characteristics) and then passes through a heat exchanger that captures the warmth or coolness for use within the building. Naturally occurring heat differentials thus provide energy to the building. In related fashion, many buildings in downtown Toronto, including the City Hall, are cooled by circulating cold water from deep in Lake Ontario through their cooling systems.[5]

"Passive houses" are perhaps the most extreme example of building energy efficiency using such strategies. These structures are so efficient that they need no active heating or cooling systems. Since they are so tightly sealed, indoor air quality can be a problem, so generally they employ heat exchangers to allow some ventilation without losing efficiency. Starting in Germany during the 1990s, there are now passive houses in most parts of the world.

To reach ZNE status, once a building passive or otherwise is as energy efficient as possible designers can add renewable energy systems to compensate for its remaining energy use. Rooftop photovoltaic (PV) panels are the most common means employed to generate electricity. To maximize a building's PV potential architects can try to create large, uncluttered flat or south-facing roof spaces. Thin-film PV technology may also allow building walls to be covered with an energy-producing layer. Municipalities can help ensure every structure can make use of photovoltaics through solar access ordinances, which keep surrounding property owners from building in such a way that would block sunlight from reaching a building.

Some buildings use small wind turbines to generate electricity, but this is often difficult since wind flow is disrupted by buildings in urban areas. Generally wind energy installations are more effective if constructed on a larger scale in the relatively small number of locations with strong year-round wind, often near coasts or offshore. Biodigesters are another emerging renewable energy source at a neighborhood or campus scale; these machines compost household garbage, yard waste, and manure to create methane gas that can be burned for heat and electricity. Solar hot water panels, not to be confused with PV, can greatly reduce use of gas or electricity for water heating, and due to their effectiveness and relatively low cost are the most common type of solar installation on residential roofs in many developing countries.

Although a few pioneering homeowners, mainly in rural areas, seek to live "off the grid," almost always buildings use the utility electrical grid as a battery, feeding excess electricity in at some times of the day and taking back current during other hours. As discussed earlier, the economic efficiency of this practice depends on utility pricing policies. Countries such as Germany have required "feed-in tariffs" through which utilities pay small-scale producers of electricity a retail rate for the energy they produce. Such measures are a strong incentive for building-scale production.

Vernacular architecture (the historical building traditions in any given place) can provide many useful techniques for energy-efficient design. On sites exposed to cold or chilling winds, such as along seacoasts or in mountainous areas, traditional building practices have often reduced the number of windows on the windward side of the building while increasing thermal mass. In hot climates, vernacular styles have often featured broad porches or partially shaded interior courtyards to provide comfortable outdoor space, as well as strategies to capture cooling breezes or allow hot air inside the building to escape out of rooftop vents during evening or night hours. For example, Arabic architecture in the Middle East frequently includes chimney-like towers to capture cooling breezes and/or release heated interior air. Traditional architecture in the hot southeastern United States makes extensive uses of wide porches or verandas to provide cool, shaded seating areas and to screen interior rooms from the sun's heat.

Reducing artificial lighting is another main concern for ecological architects. Clerestory windows (small secondary windows high up on a wall) can increase the amount of light allowed into buildings and reduce artificial heating and illumination needs. Adding skylights or light wells can also reduce needs for artificial lighting, especially within office buildings that will be occupied primarily during daylight hours. Large, open rooms with high ceilings help natural light diffuse into the interior of buildings.

GREEN BUILDING MATERIALS

Often people associate green architecture with seemingly exotic construction materials such as straw bale and rammed earth. While these represent useful strategies in certain

Figure 14.3 Simple green buildings strategies. Ceiling fans promote natural ventilation in Berkeley's renovated Civic Center building, while high ceilings and large windows let in daylight, reducing artificial lighting needs

circumstances and are discussed below, a wide variety of other alternative building materials can be used in a much broader range of cases. Essentially, any material that reduces life-cycle resource consumption or long-term social and environmental impacts is worth exploring (see Figure 14.3).

Sustainably harvested lumber is perhaps the resource with the widest applicability since much construction in North America will continue to be wood-based. (Buildings in Europe and elsewhere typically use higher proportions of non-wood materials such as precast concrete, steel, stone, and brick.) Certified sustainably harvested wood is grown within forests managed under carefully developed rules, for example, prohibiting clear-cutting, requiring buffer zones around creeks, mandating replanting, and harvesting at a rate that preserves the viability of the forest. Several associations certify wood as "sustainably harvested"; these groups include the Rainforest Alliance (with the "Smartwood" label) and the Certified Forest Products Council. Since 1993 the Forest Stewardship Council, with offices in many countries, has sought to coordinate forestry certification programs worldwide and accredit those certification organizations that comply with basic standards. Certified wood products are now becoming widely available through mainstream outlets such as some Home Depot stores. Although somewhat more expensive than conventional lumber, certified wood products have also become more price-competitive in recent years, and governments could help expand the market for them further through requirements for their use in publicly funded projects.

Reused lumber from older buildings is another sustainable source of wood products. Demolished structures frequently contain large beams and supports that can be salvaged, and such wood is often of higher quality than anything available now since it came from old-growth forests that have since vanished. To reuse such lumber, nails and other hard-ware are removed and the wood is replaned in a mill to produce finished beams, studs, and boards. Alternatively, new structural wood and fiberboard can be produced from smaller pieces of recycled wood bonded together or laminated with resins or other adhesives. Old and weathered boards can be reused directly to achieve certain architectural effects. Many unusual sources of reused lumber exist; one firm, for example, bought thousands of old railroad ties from Thailand—originally cut from beautiful tropical hardwoods—and remilled these as high-end flooring for North American homes.[6]

Steel framing may at times be an ecologically desirable alternative to wood, despite the fact that steel often costs more and may contain higher embodied energy. Metal studs and other structural building components reduce needs for logging, are relatively reusable and recyclable, and may support jobs within existing domestic industries. The replacement of wood two-by-fours with steel framing has already been happening within many construction projects owing to rising lumber prices.

Recycled concrete and asphalt is another alternative material that offers environmental benefits over new versions of these substances. Any rubble from past construction can be crushed on site to produce aggregate for new foundations or paving. Materials can also be delivered to off-site recycling facilities. In addition, new paving materials are emerging that do not have the high energy and/or petrochemical inputs of concrete or asphalt (the latter is a mix of gravel and heavy hydrocarbons left over from oil refining). One such alternative uses pine pitch to bind a paving surface that is harder than concrete but less costly.[7]

A wide variety of other materials can be recycled into green construction products. Durable, attractive, "plastic lumber" made from recycled consumer materials is ideal for decking, park benches, or other uses where wood might otherwise be subject to rot. Recycled glass collected through municipal recycling programs is frequently mixed into asphalt used for roads and parking surfaces. Recycled wood from bowling alley lanes as well as other recycled glass and plastics can be used to create hard, durable, aesthetically pleasing kitchen counter-tops. Such reused or recycled building materials represent a promising avenue to greener buildings.

Rammed earth represents a new variation on one of the oldest building materials—local soil, used for thousands of years within adobe buildings in many parts of the world. In the simplest versions of this technique, earth is simply compacted within wooden forms that are removed after the wall is created. Usually a small amount of cement is mixed in with the earth to provide increased strength. In the approach known as PISE ("pneumatically impacted stabilized earth") construction—based on earth-building techniques traditionally called "pise" in France—a mixture of soil and concrete is sprayed into forms to create a stabilized earth wall.[8] Steel reinforcing or sections of reinforced concrete can be added for earthquake safety. Use of rammed earth typically creates building walls of 12 to 18 inches in thickness whose thermal mass is extremely good at retaining heat to warm the building at night and coldness to cool it during the day. The result can be a comfortable and energy-efficient dwelling whose walls are also fireproof and resistant to rot or insect damage.

Straw bales have been widely used in recent years in a variety of buildings and provide exceptional thermal insulation as well as utilizing local agricultural waste. Typically bales are used to fill in a wall that is supported by post-and-beam construction; the wall is then plastered and painted. But load-bearing straw bale walls can also be created that are "pinned" with iron rebar or wood. If properly dried and insulated, straw bales are rot and fire-resistant. Building codes in many western US states have now been changed to allow use of this material. In northern California surplus rice straw from the Central Valley, often burned as an agricultural byproduct, has been used in a number of homes, helping to solve a major air pollution problem.

Bamboo is an alternative construction material traditionally used in Asia and Latin America which has great potential applications within North America as well. Incredibly strong and light, it can be used for framing small buildings or as a material for flooring, furniture, or fences. With a tensile strength greater than steel, it can be used to replace iron rebar in reinforced concrete. This extremely fast-growing resource could be cultivated sustainably in tropical or subtropical countries as a substitute for wood from temperate forests.

"Cob" construction, which mixes earth with straw or other fibrous materials, is also enjoying a comeback in England, where builders used it for cottages that are still standing after 500 years. Other traditional practices such as thatched roofs are being explored again as well. Roofs made from a variety of natural materials—clay tiles, wood shingles, thatch, or sod—represent a green alternative to the use of asphalt shingles that tends to dominate current construction. Some of these other materials may also be far longer-lasting than flimsy asphalt-based roofing, which is typically intended to be replaced every 20 years (see Figures 14.4 and 14.5).

Figure 14.4 Earthships. The Earthships community near Taos, New Mexico, was a pioneer of off-the-grid, zero-net-energy housing using earthen construction and materials such as used tires and glass bottles

Figure 14.5 Passive solar design. Commissioned by the Dutch Environment Agency, the community of Ecolonia uses passive solar architecture in many of its 101 homes, as well as solar hot water heating and thick, heat-retaining walls

One further area that designers of green buildings must consider is indoor air quality. Paradoxically, highly energy-efficient buildings which allow very little leakage of heated or cooled interior air to the outside are also at greatest risk of air quality problems, as any outgassing from construction materials, carpeting, paints, and furniture will remain within the building. But other, more conventional buildings are also at risk of "sick building syndrome," especially if they are entirely climate-controlled and lack openable windows, if they use carpets or particleboard rich in chemicals like formaldehyde, or if they have vents, ducts, insulation, or inadequately drained foundations that contain molds or other irritants. As buildings have grown more hermetically sealed and full of synthetic materials, this risk has grown. The solution is to design buildings with great care to provide sufficient ventilation, user-adjustable climate control mechanisms, and healthy materials.

Inside buildings, paints and finishes that contain low quantities of volatile organic compounds (VOCs) can also improve indoor air quality. Carpeting that is recycled and/or recyclable—not to mention made of natural materials such as wool or hemp—is preferable to synthetics. Products that outgas dangerous chemicals such as formaldehyde can be avoided. Such measures, along with natural ventilation and openable windows, can help create far healthier buildings.

For most urban and suburban projects there will be little alternative to connecting to a municipal sewer system. Municipal ordinances usually require it, and sufficient land may not be available for either septic systems or more innovative ecological sewage treatment methods. But on large sites and in more rural areas alternative systems are possible, such as constructed wetlands or "living machines" in which sewage is filtered through a series of tanks, usually in a greenhouse, that contain water hyacinths or other plants to purify the water. Other ecological wastewater infrastructure may include gray water systems in buildings to use shower and sink water for toilet flushing or landscape irrigation, and rainwater catchment systems to hold storm-water running off roofs or pavement. Such reusable water can be routed to cisterns, stored, and then applied to secondary uses.

USE OF NATURAL FORMS, FLOWS, AND MATERIALS

In various ways ecological site design can take into account natural flows of materials and energy in a particular place. Historically, many cultures have sought to do this through disciplines such as the Chinese practice of feng shui, which considers the many ways in which chi, or energy, flows through the landscape. Some feng shui principles directly reflect ecological factors, for example by taking into account the flow of air down a valley or the relation of a house to neighboring watercourses. In Western culture taking natural air, water, and energy flows into account may involve a more systematic process of studying site characteristics and determining how best to mesh design with sun, wind, slope, hydrology, soils, and vegetation. Landscape architects might, for example, plant windrows of trees or shrubs to shield buildings from cold winds, or call for deciduous trees to shade roofs as well as southern and western building faces during the summer while allowing sunshine to enter south-facing windows in the winter. They might also map flows of groundwater and site buildings well away from them to minimize flooding and mold. Architects might analyze the energy used to make building materials ("embodied energy"), and reduce the total energy brought to the site during the process of development by choosing appropriate materials, such as locally produced brick, stone, rammed earth, straw, or remilled lumber.

"Making nature visible," as ecological architect Sim Van der Ryn and designer Stuart Cowan put it, can help the public understand the unique characteristics of each site or place.[9] There are many ways to do this within site planning. Creeks, drainage swales, rock outcroppings, or small patches of native vegetation can be made centerpieces of landscape design. Flowing water in any form is a pleasing and intriguing natural element on a site. Runoff from roofs can be channeled into "rain fountains" that provide a delightful design element. Edible landscaping helps residents and visitors understand how food can be grown locally. Composting facilities and gardens also help demonstrate connection to the earth. Even within buildings green designers can let natural water, air, and energy flows influence design details. Water flows and gardens within buildings, for example, are a delightful amenity for occupants, while natural ventilation and daylighting are likely to improve productivity.

Preserving the traditional forms and characteristics of the local landscape is a further common theme in ecological site design. Careful study of the natural landscape and historical building patterns is required to develop strategies for doing this locally. One example is Sea Ranch, a vacation community on the northern California coast near Mendocino that was designed in the 1960s by architects Charles Moore and Donlyn Lyndon working with landscape architect Lawrence Halperin. At Sea Ranch both the buildings and the site plan have been designed to reflect the pre-existing landscape of wind-swept meadows along coastal bluffs, cypress windbreaks, and forested hills. Houses are made of weathered wood, rooflines are pitched to emulate the line of the wind-sculpted cypress, and parked cars and driveways are hidden behind weathered wooden fences. Owners are required to maintain naturalistic landscaping and forbidden from erecting fences, so that the overall effect is of homes sprinkled here and there among meadows and forests, rather than a patchwork of individual yards. A network of trails next to the sea and throughout the development provides public access. Sea Ranch does not necessarily represent a sustainable community overall—it is otherwise a low-density, motor vehicle-dependent vacation development for the well-to-do—but its design illustrates how the form and character of a natural landscape may be preserved within development.

Following design critic Kenneth Frampton, architect Douglas Kelbaugh argues for a "critical regionalism" in which ecological design incorporates the vernacular architecture of a particular region.[10] For example, a regional design ethic for the Pacific Northwest of the US and Canada might emphasize use of wood (dense forests dominate the landscape), wide roof overhangs to keep off rain, and generously sized windows to let in the soft, slanting light of this high latitude. Vernacular building design in Mediterranean, Latin American, and southwestern US locales, in contrast, is quite different. Construction materials are often stone or adobe (bricks made from local clay or mud), colors reflect the warm earth tones of these regions, and buildings are often designed to protect against heat, with thick walls, small windows, and shaded interior courtyards. Using regionally based design strategies in this way helps preserve local tradition and culture while offering practical advantages in terms of energy and materials use.

GREEN BUILDING PROGRAMS AND STANDARDS

In recent decades many levels of government have adopted requirements for energy- and water-efficient construction that have greatly improved the efficiency of new buildings. California, for example, first adopted its Title 24 Energy Efficiency Standards for Residential and Nonresidential Buildings in 1978, covering everything from insulation and lighting to water heating, air conditioning, fireplace construction, clothes dryers, and pool heaters. As a result of this improved regulation, residents of the state have saved an estimated $65 billion in electricity and natural gas costs. The 2013 version of this code by itself is expected to save 300,000 tons of CO_2 in the first year, and eliminate the need for eight large 500 MW power plants over 30 years.[11]

Over the years a number of individual cities have started Green Builder programs to promote ecological building practices. The city of Austin, Texas, for example, began with an energy rating system for new buildings during the 1980s. In 1991 the city expanded this into a more comprehensive Green Builder program. A central component of this system is a four-star rating scale assessing whether buildings meet criteria in areas of energy efficiency, water efficiency, materials efficiency, health and safety, and community. Municipal guidelines provide developers and architects with details about how these criteria can be met, and the city offers a variety of technical-support services and cash incentives for green buildings. In the 2000s Austin took the further step of requiring all new residential construction to be zero net energy by 2015. Although such programs currently affect only a small percentage of the US building stock, their influence is growing.

Beyond such initiatives, many other cities have been reviewing their building codes to ensure that they allow and encourage practices such as gray water recycling, passive solar architecture, and straw bale construction. San Francisco, for example, adopted an ordinance requiring dual plumbing systems, including one for gray water, on new commercial buildings in the late 1990s, with the water to be used to irrigate public landscaping. Municipalities can also issue nonbinding green building guidelines as a simple and basic step to promote more sustainable building practices. Such guidelines might spell out recommended practices for energy-efficient design, use of sustainable construction materials, recycling of construction debris, and so forth. Some cities are now encouraging or requiring energy "benchmarking," whereby large commercial or residential buildings are required to report actual building energy-use intensity (a measure of efficiency) as well as carbon emissions. Planning commissions, zoning boards, and design review committees might be instructed to look favorably on projects meeting these guidelines, or even to give them density bonuses or other incentives.

Since the late 1990s professional coalitions have developed green building rating systems in a number of countries. Britain emphasizes the BREEAM (Building Research Establishment Environmental Assessment Method) system, Canada uses the Green Globes, and the LEED (Leadership in Energy and Environmental Design) system dominates in the US. The US Green Building Council developed the LEED Green Building Rating System in the late 1990s to establish uniform, comprehensive green building guidelines that could be applied across the country. Applied initially to industrial and commercial buildings as well as to residential buildings higher than four stories, the LEED rating for new construction requires developers to complete a detailed checklist and awards 64 possible points for green building practices. Basic certification requires 26 points, while silver, gold, and platinum LEED certificates are awarded to buildings meeting higher standards. Main categories of the checklist include sustainable sites (site selection, alternative transportation, stormwater management, etc.), water efficiency, energy and atmosphere, materials and resources, indoor environmental quality, and innovation in the design process. In addition to LEED-NC, the USGBC has developed subsequent versions including LEED for existing buildings, core and shell, neighborhood development, schools, homes, and health care. Although open to criticism on some fronts, in particular for its "design by checklist" approach, which reduces the incentive for developers to greatly exceed the standard on any particular item, LEED nevertheless provides a valuable tool that city governments and public agencies in particular can use to specify particular levels of environmental performance (see Box 14.1).

MEETING SOCIAL NEEDS AT THE BUILDING SCALE

Although sustainable architecture is often seen primarily in terms of ecological factors, many social factors and equity issues are equally vital in the long run. Particularly important are building and site designs that meet the needs of a wide range of users, that will be flexible and adaptable over time, and that will promote community and social interaction.

In their classic book of design guidelines *Housing as if People Mattered*, Clare Cooper Marcus and Wendy Sarkissian point out how design of medium-density housing has often ignored the needs of users, and how simple steps such as adding porches, walkways, a range of private and semi-private outdoor spaces, and context-appropriate building design

Box 14.1 LEED FOR NEW CONSTRUCTION AND MAJOR RENOVATIONS PROJECT CHECKLIST

Sustainable Sites   26 Possible Points

☑ Prerequisite 1	Construction Activity Pollution Prevention	Required
☐ Credit 1	Site Selection	1
☐ Credit 2	Development Density and Community Connectivity	5
☐ Credit 3	Brownfield Redevelopment	1
☐ Credit 4.1	Alternative Transportation—Public Transportation Access	6
☐ Credit 4.2	Alternative Transportation—Bicycle Storage and Changing Rooms	1
☐ Credit 4.3	Alternative Transportation—Low-Emitting and Fuel-Efficient Vehicles	3
☐ Credit 4.4	Alternative Transportation—Parking Capacity	2
☐ Credit 5.1	Site Development—Protect or Restore Habitat	1
☐ Credit 5.2	Site Development—Maximize Open Space	1
☐ Credit 6.1	Stormwater Design—Quantity Control	1
☐ Credit 6.2	Stormwater Design—Quality Control	1
☐ Credit 7.1	Heat Island Effect—Nonroof	1
☐ Credit 7.2	Heat Island Effect—Roof	1
☐ Credit 8	Light Pollution Reduction	1

Water Efficiency   10 Possible Points

☑ Prerequisite 1	Water Use Reduction	Required
☐ Credit 1	Water Efficient Landscaping	2–4
☐ Credit 2	Innovative Wastewater Technologies	2
☐ Credit 3	Water Use Reduction	2–4

Energy and Atmosphere   35 Possible Points

☑ Prerequisite 1	Fundamental Commissioning of Building Energy Systems	Required
☑ Prerequisite 2	Minimum Energy Performance	Required
☑ Prerequisite 3	Fundamental Refrigerant Management	Required
☐ Credit 1	Optimize Energy Performance	1–19
☐ Credit 2	On-site Renewable Energy	1–7
☐ Credit 3	Enhanced Commissioning	2
☐ Credit 4	Enhanced Refrigerant Management	2
☐ Credit 5	Measurement and Verification	3
☐ Credit 6	Green Power	2

Materials and Resources   14 Possible Points

☑ Prerequisite 1	Storage and Collection of Recyclables	Required
☐ Credit 1.1	Building Reuse—Maintain Existing Walls, Floors and Roof	1–3
☐ Credit 1.2	Building Reuse—Maintain Existing Interior Nonstructural Elements	1
☐ Credit 2	Construction Waste Management	1–2
☐ Credit 3	Materials Reuse	1–2
☐ Credit 4	Recycled Content	1–2
☐ Credit 5	Regional Materials	1–2
☐ Credit 6	Rapidly Renewable Materials	1
☐ Credit 7	Certified Wood	1

(Continued)

Indoor Environmental Quality		15 Possible Points
☑ Prerequisite 1	Minimum Indoor Air Quality Performance	Required
☑ Prerequisite 2	Environmental Tobacco Smoke (ETS) Control	Required
☐ Credit 1	Outdoor Air Delivery Monitoring	1
☐ Credit 2	Increased Ventilation	1
☐ Credit 3.1	Construction Indoor Air Quality Management Plan—During Construction	1
☐ Credit 3.2	Construction Indoor Air Quality Management Plan—Before Occupancy	1
☐ Credit 4.1	Low-Emitting Materials—Adhesives and Sealants	1
☐ Credit 4.2	Low-Emitting Materials—Paints and Coatings	1
☐ Credit 4.3	Low-Emitting Materials—Flooring Systems	1
☐ Credit 4.4	Low-Emitting Materials—Composite Wood and Agrifiber Products	1
☐ Credit 5	Indoor Chemical and Pollutant Source Control	1
☐ Credit 6.1	Controllability of Systems—Lighting	1
☐ Credit 6.2	Controllability of Systems—Thermal Comfort	1
☐ Credit 7.1	Thermal Comfort—Design	1
☐ Credit 7.2	Thermal Comfort—Verification	1
☐ Credit 8.1	Daylight and Views—Daylight	1
☐ Credit 8.2	Daylight and Views—Views	1

Innovation in Design		6 Possible Points
☐ Credit 1	Innovation in Design	1–5
☐ Credit 2	LEED Accredited Professional	1

Regional Priority		4 Possible Points
☐ Credit 1	Regional Priority	1–4

LEED 2009 for New Construction and Major Renovations

100 base points; 6 possible Innovation in Design and 4 Regional Priority points

Certified	40–49 points
Silver	50–59 points
Gold	60–79 points
Platinum	80 points and above

could greatly increase the livability of housing. They also champion the method of "post-occupancy evaluation" in which the actual experience of building users—rather than abstract architectural theory—is used to help design more user-friendly buildings in the future. The growing discipline of environmental design research has adopted such methods to explore the physical and psychological impacts of a wide range of urban spaces.

New forms of housing with shared facilities, gardens, courtyards, or other spaces may help provide residents with a wide variety of amenities. Apartment buildings or condominium complexes can feature interior courtyards, pools, gardens, laundries, game rooms, exercise facilities, and common rooms. Such facilities can provide shared cars that can be signed out on an hourly basis, meaning that residents who don't need to drive every day do not need to own a car, or can get rid of a second car. Amenities like these can help make urban living an attractive alternative to the large-lot suburban house.

There has been a long history of intentionally designed cooperative living communities that have sought to provide more community-oriented and equitable living environments. These include utopian communities of the nineteenth-century, cooperative housing projects in the America of the 1930s, the communes of the 1960s and later, and various more recent eco-villages. Intentional communities such as these offer one way to provide more supportive living situations for people—especially families with children, the elderly, or the disabled. Cohousing in particular is a recent movement that has sought to create alternative living environments that promote community. Beginning in Denmark in the 1970s, the movement spread to the United States around 1990 and has resulted in hundreds of new developments in which a few dozen residents usually have their own private units (with their own kitchens) while sharing kitchen and dining facilities in a "common house."[12] Cohousing residents typically design their own living environment from scratch through several years of meetings. However, a wide variety of much less formally organized shared-living situations has been present in cities historically, and is being expanded today.[13] Cooperative houses, blocks in which neighbors take down fences and share backyards, and informal sharing of laundry and other facilities represent ways for residents to minimize resource consumption and improve urban livability.

One essential consideration in sustainability-oriented architecture is how buildings can evolve and be adapted over time to the needs of various sorts of users, often with different ages, family configurations, or cultural backgrounds than the initial residents. In particular, much of the housing built in recent decades in North America has been in the form of single-family detached units appropriate for a family with young children. Typically this housing features large unit sizes, open floor plans, and suburban locations that may be less useful, for example, for elderly or single individuals or for shared homes of unrelated adults. Retirees might wish they could close off part of the house and rent it as a separate apartment for additional retirement income. Self-employed residents might prefer to have much of the floor space devoted to an enclosed home office or workshop with a separate entrance for clients. Specific details of floor plan and design can make units more or less adaptable to these different uses over time.

Adaptable housing and office space can in turn help promote neighborhood diversity and vitality by accommodating different family sizes and configurations, by providing a diversity of unit sizes and prices for varying income levels, by allowing home businesses and small-scale entrepreneurial activity to help unemployed or lower-income individuals increase their income, and by allowing unemployed, elderly, or lower-income households to easily rent rooms or apartments for supplementary income. Such flexibility helps buildings serve human needs well in the long term. By increasing the variety of residents and activities in a neighborhood, such building design can also avoid the sterile, homogeneous character of many current subdivisions and meet the needs of diverse households.

Overall, we have a long way to go in order to create building and site designs that truly promote sustainability. We have made a lot of progress in the past couple of decades toward creating model buildings that are low energy or zero net energy, that use sustainably harvested wood and recycled materials, that are healthy for inhabitants, and that help create vibrant neighborhoods and human community. But such buildings are still rare, and many incentives favor less sustainable construction. So one of the core tasks for practitioners is to help every increment of building bring about more sustainable communities. This will take a great deal of work, but will be a creative and one hopes rewarding task for the next generation of planners and designers.

15

SOCIAL EQUITY AND ENVIRONMENTAL JUSTICE

One of the most disturbing trends both globally and within individual countries is the widening of gaps between rich and poor, and the continued existence of widespread poverty and discrimination within affluent and supposedly democratic societies. These inequities have many direct and indirect effects that weaken social capital and harm the environment. Poverty-stricken families seeking fuel deforest countrysides in the developing world. Potential engineers and designers who might have created more sustainable cities instead drop out of school to support their families. The richest 1 percent of the population appropriate resources that societies instead might have devoted to addressing social and environmental problems. And elites and the corporations they control shape the politics of countries towards their own ends rather than any sort of common good. These are huge problems standing in the way of progress toward sustainability.

At this point many if not most countries have adopted civil rights legislation and regulations against discrimination. Although much more work remains to be done, societies are getting more tolerant of racial and ethnic minorities, women, gays and lesbians, and other non-dominant groups. But still discrimination lingers, often in less overt ways. Moreover, despite occasional protests such as the 2011 Occupy sit-ins economic inequalities continue to widen.

Since the neoconservative resurgence of the 1980s, with the election of Ronald Reagan in the United States, Margaret Thatcher in Britain, and Helmut Kohl in Germany, the concept that government should work actively to promote economic equality has largely disappeared from public discourse. Ruling elites have weakened progressive tax structures that helped equalize resources and shredded government safety nets for the least well-off, while increasing handouts of public money to the upper classes through tax breaks, industrial subsidies, and giveaways of public resources. As Kevin Phillips has shown, such policies have been typical throughout the history of the United States, but are now leading to an unprecedented concentration of wealth and power and serious questions about the future of American democracy.[1]

Developing countries have their own problems with inequality, often as a newly wealthy professional class engages in rapidly expanding consumption while millions of their peers continue to live in poverty. Rising affluence in China, for example, has led to a whole generation of "princelings" who because of their family connections with government often feel themselves above the law, and therefore have appropriated billions through corrupt government contracts and favorable government treatment of their business ventures. The situation in India is similar, with newly wealthy individuals able to buy favorable treatment from an underpaid and highly corrupt civil service. In the Middle East fantastic wealth in the oil-rich countries has helped reinforce conservatism elites who amass fortunes while oppressing women and foreign guest-workers. In Africa, countries are still struggling to shake the legacy of the "kleptocracies" that followed independence from colonial powers in the 1960s.

Worldwide, affluence too often brings inequality and injustice. Wealth and power lead to corruption, increasingly stratified societies, and misuse of the public sector to help the rich get richer while the poor benefit little. There has been very little countervailing force for social equity, and so inequality has grown while various forms of discrimination and injustice persist. Even societies that are becoming more tolerant and diverse in terms of civil rights for women and minority groups still countenance other entrenched forms of injustice, especially in the form of economic inequality.

A JUST SUSTAINABILITY

As Julian Agyeman has argued, efforts to promote sustainability must take social equity seriously, producing a "just sustainability."[2] Without this, these initiatives are unlikely to be effective and become instead the agenda of an upper-middle-class environmental constituency that has very little in common with the majority of the world's people.

Since the origins of the sustainability discourse in the 1970s, social justice advocates and grassroots organizations in the developing world have prioritized equity while more mainstream institutions in affluent countries often focus on environmental topics.[3] The two groups have seen sustainability problems through radically different lenses—as issues of maldistribution and First World overconsumption by those in the developing world, and as issues of environmental management, regulation, and technology by those in developed countries. In particular it has been too easy for policy-makers in affluent countries to focus on technocratic solutions to sustainability problems, ignoring their structural roots and the inherently unsustainable nature of consumption-oriented lifestyles. Substantial progress toward global sustainability is unlikely unless such constituencies take equity more seriously, acknowledging the legitimacy of equity demands, the exploitative nature of much current economic development, and the need to move towards standards of living that can be shared sustainably worldwide.

By "social equity" we generally mean several related things: equality of opportunity, equality of conditions, and justice in the sense of fair and equal treatment within the procedures and institutions of society. All are important to sustainability. But each implies different types of actions.

Equality of opportunity demands that all members of the public have available a wide range of educational opportunities, economic pathways, and social programs enabling them to live a full and productive life. Freedom from discrimination is essential to this goal so that individuals have mobility within society. This is the dimension of equity that many elite groups would prefer to focus on, since it is the easiest to put in place and doesn't fundamentally threaten existing imbalances of resources. Free public education has long been the cornerstone of such opportunity. Many forms of discrimination have been formally outlawed, and social safety nets established so that in principle individuals need not fear destitution or hunger. Governments have also made efforts to place parks, health clinics, job training programs, and good public transit service in lower-income areas, in large part to improve opportunity. Summer jobs programs, after-school programs, and day care and early childhood development programs have extended many opportunities to poor children and families.

All of this is great. However, these opportunities have been unevenly distributed, the process of gaining them has rarely been easy, and even these baseline changes have been resisted in many countries by right-wing forces. Civil rights, feminist, gay rights, and disability rights movements have had to fight hard to pass legislation against discrimination. Social safety nets are minimal in many places.

Although free public education is one of the cornerstones of equal opportunity, in recent years most American states have starved the public schools of funding. Conservatives seek instead a voucher system for education that would reallocate much public money to privately run charter schools. Large class sizes, low teacher salaries, and an explosion of standardized testing (which consumes class time for both preparation and testing) have all worked against public education. Meanwhile, decades of state budget cutbacks have vastly raised tuition fees at US public universities, essentially privatizing these higher educational systems and making them difficult for low- and middle-class students to attend. The virtually free university systems in Britain and European countries are also under assault, with many initiating or raising tuition fees. Australia had free public universities until 1988, but no more. So basic steps to provide equality of opportunity cannot be taken for granted.

Affirmative action—the principle that members of historically disadvantaged groups should be given preference in admissions to universities and other institutional processes—is a further step to try to equalize opportunities between differently advantaged groups. Now illegal for many purposes in the US, affirmative action is nonetheless desirable to spread opportunity to those who might not otherwise have access to it. Targeted education and training programs for historically disadvantaged groups are important as well, to improve the situation of the least well-off to where they can compete on an equal basis with more privileged groups.

Conservative politicians have attacked public programs aimed at increasing opportunity for supposedly reducing individual initiative and "coddling the poor." But such criticisms ignore the large entrenched subsidies that the upper class benefits from, while in effect blaming the poor for their poverty. The conservative worldview tends to see society as a collection of autonomous individuals all able to compete freely for advancement, rather than an interdependent world in which some people are born with much better opportunities than others, and in which everyone benefits from mutual assistance and public guarantees of opportunity. The former entrenched but unrealistic vision is a major obstacle to social equity.

Equality of conditions is the second main pillar of social equity, but an objective that few current world leaders want to take on. This ideal requires redistribution of resources and strong efforts to improve living conditions for the world's poor; privileged groups typically resist these things. However, some societies have historically had success with such redistribution. One main approach has been through tax policy. Traditionally most countries and states have used progressive taxation (higher rates at higher income levels) to reduce disparities in terms of income. In the post-Second World War era marginal rates of 90 percent or more on the highest income brackets were the norm in the UK, Denmark, and even the US (under the Truman, Eisenhower, and Kennedy administrations). Now, such rates hover around 35 percent in the US and 45 percent in the UK for amounts over 150,000 pounds. For decades the chief conservative political goal has consisted of lowering tax rates even more as well as reducing or eliminating capital gains and inheritance taxes, the main ways by which the rich get richer and then pass their wealth on to their children. This strategy has been very successful at concentrating wealth. The US, for example, taxes capital gains (produced, for example, by trading stocks) at 15 percent, but income (i.e. wages, the main way less affluent people make money) at rates of up to 35 percent. Such a system quite blatantly helps the rich to get richer. The most direct way to counter such growing inequality would be a reversion to the forms of progressive taxation common 50 years ago. The rising reliance on sales taxes in the US to fund state and local services is also highly regressive, since purchases represent a higher percentage of poor people's income than rich ones'. Minimizing sales taxes or the value added taxes (VAT) used in many other countries is desirable in terms of promoting equity.

Minimum wage laws (usually adopted nationally or by states) or living wage ordinances (usually adopted by cities to reflect local cost of living) are other strategies aimed towards improving conditions for those at the bottom of the economic hierarchy. Municipal living wage ordinances are increasingly popular given the disproportionately high cost of living in many cities. More than 30 local California jurisdictions, including the cities of Los Angeles, San Francisco, San Diego, and Sacramento, have passed such legislation.[4] Requirements for employers to provide benefits such as health care and disability insurance for workers help as well to ensure that anyone working full time will not live in poverty or be without coverage if he or she has a medical problem.

Affordable housing programs are another step toward equality of conditions, in that they attempt to ensure that the poor are decently housed and not threatened by homelessness. Local governments can meet affordable housing goals through direct subsidy of nonprofit affordable housing builders, inclusionary zoning requiring that for-profit builders include a certain percentage of affordable units within their projects, other zoning changes to ensure that sufficient land is available for relatively dense multifamily buildings likely to be more affordable for renters, or (usually outside the US) direct public construction of affordable housing units. An "affordable" unit in the US is generally defined as one that costs no more than a third of a household's monthly income. Cities typically try to ensure a supply of housing affordable to low-income households making 80 percent of the local median income. However, they devote relatively little attention to housing "very low income" residents making 50 percent or 30 percent of the area median. An emphasis on homeownership as opposed to rental housing leaves these people out, since individuals or families at the lower end of the income distribution are unlikely to be able to buy their homes. Homelessness and overcrowding are the frequent result of the public and private sectors' failure to provide adequate low-income housing.

Economic development strategies aimed at promoting decent-wage jobs and training for low-skill workers are a final main way to promote equality of conditions. The US and other industrialized countries have struggled to maintain decent-paying jobs as manufacturing has moved to developing countries, leaving an economy featuring many poorly paid service jobs without benefits. Meanwhile, workers in developing countries frequently have few bargaining rights or health and safety protections within new industries, and their pay is often insufficient to keep up with rising standards of living. In both cases the public sector may need to actively intervene to ensure decent wages and reasonable working conditions. Through education and training it can also try to ensure that workers' skills match changing economic conditions, avoiding structural mismatches in the economy that can lead to unemployment and poverty.

The third main goal of social equity advocates—justice—can be attained through yet another set of strategies. Once civil rights have been established they must be made reality through the courts, through regulation, and through enforcement of laws, ordinances, and procedures. This is often a long process, with foot-dragging by many different parties. A century after slavery ended in the United States, the US housing market was still highly segregated. Realtors would not show houses in many neighborhoods to black buyers, homeowners would not sell to them, and banks would not give them loans. The public sector had to take specific regulatory steps to ensure fair and equal treatment of black homebuyers, and even today, some 30 years on, discriminatory practices have not ended in many places. During the housing boom of the 2000s, for example, banks disproportionately made risky home loans to minority residents who couldn't afford them (ironic in light of their previous resistance to lending them money), leading these populations to suffer greatly in the post-2008 housing collapse.

Fair and equal treatment by law enforcement and the courts is particularly important in terms of providing justice. Here too, many problems exist. Police departments are often prejudiced against particular racial or ethnic groups, and have only slowly begun to appreciate, for example, the need to treat domestic violence against women seriously. Once in the court system, poor individuals often don't have access to good legal representation in the US. Public defenders are often overloaded with an excessive number of cases, and are often in any case relatively young and inexperienced lawyers. The criminal code and sentencing provisions often unfairly penalize the poor and minorities, locking up individuals for long periods of time for relatively minor drug or theft offenses when white collar criminals responsible for illegally misappropriating much larger amounts of money are treated lightly. The treatment of Wall Street executives after the 2008 mortgage derivatives fiasco is especially indicative of such preferences—despite having helped bring about a major economic recession that caused untold suffering for millions of people, very few executives were prosecuted and most wound up keeping their ill-gotten profits.

Prisons are one of the great tests of any society's commitment to justice. Hidden away out of sight and mind for most people, they house large numbers of people in conditions often ranging from barely tolerable to inhuman. Some countries lock up political prisoners whose only crime is a desire to see a more democratic or just country. Other countries imprison minorities at greatly disproportionate rates, a sign that the social system has been stacked against them. Rates of imprisonment are vastly higher in some places than in others. The US federal and state prison population, for example, more than doubled between 1990 and 2009 to 1.6 million people. When individuals in local jails or on probation or parole are added, the total number of people under correctional supervision was

7.2 million, or 3.1 percent of the population.[5] More than 60 percent of those jailed were members of a racial or ethnic minority group, and on any given day 10 percent of African American men in their thirties were imprisoned.[6] Those convicted of felonies are permanently barred from voting in many states, meaning that about 5.9 million people, or 2.5 percent of the US voting population, are systematically disenfranchised through the prison system.[7] These figures speak to great social injustice that is rarely mentioned in politics or public policy circles. Conditions within prisons are a separate but related injustice. Many of these institutions are grievously overcrowded, with inmates at the mercy of prison gangs and callous guards. Inadequate medical care is a related problem in many places.

ENVIRONMENTAL JUSTICE

Environmental justice, the principle that disadvantaged communities should not be subject to disproportionate environmental impacts, is one key element of a just sustainability. As mentioned in Chapter 4, the environmental justice movement is the product of the realization that communities of color and lower-income neighborhoods are primary sites for waste incinerators, landfills, and other locally unwanted land uses (LULUs). African American leaders in Warren County, North Carolina, for example, organized to fight a PCB landfill in 1982 and eventually convinced the state to clean up the site.[8] Follow-up studies by the US General Accounting Office and the United Church of Christ confirmed that hazardous materials landfills were disproportionately located in African American communities in the American South. Advocates also called attention to the extreme whiteness of the mainstream environmental movement, and demanded greater representation. Since that time environmental justice concerns have become widespread within social equity organizing campaigns, and a topic of frequent study within academia as well.

Planners and public officials can address environmental justice issues in many ways. These include procedural steps to increase the transparency of public decision-making and the involvement of under-represented constituencies in governmental processes, as well as substantive actions to reduce environmental threats and clean up sites within disadvantaged communities. Particular attention has focused on reducing emissions and contamination from large industrial facilities near communities of color, more sensitively siting landfills and other facilities, and monitoring international flows of wastes and toxic materials to ensure that they do not harm disadvantaged communities in developing countries.

However, as long as industrial societies produce large amounts of waste and pollution, these things are likely to harm some communities somewhere. And chances are that the damage will affect communities that are less politically powerful than others, whether they are minority neighborhoods in the US or Europe, or poor workers and factory towns in Asia. Therefore the best way to address many environmental justice concerns in the long run is to reduce the overall threats. Moving society away from reliance on industries that produce harmful byproducts is therefore an important part of the environmental justice agenda. Reducing excessive consumption of material goods—which in turn leads to waste and pollution—can be seen as part of this agenda as well.

CLIMATE JUSTICE

Climate change is perhaps the ultimate environmental justice issue. The tragic irony is that the effects of global warming will hit hardest those in the world who are least responsible for it—low-income people on all continents who often live in the locations most vulnerable to flooding, drought, storms, or food insecurity. By definition those with the fewest resources will be least able to adapt as the climate changes. They will have most difficulty relocating away from low-lying areas, buying more expensive imported food, or irrigating their farmland if drought strikes. Thus, reducing greenhouse gas emissions and eventual climate change impacts may be the single most important environmental justice initiative of the next century.

Since the term "climate justice" arose in the early 2000s, advocates have advanced various different arguments. Chief among these are the need to address the vulnerability of poor neighborhoods and communities of color to climate change, dramatized by the damage Hurricane Katrina did to New Orleans's lower-income, heavily African American Ninth Ward. Another main emphasis is on the need of affluent countries and populations to take responsibility for their current and past emissions, and to assist less well-off areas in preparing for the worst. A set of Bali Principles for Climate Justice, agreed by international NGOs in 2002, also calls for a moratorium on fossil fuel exploration, advocates a principle of "ecological debt" that industrialized countries and transnational corporations owe the rest of the world, and emphasizes the need for disadvantaged populations to be involved in international climate negotiations.[9]

Social justice considerations may arise within many types of planning to address global warming. For example, carbon taxes or cap-and-trade systems to reduce GHG emissions may place disproportionate financial burdens on the poor. Individuals on lower incomes tend to be more heavily impacted by increases in the cost of gas or utilities, since these things take a larger share of their income and they have less ability to buy more fuel-efficient cars or houses. Possible steps to offset these impacts include subsidized programs for home energy efficiency retrofits or fuel-efficient vehicles, steps to make public transportation better and cheaper, and tax credits to offset rising energy expenditures for lower-income taxpayers.

Likewise, any steps to develop biofuels need to take into account effects on food systems, and thus on the world's poor. During the mid-2000s it quickly became apparent that government subsidies and mandates for biofuels in North America and Europe were driving worldwide grain prices higher, leading to hardship for many in the developing world. Although farmers and the politicians representing them were happy with the higher prices, environmentalists quickly backed away from promoting biofuels derived from crops that might otherwise be used for human consumption, seeing the social justice and humanitarian impacts of this policy.

Climate change illustrates just how profoundly social equity concerns are interwoven with sustainability planning. Virtually any initiative to address global warming has equity implications in terms of how it affects vulnerable populations and how it allocates costs and benefits among demographic groups. However, not addressing global warming has far larger equity impacts on disadvantaged populations and future generations. All equity dimensions of the climate problem cannot be foreseen at this point. But we do know that there will be many, and that an intensive effort will be needed in the future to identify and address them.

16

ECONOMIC DEVELOPMENT

Public, private, and civil society sectors can take a variety of different steps to redefine the economic context around us in ways that promote sustainability. The needs in this regard are enormous in order to create economies that use far less energy and fewer resources, create less pollution, produce goods and services that are really needed, and provide more meaningful, decently paid work for people.

At a local scale there are needs for a better balance of meaningful, decent-paying jobs in each community, preferably "green" jobs that help build an economy with minimal environmental impacts, high social equity, and maximum contributions to human quality of life. At regional, state, or provincial scales there is a need for broader economic coordination that maximizes the web of sustainable economic interactions within that region. At a national scale there is a need to establish a better balance between public and private sectors, and to guide economies away from the fixation on continual quantitative increases in goods production. Other needs such as for a zero-net-energy economy can perhaps be best addressed at the national scale as well, since national governments have an ability to adopt more comprehensive policy than most lower levels of government. And at the international level there is a need to allow a baseline level of global trade without allowing multinational corporations to despoil natural resources, exploit people, undermine local businesses and markets, or destroy the planet.

This array of economic challenges can be bewildering, and the subject of many books in itself.[1] Given the entrenched structural forces promoting unsustainable economic activity such as the fossil fuel industry, the process of moving to a more sustainable economy might be equated to turning around the Titanic. But many good ideas exist on how this can be done, and some progress is being made. I will attempt here simply to outline some main aspects of economic strategy particularly important for sustainability planning.

RETHINKING THE GROWTH ECONOMY

The traditional goal of governments is economic growth, measured in terms of the production of goods and services. Politicians, economists, and the media assume that such growth produces more jobs, wealth, and happiness in a society. However, this is manifestly not the case in many respects.

When high degrees of inequality exist, growth results in more wealth for a few without corresponding jobs and happiness for others. The jobs that are created may offer workers little pay and meaning. Traditional forms of economic growth may also result in environmental pollution and resource depletion, and may depend on the exploitation of workers and communities, as Wal-Mart has depended on a poorly paid, often part-time labor force receiving few benefits. Many industries create products and services that people don't really need, manufacturing demand for these through advertising while basic human needs for decent housing, education, health care, community, and quality of life go unfulfilled. Most worrisomely, a fixation on economic growth for its own sake can warp countries' values and politics, resulting in control by corporations, their lobbyists, and allied politicians rather than a more balanced and democratic collection of forces devoted to other values besides economic growth. The term "growth coalition" has been used historically to refer to boosterish pro-development local government politics,[2] but there is no reason it couldn't be applied to national and international political forces as well.

As writers such as Herman Daly and John Cobb have pointed out,[3] the sheer volume of production of goods and services in a society has relatively little to do with human happiness, quality of life, or overall social and ecological well-being. Yet growth in quantitative output remains the bottom line of economic progress. Governments and news media repeatedly trot out small changes in gross domestic product (GDP) as the measure of social progress. Yet this indicator is purely a measure of quantity of economic activity, not quality of life. Environmental disasters such as the 2010 Gulf of Mexico oil spill are "good" for GDP, in that the cleanups lead to more economic activity. Meanwhile, improvements in human health, education, ecological sustainability, or energy efficiency are not captured at all by this statistic. Indeed, energy efficiency improvements probably decrease GDP, since fewer fossil fuels are then purchased.

A number of groups have proposed different measures of success. Daly and Cobb developed an Index of Sustainable Economic Welfare (ISEW) that would take into account the costs of pollution, commuting, resource depletion, and environmental damage while counting as positives public spending on health and education. The Redefining Progress organization created a Genuine Progress Indicator that includes the economic contributions of household and volunteer work, but subtracts factors such as crime, pollution, and family breakdown.[4] Marilyn Waring proposed rewriting national accounting systems to include the value of unpaid work that women do, such as child care, elder care, and the running of households.[5]

On a more general level, writers such as Paul Hawken, Amory and Hunter Lovins, E.F. Schumacher, and Hazel Henderson have argued for revisions to current economic systems so as to better reflect values underlying sustainable development. Hawken's "restorative economics" calls for revised incentives to put the energies of entrepreneurship and economic creativity to work restoring environments rather than degrading them.[6] This approach favors, for example, taxing "bads" such as pollution and resource depletion instead of "goods" produced by businesses. Lovins has argued that emphasizing energy efficiency, renewable energy, and local energy production can actually be more profitable for businesses than large-scale, high-tech energy solutions.[7] Henderson has argued for the use of socially responsible investing practices and improved corporate citizenship to help economies move to a renewable resource base.[8] In his influential books several decades ago, Schumacher like Lovins called for "appropriate technology," especially for development within developing countries, and for small-scale, locally based solutions to development problems.[9]

Rethinking economics in such ways to move away from what Daly once termed "growth-mania"[10] is essential for sustainable development. Governments can help the process through many initiatives that refocus economies on growth in quality of life and ecological welfare rather than material production. One step is to adopt regulations and incentives to promote forms of economic activity that markets just don't prioritize on their own, such as constructing high-quality affordable housing, providing decent health care to all, replacing fossil fuels with renewable energy, and cleaning up past environmental damage. Another step as mentioned in Chapter 15 is to enact laws requiring decent pay for workers (including livable minimum wages) and humane working conditions. A third type of action is to promote strategic industries (such as renewable energy) important to a sustainable future. A final set of efforts would focus on improving corporate governance and accountability, for example by requiring explicit study of social and environmental impacts of products and actions, requiring companies to notify and negotiate with local communities before plant closures, and businesses to involve workers and community members in their decision-making. Most such initiatives have been pursued to some extent in many countries, especially the social democratic countries of western Europe. But all could be taken further, especially within more laissez-faire capitalist systems such as in the US.

At the local level, municipal governments often feel compelled to plan and zone for rapid, unsustainable growth in jobs, population, and/or commercial development. Growth at this level often includes the physical expansion of communities as well as increase in economic activity. Such as juggernaut of "progress" is taken for granted by many local officials. Politicians and planners often feel that this model is the only way to ensure jobs for local residents and raise municipal tax revenues. They may see growth as the way to upgrade deteriorating urban infrastructure, fund services, and pay off past municipal debt. Or as previously mentioned they may seek growth simply because of a boosterish political climate in which civic and business interests seem to coincide.[11]

In many places state tax structures provide powerful incentives for local governments to embrace traditional forms of economic development through zoning for big-box commercial development, suburban strips, office parks, automobile dealerships, and other forms of suburban sprawl—the phenomenon known as "fiscalization of land use." These land uses generate sales and property tax revenues for cash-starved municipalities, especially when anti-tax politics or legal restrictions prevent cities and counties from raising property tax rates or selling bonds. The fact that taxes for important services like schools are raised primarily at the local level—where land use decisions are also made—puts in place a structural incentive for pro-growth planning. If taxes were shared across metropolitan regions, or were collected primarily by state or federal government and then channeled back to local governments on the basis of population or need, then such incentives for bad local land use planning and rapid sprawl-style growth would be dramatically lessened.

The idea that rapid growth is the solution to local government problems must always be questioned. Too often it simply produces short-term benefits while creating longer-term problems. This does not mean that local government should be in favor of no-growth or even slow-growth policies, especially in terms of housing development—such positions often camouflage exclusionary attitudes towards lower-income and minority populations, attempting to keep out such populations. Rather, it means to seek Smart Growth which creates a better balance of economic, environmental, social, and fiscal well-being within the community. This may entail providing more affordable housing to correct an existing deficiency, adding local shops and parks to improve quality of life, nurturing locally owned businesses that employ local residents and meet local needs, or creating a revitalized downtown and neighborhood centers that can provide stable centers of community and economic activity. Jobs and tax revenue therefore come from within the community, rather than from attracting large new businesses from elsewhere.

A LOCAL AND REGIONAL FOCUS

Many sustainability advocates for decades have called for a more locally and regionally oriented economy. This model of economic development has the advantages of greater local control and accountability, more extensive use of local materials and labor, more human-scaled and personal businesses, and the retention of expenditures and earnings regionally. Such a "small is beautiful" strategy has intuitive virtues in an age of corporate bigness. It supports local character, culture, and sense of place, while avoiding fueling a global economy of large multinational corporations with little local accountability and a strong tradition of playing localities off against one another to secure the lowest possible labor costs and regulation. Thinkers such as Jane Jacobs and Michael Shuman have argued for building locally oriented economies through "import replacement" strategies helping local businesses produce products and services that were formerly imported.[12] Traditions such as micro-enterprise lending in the developing world also emphasize economic development through the growth of small, local businesses whose ownership is spread throughout communities.

The idea is not to have a totally self-sufficient local economy. This is virtually impossible given the present state of global interconnection, and not particularly desirable either, since we all gain much from goods and services created elsewhere. Rather, the aim is to have products and services produced locally whenever this can reasonably be done, and to incorporate the costs of long-distance transport and distant environmental impacts into economic decision-making. Smaller, local businesses tend to be more closely linked to local resources and cultures, exert less of a homogenizing influence on local societies, and produce less of an overall concentration of economic power and wealth. With a combination of municipal support and appropriate regulation (for example living wage ordinances), they would ideally provide stable, decent-paying, meaningful jobs for the community which global corporations, with their highly routinized job specializations and constant threats of layoffs or plant relocations, often fail to do.

A sustainable, locally oriented economic development strategy will take advantage of local resources, history, and skills to build these businesses. If firms are relatively diverse and small scale, the local economy is likely to be more resilient in the face of recession or economic disruption than if jobs are concentrated in one or two large firms that could go bankrupt or depart overnight. If businesses are locally owned, they are also likely to be more active corporate citizens and more extensively involved in the community than if they are branches of multinational corporations.[13] And if local governments emphasize forms of business that don't pollute, don't produce toxic wastes, specialize in environmental cleanup, or produce products and services really needed by people, this will clearly bring other long-term benefits for the community as well.

Many local governments can safely reassess the amount of land they have zoned for new, large-scale commercial or industrial development. Following the growth ideology, cities typically over-zone for potential new employers, designating vast tracts of land for new office parks or malls. But these types of business may simply attract new residents that will require additional services in the long term, drive up local housing costs, congest local roads, and lead to many other secondary growth problems. Instead, a focus on a stable, high-quality employment base for existing residents seems more appropriate. If new jobs are seen as desirable, for example to achieve a better balance of employment and housing, then the type and location of these employers need to be carefully considered. The new jobs should match the skills of existing workers and pay decent wages. And companies should be encouraged to locate in existing developed areas rather than new, sprawling office parks. Their workers will then be more likely to patronize existing local businesses and help create vibrant downtowns, and will have better access to public transportation and existing residential neighborhoods.

PUBLIC CONTROL OF CAPITALISM

A basic question for societies is the desired balance between public, private, and civil (nongovernmental) sectors of society. Getting this framework right is essential for sustainable development so that economic objectives, policies, and values are balanced with social and environmental goals.

The twentieth century saw extremes of both public and private sector control. Former socialist countries such as the Union of Soviet Socialist Republics and China (until the late 1970s) had virtually no private sector, just an authoritarian public sector. These countries did not do so well economically, lacking the ability to efficiently produce and distribute necessary products often including foodstuffs. They didn't do well environmentally either, since environmental goals weren't part of their public sector's agenda. Meanwhile, the US and to a lesser extent other countries such as Canada and Australia pioneered forms of laissez-faire capitalism in which the private sector operated within a weak regulatory framework and an environment of weakening labor union power. Corporations and the wealthy also enjoyed substantial ability to influence elections and legislative policy-making, now increased further by events such as the 2010 *Citizens United* Supreme Court decision in the US striking down limits on financial contributions to independent political organizations. Although such countries did manage to adopt environmental and civil rights legislation, the political strength of the private sector kept more progressive reforms at bay. Rethinking the growth economy, rising social inequality, or dependency on fossil fuels was out of the question at century's end.

Neither model—socialism or laissez-faire capitalism—produced a remotely sustainable society. Most socialist systems eventually collapsed under their own weight, with those societies becoming capitalist democracies or dictatorships. In large part due to the influence of civil society, capitalist countries were able to partially address some environmental and social issues. However, most have so far proved incapable of planning for a transition away from fossil fuels or other fundamental forms of sustainability planning. Private sector-dominated economic systems are also busily spreading materialistic values around the world, creating desires for consumption among the world's people that will be very difficult to satisfy in a sustainable manner. And they tend to generate increasing inequality, which also does not bode well for sustainability.

So, planning for a third way seems necessary, but something much stronger than the relatively neoconservative Third Way of Tony Blair and Bill Clinton during the 1990s. Such a system would place capitalism within a strong framework of social control, prohibiting, for example, the types of speculation that led to the late 2000s global recession. It would require corporations to adhere to much stricter licensing and public accountability requirements, making them citizens of society rather than free-floating power centers. It would maintain a strong public sector role in meeting basic human needs related to health care, housing, education, and economic security, and would insulate political systems from the influence of private sector money. It would nurture civil society, which serves in many ways as the conscience of society, and would probably rely on NGOs to perform many service functions, in light of the relative nimbleness and efficiency of many NGOs and QUANGOs.

Such a revised balance between public and private sectors probably comes closest to the social democracy practiced in European countries after the Second World War. Interestingly enough, former Norwegian Prime Minister Gro Harlem Brundtland, whose name is associated with the most commonly used definition of sustainable development, is a social democrat within that country's Labour Party. But many improvements to the European social democratic model are possible, for example to downplay the role of large corporations and promote more locally oriented economic development.

Within such a structure, revised forms of corporate governance could help make economic activity more socially and environmentally responsive. In the United States, the few small corporations that existed at the time of the nation's founding received limited licenses from states to perform a specific task such as operating a toll road or a canal for a limited period of time. Gradually, licensing requirements have been loosened to the point of near-meaninglessness. Often corporations play one state off against the other to locate their mailing addresses in those, such as Delaware, that allow the weakest charters. A related problem has been that the US Supreme Court granted corporations the same rights as persons in its 1889 Santa Clara decision, with few of the responsibilities. Addressing this imbalance through revised chartering procedures spelling out requirements for social and environmental responsibility is likely to be an eventual part of developing a more sustainable economy.

A green social democracy might promote alternatives to the private, for-profit corporation. Nonprofit corporations, organized to perform purposes other than making money for shareholders, are one alternative. Cooperative businesses, which are jointly directed either by producers or consumers, represent another. There has been a rich tradition of cooperatives within most industrialized countries. In the United States, household products such as Sun-Kist oranges and Land-o-Lakes butter are marketed by producer coops, while cooperative grocery stores function as consumer coops. Japan also has a very active cooperative movement, including 619 consumer coops with some 26 million members nationwide.[14] The general rationale behind coops is that they spread profits and responsibility broadly among community members, promoting a more equitable and responsive society. They also lack the often manipulative, profit-driven character of large corporations.

A more far-reaching discussion of new, more sustainable economic systems is well beyond the scope of this book. However, clearly major revisions are needed to the forms of laissez-faire or state-sponsored capitalism that dominated worldwide at the beginning of the twenty-first century. Economic reform can be seen as a planning activity just as much as any other, and perhaps more important than most. Some steps can be taken at national or international scales; others can be taken regionally or locally. Whatever the level, the point must be to better incorporate long-term social and ecological welfare into economic systems, and to defuse and eventually eliminate the forms of economic power that currently resist sustainability initiatives, for example as the fossil fuel industry currently resists action on global warming. Such reforms are urgent and essential. Though they will be fiercely resisted, they will produce multiple benefits for societies in the long run.

17

POPULATION

Few of the world's political leaders, environmental advocates, or citizens want to touch the subject of population, and so, except for a relatively brief period in the late 1960s and 1970s following publication of Paul and Anne Ehrlich's book *The Population Bomb*, it has been little discussed. Those who do bring up this topic run the risk of being accused of being elitist or anti-immigrant, since much of the most rapid population growth is occurring in the developing world or among immigrant groups within industrialized countries. At both local and national levels it is assumed that populations will just keep growing until rising worldwide affluence leads to smaller family sizes and natural population stabilization at 9 or 10 billion around the year 2100, and that attention should focus instead on accommodating this increase.

Yet population growth globally and within individual countries is an issue with enormous implications for resource use, land use, environmental protection, social equity, and quality of life. No matter how well we plan for growth, population expansion will have unavoidable sustainability impacts. Globally, very serious questions exist regarding what level of human population can be sustained far into the future, given that even with better technology our per capita impacts are unlikely to be zero. So, like it or not, population must be considered within discussions of sustainability planning, and ways must be found to help bring about a sustainable human population on the planet.

IPAT REVISITED

One way to come to grips with population is to return to a famous formula developed in 1972 by Paul Ehrlich and his then-graduate student (and later presidential science adviser) John Holdren.[1] Considering the data on rising global environmental problems at that time, they proposed a simple equation: global environmental impact (I) = population (P) times affluence (A) times technology (T). Rising affluence in their view equated with higher consumption, pollution, and resource depletion. This could be ameliorated to some extent by improved technology—more efficient machines and production processes, better pollution control devices, and more creative substitution for scarce resources. But the future of the global environment in their view was extremely dependent on the relation between these factors.

In the years since then, the I=PAT formula fell out of the public eye, and also out of favor among global development experts. Certainly it was more useful in some contexts than others, and did not fit well to certain more manageable environmental problems. For example, strong public policies to set up national parks and wildlife preserves could potentially help preserve biodiversity within a given country or globally even if population and consumption grew and technology remained the same. Thus, I=PAT would not fit in this instance. (In practice, such policies have been successful within particular countries but not globally, where impacts from rising population, consumption, and pollution are greatly threatening biodiversity.)

However, at a worldwide scale in terms of problems such as global warming, I=PAT still holds much relevance. Global impact in terms of greenhouse gas emissions is very much a function of population, technology, and affluence. The equation can even be refined, since equity (E) plays a role as well. If affluence is not equally distributed—i.e. if a small minority of the world's population produces a large share of GHGs, as is currently the case—then it may be possible to have a relatively large population without devastating impact if technology is also quite clean. However, if equity rapidly increases as billions of people in developing countries enter the global middle class, which is currently happening, then impact quickly becomes intolerable. Even with cleaner technology each of the world's residents will still cause significant GHG emissions, for example, and the collective impact will be unsustainable. The most relevant formulation may thus be I=PATE. The bottom line is that population matters a lot. This is especially true if technology doesn't improve immediately, the dream of materialistic lifestyles continues, and an increasing percentage of the world's population comes to share in them.

A SUSTAINABLE LEVEL

What would be a sustainable level of human population? Deep Ecologists Bill Devall and George Sessions, writing in the 1980s, argued that a sustainable population would be only a few hundred million.[2] That point of view seems unnecessarily draconian, and was perhaps more a rhetorical stance than a realistic proposal. On the other end of the spectrum we have United Nations estimates for eventual twenty-first-century population growth, which show the number of people globally growing to about 10 billion in the mid- to late twenty-first century before declining slightly thereafter.[3] The presumption is of course that somehow or other we will find a way to feed such an increase.

At the moment GHG emissions appear to be the limiting factor in terms of calculating the Earth's sustainable human population level. As we have seen, if we want to avoid a catastrophic runaway greenhouse effect, it is essential to lower overall emissions to almost zero as soon as possible. This is a tall order, to say the least. Even with much better technology—an all-electric vehicle fleet, most electricity produced by renewable sources, massive retrofits to make buildings more energy efficient, a complete overhaul of industry, and so forth —it is hard to imagine reaching zero net emissions through technical improvements alone.

It is also hard to imagine levels of per capita consumption dropping globally to, say, those of the current population of India. Too many millions of people worldwide want a better life, which they define in terms of the consumption-oriented images they see from corporate advertising and entertainment from the currently industrialized countries.

The average American household produces about 50 tons of CO_2-equivalent annually, the average European household 20 tons, the average Chinese household 8 tons, and the average Indian household 4 tons.[4] With efficiency improvements and renewable energy, by the middle of the century these numbers are likely to converge on average emissions of perhaps 10–15 tons. This would be a big step forward for many countries—an 80 percent reduction of US emissions, for example—but still a large quantity of climate-warming gases added to the atmosphere each year, on the order of 40 gigatons annually, roughly the current level. Improving global equity, in other words, will more or less cancel out the benefits of improved technology. Help is going to have to come from somewhere else besides technology, meaning a reduced population as well.

Estimates of a sustainable global population are necessarily conjectural, since this number will depend on consumption levels and technology. But given the desire of people worldwide for greater material comfort, I believe that this will be in the 2–3 billion range, less than half the current level. Climate impacts then would still be significant, but with continual improvements in technology and simpler lifestyles might be manageable. Other global impacts on biodiversity, resources, and ocean health might be manageable as well.

PLANNING TO STABILIZE POPULATION

How would we ever get to such a sustainable population level? Luckily, the answers to that question are fairly well known. They have been developed over many decades through the work of academic researchers, international NGOs, United Nations agencies, and events such as the World Summit on Population held in Cairo in 1995.

The world's population can be stabilized and reduced without any coercive measures by focusing on six things: family planning, poverty, economic security, education, health care, and the status of women. As populations become more secure economically and as women gain education, opportunities, and access to family planning services, birth rates naturally come down. This has been the experience of Europe, of Asian countries such as Japan and Korea, and of longer-term resident populations within North America, New Zealand, and Australia. (Immigration and birth rates within recent immigrant populations have inflated the population projections within some of these countries.) Birth rates in European countries such as Italy are well below the replacement level, and they would be losing population steadily if it were not for immigration. As it is, their populations are more or less stable despite substantial inflows of migrants.

In every society large families have been a hedge against an uncertain future. If some children may die in infancy or childhood, and if health care, income, and housing are highly uncertain in old age, people tend to have larger families. Also, if women have no life choices except to stay at home and have children, then larger families are likely to result. Societies that don't provide educational and economic opportunities to families also condemn their people to the same uncertain existence that has led to overpopulation in the first place, and the pattern of rapid global population growth will continue.

Here again equity comes into play. The economic situation of hundreds of millions of people worldwide would already be much improved if the global elites, corporations, despotic national leaders, and local upper classes had not siphoned off much of the world's wealth. Global population might already have stabilized at lower levels. But neither equity nor population issues have so far been dealt with, so more concerted initiatives will be necessary in the future, both within individual countries and by the international community.

One main strategy would be for countries worldwide to take more seriously the UN's own Millennium Development Goals (more about those in Chapter 19). If even a fraction of military budgets, especially that of the US, were allocated to this purpose enormous progress could be made. Coordinated diplomatic initiatives would help as well. Much better public health services could be established in developing countries. Primary and secondary education could become universal. Equal rights and opportunities could be guaranteed to women in many places where they aren't now, and the means to family planning made available worldwide. Hunger and starvation could be virtually eliminated. Such steps would take leadership and a concerted educational campaign to inform global publics of these needs. Virtually none of the American population, for example, has ever heard of the Millennium Development Goals, and they are almost never mentioned in the mainstream US media. Leadership by politicians, NGOs, educators, and the news media, however, could start public discussion of such topics, perhaps laying the foundation for national and international action to once and for all deal with the overpopulation threat that has hung over humanity for so long.

18

GOVERNANCE AND SOCIAL ECOLOGY

Most of us think of sustainability planning primarily in terms of planning for energy efficiency, reduced GHG emissions, green buildings, bicycle and pedestrian travel, socially responsible economies, environmental justice, and other similar initiatives. Those goals are certainly excellent. However, most societies haven't made nearly enough progress toward them because of poorly functioning political systems. Their systems of governance simply aren't up to the task of planning for sustainability; they avoid dealing with environmental and social problems or actively assist unsustainable forms of development. The need is therefore to plan for improved governance, and to shape social evolution so that societies become more capable of making positive long-term choices for themselves and the planet.

Many types of governance problem exist. Some countries are ruled by dictators or one-party elites with little interest in sustainability. Their agenda is simply to perpetuate their own power, often by whatever means necessary. Other societies are mired in corruption, with leaders diverting large amounts of public funds into Swiss bank accounts and/or low-level officials preoccupied with acquiring income through bribes. In the latter case government salaries are often not enough to live on, but the question then is why political leaders allowed that situation to happen. Still other cultures are nominally democratic but have political systems dominated by wealthy elites and large corporations who use government for their own ends, with insufficient social capital within civil society to balance these powerful forces. Such plutocracies include many developing countries and, arguably, industrialized countries such as the United States.

The types of dysfunctional governance around the world are many, but confronting these shortcomings and improving collective decision-making is an essential element of planning for sustainability. In many ways professionals and citizens can plan for better-functioning political systems that are capable of moving towards long-term social and ecological well-being. Progress toward sustainability most likely will not depend on monolithic government institutions to bring about change by themselves; indeed establishing such structures is often counterproductive, as they too easily become rigid and bureaucratic. Rather than emphasizing Government with a capital "G," the focus would be on small-g governance, a flexible, adaptive network of institutions throughout society that can guide progress towards sustainability while operating in a transparent, resilient, and flexible manner. Given the near-global consensus expressed through UN treaties and declarations that people should be actively involved in governance, improved mechanisms of public input will also be needed to guide these political frameworks.

FUNCTIONAL DEMOCRACY

Political theorist Benjamin Barber has long argued for "strong democracy" that is based on civic commitment, public participation, and a focus on mutual responsibilities rather than just individual rights.[1] Herman Daly, Thomas Prugh, and Robert Costanza have also argued for a strong democracy in which political participation is a way of life.[2] These visions of greater democratic commitment are an important starting point for a sustainable society. However, more structural changes appear necessary as well. At its foundation, strong democracy requires four things: a clean system, real choices, full participation, and an enlightened electorate. Practical things can be done to help bring about each of these to a much greater extent than at present.

A clean political system is one that is not corrupted by money, wealthy elites, corporations, or other special interests. Individuals are not allowed to buy elections or political favors. Campaign donors are not allowed special access to decision-makers. Elected officials are required to declare potential conflicts of interest and to recuse themselves from decisions in which they have a conflict. Elections are held fairly, with all adult citizens able to vote and not prevented from doing so by intimidation, poll taxes, or logistical difficulties (such as having polling places in inconvenient locations or elections held during inconvenient times such as work hours). It's important also to prevent political parties from stealing elections through ballot-box stuffing, fraudulent voting machines, or selective vote-counting.

This first criteria of strong democracy is often not met either within developing countries or developed countries such as the US. But the practical steps toward meeting it are relatively simple and well known. These include public funding of elections, strict limits on private political financing and lobbying, transparency laws for government meetings, conflict-of-interest disclosure, and basic repairs to the machinery of elections.

Real choices for voters is a second main criteria for democracy. Generally this means multiple political parties presenting a range of points of view for voters to choose from. Although candidates may not always be able to articulate these viewpoints convincingly, at least voters will be presented with some degree of choice. One-party systems such as in Turkmenistan, Vietnam, and Cuba clearly do not satisfy this requirement. Two-party systems may not either if both parties hold similar positions, or if important points of view are not represented within election discourses. The US again fails on this account. The Democratic and Republican parties have made sure that no other political party is allowed to seriously challenge them, for example by preventing other parties from participating in televised debates. Within the 2012 presidential election, for example, this led to a situation in which there was no viable candidate speaking about many of the most important issues of the day, including climate change, environmental protection, poverty, election financing, or the state of American democracy itself.

Again, steps to improve democracy in this regard are relatively straightforward. In particular, allowing multiple political parties to participate and providing public funding to any that attract a significant following is essential. Germany has long been an example of how to do this, and provides public funding to any party achieving 0.5 percent of the vote in a national or European election, or 1 percent of the vote in a state election.[3] Such funding helped the German Green Party become established as a political force in the 1980s, and arguably is one reason why Germany leads the world in many aspects of sustainability planning today.

The third criteria for strong democracy is full participation by the electorate. If leaders are only elected by a subset of adult citizens not reflecting the demographics or political views of the whole population, that can hardly be said to be real democracy. Once again the US fails in this regard. Off-year and primary elections frequently only see turnouts in the 30–40 percent range, while even the most heavily contested elections in recent decades, such as the 2008 presidential contest, only involve about 65 percent of adult citizens.[4] Those who don't vote are primarily on the liberal end of the political spectrum; if everyone voted, the Democratic Party would have won landslides in most recent elections. Other countries of course have their own difficulties. In Britain, for example, the percentage of registered voters actually voting in 2001, 2005, and 2010 national elections ranged from 59 to 65,[5] hardly an example of healthy democracy.

A variety of steps could improve voter participation in most places. The starting point is to outlaw intimidation and tampering. Another strategy is to make voting compulsory, as done by Australia and 22 other countries, with lack of voting punishable by fines or community service.[6] Granted, there is little enforcement in many of these countries and citizens can easily ask to be excused from voting. However, turnouts are nonetheless relatively high, often around 90 percent. Election day can be made a national holiday, making it easier for many people to vote. Advance voting or voting by mail can be allowed. Voting registration can be made far easier, with same-day registration allowed at the polls. Prisoners and persons with past felonies can be allowed to vote. Finally, since education is the factor most likely to determine whether someone will vote, improving education for society as a whole can also help improve democratic participation. Such steps could almost certainly get participation up into the 80–90 percent range in most countries.

The fourth and perhaps most important criteria for strong democracy is an enlightened electorate—one that can think critically, see through flimsy hypocrisies and rationalizations, understand the important issues of the time, and bring ethical understanding to bear on current topics. This is a tall order in any country. However, societies can move towards this ideal in a number of ways, such as by focusing on critical thinking and civics within K-12 education, by requiring young people to engage in public service that can help them understand situations in communities different from their own while contributing to social welfare, and by banning manipulative forms of political advertising such as television commercials in favor of debates or direct candidate appeals to voters. Building social capital is part of achieving this enlightened electorate. Having strong social ties, networks, and traditions of trust and cooperation within a society reinforces the values and cognitive skills necessary for constructive political participation. The sorts of education, service, and media regulation mentioned previously can help promote such social capital.

Simply acknowledging existing problems with governmental institutions is the starting point for change. Too often countries are self-congratulatory about their own internal workings, and eager to have other countries copy their model. Britain, the United States, the former Soviet Union, and other imperialistic powers have been particularly blatant in this regard. To be sure, such societies have often had good things to offer to the rest of the world. But nationalistic fervor also tends to blind countries to their own shortcomings and needs for improvement. Admitting the need to constantly improve governance is essential in order to actually do that.

Civil society can assist with this process, nudging government along, providing watchdogs, organizing input, calling attention to important issues, and at times protesting, lobbying, or litigating to force action. The media might more prominently cover existing NGO work monitoring democratic participation and government transparency worldwide. International election-monitoring organizations, perhaps affiliated with the UN, might play a larger role in monitoring and comparing democratic processes, with no country exempted since improvements are needed almost everywhere. Funding from national governments or the UN could help civil society play this role more consistently.

VISION

One basic problem with governance at many levels has been a lack of vision about how to move towards sustainability. Politicians seize upon lesser issues for election fodder, and government agencies have a vested interest in continuing their operations in the status quo rather than questioning the business-as-usual (BAU) policy. Long-term strategic thinking, when it occurs, often comes from NGOs or challenger politicians outside the existing system.

One strategy to promote vision is to ensure that a variety of political parties are involved in politics, some of which will articulate strong points of view that encourage others to be bolder as well. Another strategy is to reduce barriers to entry into politics, so that good potential candidates are not dissuaded from running by the enormous amount of money that must be raised in many countries or political party hurdles that must be overcome. Yet another potential step is to promote long-range thinking through formal planning processes, public debates about future directions, or required reports on progress toward sustainability.

Many governments have developed visionary plans in the past, and this role can be strongly encouraged in the future. Green plans—national planning frameworks emphasizing sustainability goals—have for more than two decades been adopted by countries such as the Netherlands and New Zealand as described further in Chapter 20. These documents represent a formalized, comprehensive way of thinking about the future, potentially coordinating policies and programs to improve sustainability. Governments can also take the lead in developing and updating sustainability indicators and in scrapping current measures of progress such as GDP in favor of much broader and more meaningful indicators of social and ecological health.

Most importantly, governments can develop a range of alternative strategies for any proposed action. Analyzing these alternatives can aid in developing more visionary approaches. Environmental impact review legislation currently requires this for qualifying projects in the US; indeed, such consideration of more sustainable alternatives was one of the breakthroughs of environmental institutional reform beginning in the 1970s. But the "alternatives" within official environmental impact review documents are often pallid straw men constructed by the agency in question just to satisfy requirements of the law. Much more dramatic sustainability scenarios need to be developed in many cases, and could be required by legislation.

One technique to improve sustainability scenarios is known as backcasting, and means working backward from desired end states rather than muddling forward toward vague ideals. If a goal is to reduce GHG emissions 80 percent by 2050, planners could model alternative strategies to actually get there, and then distill these into an official policy alternative to be adopted by the jurisdiction in question. Analysts could then evaluate year by year whether sufficient progress was being made, and leaders could adjust policies if not. The point is to develop a stronger and more realistic vision about how goals such as this one could actually be reached, and to build political commitment for such action.

PUBLIC UNDERSTANDING AND PARTICIPATION

In the end planning for sustainability must consider how to plan for a healthier social ecology—a social system that is capable of addressing important sustainability issues. Governance changes such as those mentioned above are a big part of this process. But also important are changes in values, lifestyles, and institutions. Changing incentive structures for different occupations, for example, would be one step to help many people reconsider their priorities—paying teachers more and investment bankers less (for example by raising taxes on the top 1 percent of earners). Changing ways the societies measure progress would be another step, eliminating GDP and instead adopting broader-based sustainability indicators.

Many likely steps to plan for a healthier social ecology go beyond the bounds of what is currently considered politically feasible, but it is important to discuss them anyway. If we don't do that, then planning for sustainability in any meaningful long-term sense becomes impossible. One key part of the challenge relates to education. School systems that are adequately funded and creatively run are an important starting point, offering education that focuses on critical thinking, active or experiential learning, and development of citizenship. Overemphasis on preparation for standardized tests, by contrast, often gets in the way of these educational goals. Not for nothing have conservatives in the US focused so heavily on promoting simplistic performance measures for public education. An enormous amount of teachers' attention is now spent in "teaching to the test." The situation is sometimes referred to as "drill and kill": drill for the tests and kill creativity or independent thought. In such an environment it is very difficult for students to develop the range of critical thinking skills and social skills that would make them good citizens, a situation that might complicate matters for politicians depending on an electorate unable to see through hypocrisy or deeply understand issues.

Another part of the social ecology challenge relates to informational environments outside of formal educational systems. Collectively we shape our minds through the environments that we allow to be created for ourselves. Television, radio, internet resources, and news media are among the main influences on our values, tastes, lifestyles, and belief systems. Many of these sources are highly commercialized and promote shallow, one-sided understandings of the world. The idea that public airwaves and communications infrastructure should be used for public good has been lost. Steps to change this situation might include a ban on advertising aimed at children, greater support of public television and radio, and stronger requirements for commercial media to serve the public interest (for example through better educational programming). It might also be wise to prohibit advertising by corporations clearly jeopardizing global sustainability, such as the fossil fuel companies, just as some countries currently prohibit advertising by corporations clearly undermining human health, such as alcohol and tobacco firms.

Other steps to promote a healthier social ecology would focus on citizen participation in governance and civic problem-solving. National service programs, through which people spent a year or two working in needy communities or on environmental restoration projects, would be one means. Such service could be particularly useful for young people, but might accommodate all ages. High schools might require students to attend and analyze local government meetings, and history, civics, and social studies might be made a more prominent part of curricula. Local governments could give residents within each community a certain percentage of local government funds to allocate within their neighborhood according to collective decision-making processes, as in the "participatory budgeting" model of Puerto Allegre, Brazil.[7] Having this real neighborhood-scale responsibility is one way to help leverage civic involvement at a very grassroots level.

Such strategies might in the end lead toward strong democracy—a rich, mutually reinforcing tradition of civic interaction, involvement, and cooperation that succeeds in sharing power broadly among members of an informed and committed public. Such a system of governance might be able to plan for sustainability directly and effectively.

Part Three

Scales of planning

19

INTERNATIONAL PLANNING

We turn now to look at possible directions for sustainability planning at many different scales of governance, from the global to the local. "Planning" is often thought of as an activity that occurs at a municipal government scale through specific land use and development decisions within cities. But countless other forms of planning occur at both larger and smaller scales as well. International agencies, national governments, states or provinces, nonprofit organizations, and individual landowners and developers all seek to shape contexts for the future. Coordination of goals starting at the broadest possible scales is essential to the process of sustainable development. Large-scale policies and programs can help smaller-scale planning happen more effectively, and can ensure that minimum standards of environmental protection and human well-being are met even if local governments are particularly resistant to providing them.

The challenge, in short, is not just to "think globally and act locally," as activists began to say in the 1960s, but to "think at many scales and act wherever possible." This chapter focuses on the broadest level of planning—the global scale—and following chapters will address increasingly smaller institutional purviews. A central question throughout is how to create an interlocking framework of sustainability planning efforts at all different levels. Since no single institution or level of government has the power to address the multitude of sustainability issues, that is the way a more sustainable society is most likely to come about.

WHO PLANS AT AN INTERNATIONAL SCALE?

Consciousness of global interdependence is one of the hallmarks of recent ecological thought and efforts for sustainable development. But institutionalized actions to coordinate development goals internationally are still relatively recent, having taken place mainly since the establishment of the United Nations at a series of events in San Francisco in 1945, and the World Bank and the International Monetary Fund at a conference in Bretton Woods, New Hampshire in 1944. (The League of Nations, founded in 1919 after the First World War, did not last long enough to accomplish much in terms of international coordination, and previous efforts by colonizing powers such as Great Britain and Spain to spread their models of public administration around the world did not necessarily take joint needs and goals into account.)

More than 50 years after the founding of these institutions, international coordination toward sustainable development is still difficult. The role of the World Bank and other development agencies established by individual countries has been problematic in terms of sustainability. Too often these agencies have funded large-scale infrastructure inappropriate to the local context and set in motion patterns of development that reinforce local elites rather than the majority of people. Moreover, the emergence since 1995 of the World Trade Organization—the non-UN international agency dedicated to liberalizing global trade—seems to be in direct conflict with UN institutions with more socially and environmentally oriented goals. On the more positive side of the equation, the rise of a global network of NGOs with a mission of promoting ecological and social welfare has helped balance to some extent the influence of corporations, the wealthy, and the institutions they work through. Overall, the gradual accumulation of global organizations and treaties as well as the rise of global consciousness are hopeful signs that our species can learn to coordinate its actions internationally to meet the challenges it faces on a small planet.

Dominant cultures have always exported their models of how to build cities, towns, and entire societies. The ancient Greeks and Romans laid out gridded cities throughout Europe and the Mediterranean, from Egypt to England, creating a standardized form of urban development that can still be seen today in many places. In similar fashion the Spanish planned cities throughout the New World starting in the 1500s, developing each new town according to principles that King Philip II codified as The Laws of the Indies in 1573. Like the Roman military towns, these laws called for a gridiron street pattern around a central plaza. Lots were reserved for a church, a hospital, and a governor's house. Subsequent urban growth was to be symmetrical and contiguous to this initial grid.[1] Meanwhile social institutions such as the Catholic Church and the Spanish army spread particular types of social order. More recently, the British in the nineteenth and twentieth centuries imposed their methods of city planning on many countries that they colonized in Africa, Asia, and North America. British administrators simply applied standards of the British Town and Country Planning Act to these foreign locations, even encouraging use of British building materials such as brick in colonial contexts. In the process a great deal of local architectural and cultural tradition was lost.

In a somewhat less direct fashion specific urban design ideas such as that of the "garden suburb"—pioneered in England and the United States in the late nineteenth and early twentieth centuries—have gained influence on all continents. More recently, American models of motor vehicle-oriented suburbs have been widely emulated, especially since governments of developing countries and developers frequently hire American or British consultants to design new communities. In a new form of cultural imperialism, First World corporations and media spread images of suburban life and material products worldwide, shaping cultural desires regarding living environments and lifestyles. International development agencies such as the World Bank have typically promoted Western models of development as well, including freeways and other large-scale infrastructure. The result is that unsustainable development models are spreading across the developing countries.

In contrast to these forces for unsustainability worldwide, in recent decades there have been more conscious and systematic attempts to build a global consensus on alternative development directions, expressed primarily through international treaties, UN conferences, some of the more progressive international development agencies, and the work of NGOs. Often downplayed or disregarded within the United States, such agreements potentially chart a global path towards sustainability. The Geneva Conventions on the rights of prisoners of war, first formalized in 1864, the Charter of the United Nations (1945), and the Universal Declaration on Human Rights (1948) were among the first of these statements of global values. Although few countries have systematically implemented all their provisions, these declarations nevertheless began the process of laying out the intrinsic rights of human beings and establishing basic norms of national behavior.

International attempts to develop consensus on principles of sustainable development date mainly from the 1970s. As mentioned previously the 1972 Stockholm Conference on the Human Environment—the original "Earth Summit"—was a catalytic event helping to develop global understanding of development crises and stimulating the first use of the term "sustainable development." The 1976 Vancouver Conference on Human Settlements ("Habitat 1") represented the first global attempt to consider desirable goals and directions of urban development. This conference approved a declaration that stressed a moral imperative to plan for poor residents of the world's cities, strongly attacked land speculation in urban areas, and called for national steps to tax the "unearned increment" of land value resulting from speculation.[2] Not surprisingly, this call was ignored in capitalist countries in which a large fraction of personal wealth has been built historically on the rapid escalation of land values.

Twenty years later a second major round of United Nations conferences was considerably more successful at focusing international attention on sustainable development directions. These conferences included:

- the 1992 Conference on Environment and Development (the "Earth Summit" held in Rio de Janeiro);
- the 1993 World Conference on Human Rights (held in Vienna);
- the 1994 International Conference on Population and Development (held in Cairo);
- the 1995 World Summit for Social Development (the "Social Summit" held in Copenhagen);
- the 1995 World Conference on Women (held in Beijing);
- the 1996 Conference on Human Settlements (the Habitat II "City Summit" held in Istanbul).

All these events developed declarations on their respective topics, usually lengthy and convoluted documents agreed to through laborious consensus processes. Yet these statements can be seen as a first step towards global principles concerning many development issues. These conferences were also notable in that they gave a major boost to the global growth of civil society—the nongovernmental sector characterized by nonprofit organizations that try to provide services or shape public policy, generally in humanistic or environmental directions.

The Rio Earth Summit is the best known of the 1990s UN conferences. Although it did not address urban development per se, this event nevertheless helped initiate much planning for sustainable communities. Chapter 28 of the Agenda 21 document adopted in Rio requires each local authority to enter into a dialogue with its citizens, local organizations and private enterprises and adopt "a local Agenda 21." Through consultation and consensus-building, local authorities would learn from citizens and from local, civic, community, business, and industrial organizations and acquire the information needed for formulating the best strategies. The process of consultation would increase household awareness of sustainable development issues. Strategies could also be used in supporting proposals for local, national, regional, and international funding.[3]

Following these guidelines, many communities worldwide began "Local Agenda 21" (LA21) planning processes, often assisted by the NGO ICLEI: Local Governments for Sustainability (formerly the International Council on Local Environmental Initiatives). In countries such as Australia, Bolivia, Denmark, Finland, Japan, the Netherlands, Norway, the Republic of Korea, Sweden, and the United Kingdom national governments helped coordinate LA21 efforts. In other countries, local governments took initiative by themselves. By 1996 more than 1800 local agencies in 64 countries reported being involved in LA21 activities.[4] By 2002 that number had grown to 6,200 LA21 efforts worldwide, although 5,000 of these were in Europe.[5] Later in the 2000s ICLEI rebranded this initiative as Local Action 21, with four main areas of focus: resilient communities and cities, just and peaceful communities, viable local economies, and eco-efficient cities.

UN agencies, bilateral development agencies such as those run by Canada, the Netherlands, and Denmark, and NGOs have sought to support such local sustainability planning. Nonprofit groups such as ICLEI have been particularly active in working to see LA21 documents implemented (indeed, ICLEI originated the LA21 idea the year before the Rio Summit).

The follow-up World Summit on Sustainable Development, held in Johannesburg, South Africa, in 2002 also under UN auspices, gave additional impetus to local efforts to implement Agenda 21 and share best practices of local sustainability planning. However, the 2012 conference in Rio—the 20th Anniversary of the 1992 Earth Summit and 40th Anniversary of the Stockholm conference—failed to achieve significant agreements. Sadly, Agenda 21 has primarily received attention in the United States as the object of right-wing attacks claiming that the UN is trying to impose an international agenda on American local governments, despite the fact that most US local governments have never heard of Agenda 21.

The 1996 City Summit in Istanbul represented the first time that the world's countries had gathered to specifically share strategies on urban development and display best practices. The resulting charter avoided certain topics strongly opposed by powerful countries such as the US, including land use and anti-speculation taxes. But it does endorse a universal goal of "ensuring adequate shelter for all," calls for changing "unsustainable consumption and production patterns, particularly in industrialized countries," and embodies international agreement on "land-use patterns that minimize transport demands, save energy, and protect open and green spaces."[6] Although national and local governments have widely ignored these goals, the Habitat Declaration represented a first step towards setting forth global principles of sustainable urban development that may someday be linked to more effective implementation or incentives.

These 1990s events led in part to one of the largest international sustainability planning initiatives, the Millennium Development Goals. Established in 2000 by the United Nations and endorsed by all 193 UN members and 23 international agencies, the MDGs call for the world to achieve eight ambitious goals by 2015. These goals consist of the following:

- eradicate extreme poverty and hunger;
- achieve universal primary education;
- promote gender equality and empower women;
- reduce child mortality;
- improve maternal health;
- combat HIV/AIDS, malaria, and other diseases;
- ensure environmental sustainability; and
- develop a global partnership for development.

Although these objectives may sound impossibly ambitious, the world has actually made progress towards many of them. Primary school enrollment has increased, especially for girls. Many more people have access to treatment for HIV, malaria, and other diseases. More communities have safe water and decent housing. The rate of extreme poverty has been cut as many of the world's countries enter the global middle class. The proportion of people surviving on $1.25 a day or less fell from 47 to 24 percent between 1990 and 2008, although in 2015 one billion people are still expected to live on this amount.[7]

These accomplishments are very real and show that concerted effort by international institutions can make a difference in terms of certain sustainability-related issues. Alas, the world is making much less progress toward others of the Millennium Development Goals. Environmental sustainability is in great jeopardy due to climate change, loss of biodiversity, and other threats. The status of women is little improved in much of the world. Decreases in maternal mortality are far from the MDG target, and hunger remains a major problem. Much more remains to be done on these and other issues.

The establishment of an international network of World Heritage sites and Biosphere Reserve sites is yet a further way to plan on a global scale. The former list includes more than 962 sites worldwide of great natural or cultural value, with candidate sites nominated by national governments with the understanding that they will be protected under the 1972 Convention for the Protection of World Cultural and Natural Heritage.[8] The Biosphere Reserve site concept was initiated in 1974 by the United Nations Educational, Scientific, and Cultural Organization (UNESCO) Man and the Biosphere program to designate places that conserve biological diversity, promote economic development, and maintain cultural values.[9] Although such areas are designated by national governments and do not necessarily now enjoy strong protection, the concept of a consistent framework of sites that are valuable to the entire human species, or to the Earth itself, is potentially a very powerful one.

On a slightly smaller scale, the European Union and its associated organizations have been a prime forum for developing continent-wide sustainability policies. In 1992 the EU's Fifth Environmental Action Plan, entitled "Towards Sustainability," attempted to address the failure of regulatory approaches to meet environmental standards by adapting the method of the Dutch National Environmental Plan, which combined regulatory, voluntary, and market approaches.[10] The 2001 Sixth Environmental Action Plan focused action on four target areas for the 2002–2012 period: climate change, biodiversity, health and the environment, and sustainable resource use and waste management. This plan established specific environmental objectives to be achieved by 2010 as well as longer-term goals. For example, it established a target of having 12 percent of energy derived from renewables by 2010, as well as a longer-term abolition of subsidies for fossil fuels.[11] A 2011 review, in preparation for the Seventh plan, concluded that the EU had made progress in protecting natural areas, adopting a comprehensive chemicals policy, and taking action on climate change. However, it saw particular needs to combat biodiversity loss, improve soil and water quality, and decouple resource use from economic growth.[12]

The 1997 European Union Treaty of Amsterdam also called for sustainable development. This document established common policy on a wide range of subjects from immigration to the protection of animals, but within its environmental policy section emphasized sustainable development. Among other actions, the treaty called for prices to be amended to take into account social and environmental costs; for all major legislative proposals to assess environmental, economic, and social costs and benefits; and for action on climate change, public health, and other subjects.[13]

Through these and other vehicles, the EU has promoted environmental analysis of agricultural, economic, and regional policies, as well as the use of environmental impact assessments.[14] It has had the power to launch large-scale cross-border market mechanisms such as the GHG emissions trading program begun in 2005, and also facilitates continent-wide planning and the transfer of knowledge concerning environmental best practices between jurisdictions throughout Europe. At times the EU has had the power to require countries to change their laws, as with Britain, which has had to codify certain civil rights, adopt wildlife protection legislation, and improve water quality standards in order to be in compliance with European courts.[15]

The European Emissions Trading System (ETS) is particularly worth noting, and was discussed earlier in Chapter 7. This cap-and-trade framework was criticized initially for giving away allowances that large firms then sold for profit, and for creating too low a price for carbon due to excessively high caps. However, it has been improved and expanded over time, for example by adding air travel emissions in 2012. Australia is planning to join the European ETS by 2018, linking its own emissions trading system launched in 2011. Although it certainly had its flaws initially, the ETS has over time become a leading example of collaborative international action to improve sustainability.

The Council of Europe, a separate organization bringing together spatial planning ministers from many countries, developed agreement on Guiding Principles for Sustainable Spatial Development of the European Continent in 2000.[16] This document expressed agreement on topics ranging from an efficient continent-wide transport system to the need to preserve ecological and cultural landscapes. The European Council then adopted a European Union Sustainable Development Strategy at a summit meeting in Gothenburg in June 2001, revised in 2006 in light of the expansion of the EU to 25 members. This policy platform includes many specific measures related to reducing carbon dioxide emissions, pollution, and nonrenewable resource use. Although implementation of such initiatives remains slow, a 2011 review of sustainability indicators for the EU showed progress on

GHG emissions, renewable energy, and poverty, but declining sustainability in terms of resource productivity, energy consumption of transportation, conservation of fish stocks, and employment of older workers.[17]

Multinational development organizations—particularly the World Bank and the United Nations Development Programme (UNDP)—took up the subject of sustainability very actively in the early 1990s, and altered their urban development policies and programs based on it. The UNDP adopted "sustainable human development" as a primary goal, and prepares a Human Development Index that is a leading global indicator of progress toward this end. The World Bank appointed Egyptian-American architect Ismael Serageldin as Vice President for Sustainable Development in 1989, and initiated a series of annual conferences on environmentally sustainable development in the early 1990s. The World Bank also established a Global Environment Facility in 1991 that became the world's leading funder of environmental initiatives, including the UN's Framework Convention on Climate Change, during the 1990s and 2000s. However, the extent to which these initiatives affected World Bank operations is debatable. Activists continued to criticize the agency for funding large-scale, environmentally destructive dams and power plants, for requiring recipient countries to adopt "structural adjustment" policies cutting budgets and opening their economies to international investment, and for emphasizing growth in GDP as a measure of progress. Some called for outright dissolution of the Bank, believing that its projects overall work against the cause of sustainability in developing countries.

Many bilateral development agencies—coordinating aid between individual countries and developing countries—have pursued strategies more closely aligned with sustainability planning. The bilateral development agencies of Canada, Denmark, Sweden, and the Netherlands have been particularly active at working to develop sustainability planning worldwide, funding and providing technical assistance to local programs in areas such as alternative energy, public health, appropriate transportation, and local self-sufficiency.

Along with national, multinational, bilateral, and other official international agencies, much action on the international scale is undertaken by the institutions of civil society. Various environmental organizations, religious groups, trade associations, networks of local officials, and other entities have become highly active on the international scene, both in educational and program-implementation roles, and as legitimate players within decision-making. Many international networks of cities participating in sustainability planning efforts have been established. These associations include the European Sustainable Cities and Towns Campaign, the Sustainable Communities Network, the International City Management Association, the Sustainable Cities Programme (a project of the United Nations Center for Human Settlements (UNCHS) and the United Nations Environment Programme (UNEP)), and the Urban Environment Forum (a UNCHS-affiliated network of cities). Participants at the European Conference on Sustainable Cities and Towns, in Aalborg, Denmark, in 1994 approved a particularly influential document known as the Aalborg Charter. This conference also gave birth to the first organization mentioned above, which as of 2012 had been joined by more than 2500 local authorities representing more than 100 million people in 40 European countries.[18] In addition to local efforts, the Campaign worked with the European Union in the 2000s to secure commitments to using environmentally related taxes and incentives, stricter targets for greenhouse gas emissions, and planning for more sustainable European transportation systems.

NGOs are playing a major role internationally around specific issues and techniques. For example, ICLEI has conducted extensive environmental impact assessment training with staff from 17 African cities in 6 sub-Saharan countries.[19] The aim has been to increase the capacity of local officials, politicians, and members of the public to recognize and respond to environmental problems. Oxfam has done an enormous amount of work to promote sustainable agricultural practices in many countries. Doctors Without Borders has been active not just in responding to public health emergencies, but in pushing for more sustainable long-term solutions to health care, including improved human rights conditions, better access to medicines, and improved distribution systems for food. Many NGOs have also been active in pushing for stronger international treaties and in creating and promoting best practice examples of sustainable community development.

One of the most ambitious efforts to establish international consensus on sustainable development principles is represented by the Earth Charter, a document developed by an international network of NGOs and international agencies starting in the 1990s (see Box 19.1). This document aims to establish global consensus on fundamental ethical principles underlying sustainable development, stressing global interdependence and acknowledgment of shared responsibility for global well-being.[20] Its exact usefulness in the future will depend on how it is viewed and used by governments, NGOs, and ordinary citizens, but it joins the Global Declaration of Human Rights and other United Nations documents as a pioneering attempt to establish a foundation of common global values on which to base action.

Such networks and programs help counterbalance global expansion by corporations and business-oriented agencies such as the WTO. However, their power is not strong in relation to these economic interests and the national governments that frequently support them. The risk is that economic globalization by itself will bring about the opposite of sustainable development—a world in which corporations are relatively free to exploit local people and resources, circumventing local laws and regulations by appeals to international trade agreements.[21] In this world, corporations would dominate local, national, and international politics, standardized products would displace those of local cultures, values of materialism and individualism would trump those of cooperation and collective welfare, and public sector action to promote environmental protection, social equity, or local community would become extraordinarily difficult. Such troublesome dynamics might take place under cover of an atmosphere of tolerance and modest reform on the surface, as long as these did not threaten the fundamental processes of the accumulation of wealth and power.

To counter such a prospect, a sustainability planning agenda could strengthen international agreement on social and environmental goals, and back this up with agreements on specific targets and funding for implementation. Those working internationally could also increase the sharing of ideas and best practices concerning sustainability planning. Dedicated efforts by citizens, NGOs, international institutions, the media, and progressive politicians could help counterbalance the weight of business and international capital and make sure that sustainability objectives are the focus of planning.

Such an organizing effort began in the 1990s as activists connected with the International Forum on Globalization (based in San Francisco) and *The Ecologist* journal (based in

Box 19.1 THE EARTH CHARTER PRINCIPLES

I: Respect and Care for the Community of Life
1. Respect Earth and life in all its diversity.
2. Care for the community of life with understanding, compassion, and love.
3. Build democratic societies that are just, participatory, sustainable, and peaceful.
4. Secure Earth's bounty and beauty for present and future generations.

II: Ecological Integrity
5. Protect and restore the integrity of Earth's ecological systems, with special concern for biological diversity and the natural processes that sustain life.
6. Prevent harm as the best method of environmental protection and, when knowledge is limited, apply a precautionary approach.
7. Adopt patterns of production, consumption, and reproduction that safeguard Earth's regenerative capacities, human rights, and community well-being.
8. Advance the study of ecological sustainability and promote the open exchange and wide application of the knowledge acquired.

III: Social and Economic Justice
9. Eradicate poverty as an ethical, social, and environmental imperative.
10. Ensure that economic activities and institutions at all levels promote human development in an equitable and sustainable manner.
11. Affirm gender equality and equity as prerequisites to sustainable development and ensure universal access to education, health care, and economic opportunity.
12. Uphold the right of all, without discrimination, to a natural and social environment supportive of human dignity, bodily health, and spiritual well-being, with special attention to the rights of indigenous peoples and minorities.
13. Strengthen democratic institutions at all levels, and provide transparency and accountability in governance, inclusive participation in decision making, and access to justice.
14. Integrate into formal education and life-long learning the knowledge, values, and skills needed for a sustainable way of life.
15. Treat all living beings with respect and consideration.
16. Promote a culture of tolerance, nonviolence, and peace.

London) helped mobilize opposition to world trade conferences in locations such as Seattle (1998), Genoa (2001), and Cancun (2003). Public protests at such events, reported by the world's media, helped let vast numbers of people know that opposition to economic globalization does exist. Meanwhile, anti-globalization activists have organized a series of World Social Forums in cities such as Porto Alegre (2002 and 2003), Mumbai (2004), Nairobi (2007), and Dakar (2011) to bring together tens of thousands of grass-roots organizations and individuals in an explicit counter to the WTO and the elite meeting of global economic leaders every January in Davos, Switzerland.

INTERNATIONAL PLANNING ISSUES

Sustainability issues at an international scale are numerous and many are by now well known. Agreements to combat global warming are one area in which the world's countries and other institutions have tried to plan jointly, although without a great deal of success so far. As discussed in Chapter 7, the major agreement so far has been the 1997 Kyoto Protocol, which was taken seriously by some countries but not by others. Subsequent negotiating sessions in the 2000s and early 2010s produced little of significance.

Global action to address depletion of the Earth's ozone layer has been more successful. The 1987 Montreal Protocol (amended in several subsequent sessions) was eventually ratified by 183 countries. This agreement called for countries to phase out production of ozone-depleting chemicals such as chlorofluorocarbons (CFCs), halons, carbon tetrachloride, and methyl chloroform by the year 2000 (2005 for methyl chloroform). Gradual reductions in methyl bromide were subsequently negotiated as well. Industrialized countries also established a Multilateral Fund, administered by the United Nations Development Programme and United Nations Environment Programme, to help developing countries meet Montreal Protocol targets. The result has been stabilization of the size of the hole in the layer that appears each fall over Antarctica. Models predict that the ozone layer will gradually recover over the next century.[22]

Another main topic addressed by international mechanisms is loss of biodiversity. After decades of advocacy by the international NGO known as WWF (formerly the World Wildlife Fund), the 1992 Earth Summit approved a Convention on Biological Diversity which has now been signed by 193 countries. The parties refine implementation mechanisms at conferences every two years, and have expanded the original framework to cover new issues such as biotechnology, biosafety, and plant conservation.[23] The United Nations General Assembly has now declared 2011–2020 the Decade on Biodiversity. However, this framework cannot directly address many of the most fundamental threats to biodiversity such as climate change and human population increase, so whether such agreements are up to the task in the long run is questionable.

Other topics are less than global in scope, but transnational in the sense that they affect several countries or a part of the world, crossing national boundaries. These "subglobal" issues include acid rain, other forms of air pollution, water use, and ecosystem protection in transnational areas such as the Great Lakes, the Mediterranean, or the Amazon. Some treaties, cleanups, and other actions have been negotiated in these instances. For example, the US, Canada, several states and provinces, and a variety of business and environmental groups have collaborated on a series of Great Lakes Water Quality Agreements, which, despite slow and uneven implementation, have substantially reduced discharges of many pollutants and toxic chemicals into these shared bodies of water.[24]

As international institutions, trust, traditions of cooperation, and understanding become better developed, more can be done to plan for solutions to global issues. Meanwhile, some existing institutions will need to be rethought. The Bretton Woods institutions (the World Bank, International Monetary Fund, and associated agencies) have a great institutional momentum and attachment to mainstream agendas of capitalist economic expansion; much more substantial reforms are likely to be required. New institutions specifically set up to promote sustainable development are still relatively weak. New mechanisms may need to come into existence to deal with issues that are not presently being well addressed, such as climate change, population, deforestation, and rising inequity.

SUSTAINABILITY ISSUES IN INDUSTRIALIZED VS. LESS-DEVELOPED COUNTRIES

Cities and towns the world over share many problems because they are increasingly being shaped by similar forces, such as the spread of motor vehicles, technology, and materialistic lifestyles. It makes less and less sense to talk of "developed," "underdeveloped," and "developing" countries, First World and Third World, South and North. All of these distinctions, to the extent that they were ever meaningful, have become blurred by recent changes. A growing number of countries occupy a middle ground or illustrate a great range of development within themselves.

Countries such as Turkey, Brazil, South Korea, and Mexico have become highly urbanized, technologically sophisticated, and affluent in many ways, even while large segments of their populations remain poor and isolated. In Turkey, for example, cities such as Istanbul and Ankara are relatively modern and well tied in to global finance and business networks. The country's highly efficient intercity bus system is far better developed than similar ground transportation in the United States. Yet many other parts of the country remain without electricity or telephones, and literacy nationwide is only in the 40–50 percent range. Even in its large cities much of Turkey's housing is informal—built illegally by immigrants from the countryside, as occurs throughout the developing world. So this country like many others illustrates a complex mixture of characteristics that are both "developed" and " less developed."

Conversely, many communities within the "developed" United States suffer from enormous poverty, pollution, inequality, and deteriorated infrastructure. Immigrants and poor residents create a variety of informal housing types in this affluent country—most dramatically the large "colonias" developments along the Texas/Mexico borderland—and homelessness is a major problem in most large cities. Much of the US development is also clearly unsustainable in terms of its environmental or social impacts. So the "developed" label is problematic also, and does not necessarily mean that these countries are consistently or well developed.

Still, many basic challenges vary between industrialized and less-developed countries. Cities in less-developed countries face difficult problems by virtue of their extremely rapid growth, poverty, and level of technical and administrative development. In particular, such countries often feature rapidly growing "megacities"—cities of more than ten million people that have doubled, tripled, or quadrupled in size within a few decades. Cities of such size never existed until the middle of the twentieth century. In 1970 there were only 2 megacities, Tokyo and New York. In 1990 there were 10, and in 2011 23. The United Nations projects 36 in 2025, the large majority in developing countries. At least 10 will be over 20 million inhabitants (see Box 19.2 and Figures 19.1 and 19.2).

Box 19.2 THE GROWTH OF MEGACITIES (POPULATION IN MILLIONS)[25]

1970	1990	2011	2025 (expected)
Tokyo (23.3)	Tokyo (32.5)	Tokyo (37.2)	Tokyo (38.7)
New York (16.2)	New York (16.1)	Delhi (22.7)	Delhi (32.9)
	Mexico City (15.3)	Mexico City (20.4)	Shanghai (28.4)
	São Paulo (14.8)	New York (20.4)	Mumbai (26.6)
	Mumbai (12.4)	Shanghai (20.2)	Mexico City (24.6)
	Osaka-Kobe (11.0)	São Paulo (19.9)	New York (23.6)
	Kolkata (10.9)	Mumbai (19.7)	Sao Paulo (23.2)
	Los Angeles (10.9)	Beijing (15.6)	Dhaka (22.9)
	Seoul (10.5)	Dhaka (15.4)	Beijing (22.6)
	Buenos Aires (10.5)	Kolkata (14.4)	Karachi (20.2)
		Karachi (13.9)	Lagos (18.9)
		Buenos Aires (13.5)	Kolkata (18.7)
		Los Angeles (13.4)	Manila (16.3)
		Rio de Janeiro (12.0)	Los Angeles (15.7)
		Manila (11.9)	Shenzhen (15.5)
		Moscow (11.6)	Buenos Aires (15.5)
		Osaka-Kobe (11.5)	Guangzhou (15.5)
		Istanbul (11.3)	Istanbul (14.9)
		Lagos (11.2)	Cairo (14.7)
		Cairo (11.2)	Kinshasa (14.5)
		Guangzhou (10.8)	Chongqing (13.6)
		Shenzhen (10.6)	Rio de Janeiro (13.6)
		Paris (10.6)	Bangalore (13.2)
			Jakarta (12.8)
			Chennai (12.8)
			Wuhan (12.7)
			Moscow (12.6)
			Paris (12.2)
			Osaka-Kobe (12.0)
			Tianjin (11.9)
			Hyderabad (11.6)
			Lima (11.5)
			Chicago (11.4)
			Bogata (11.4)
			Bangkok (11.2)
			Lahore (11.2)
			London (10.2)

Figure 19.1 Informal settlements. Much housing in the developing world is built in "informal" settlements such as this one outside Istanbul, where residents lack title to the land and basic services are often missing. Sustainability planning in such locations must take into account basic needs for clean water, sanitation, health services, food, and employment

Figure 19.2 Green development in China. This new neighborhood in Harbin, China combines energy-efficient buildings with a network of pedestrian-only outdoor green spaces. However, it is a gated enclave for the wealthy, and so doesn't meet social equity goals

These gargantuan urban places, most of them in the developing countries, must cope with the planning challenges of providing basic infrastructure and services such as sewage, water, electricity, roads, public transit, public health, and education. They are also frequently faced with massive transportation congestion and growing automobile use within street systems that are often totally unprepared for such needs. Bangkok is a good example; its streets and alleys were developed primarily for foot traffic and do not connect in ways that help diffuse automobile traffic across the city. Basic emissions controls and other pollution regulation are also frequently lacking in developing countries cities, as is efficient government and basic information on land tenure (legal ownership of land). Poverty, hunger, and health care are pressing needs.

These challenges—in many cases faced by industrialized countries in the nineteenth century—are often overwhelming in the age of megacities. Ensuring that the population has access to food, water, shelter, and sanitation often takes precedence over questions of how to reduce pollution and restore urban ecosystems. The pressing need for action and limited governmental resources can sometimes lead to highly creative and sweeping innovations, such as Mexico City's system of allowing cars to drive only on odd- or even-numbered days, depending on their license plate number, or the dramatic land use planning and social programs found in Curitiba, Brazil. In this latter city, widely trumpeted as a best practice of urban development, planners unable to afford a subway system organized the city's land development around a highly efficient bus network instead (as well as undertaking many other creative planning endeavors). They provided the private bus companies with dedicated lanes, elevated tube-like loading platforms to speed boarding, and rapid transfer facilities. Meanwhile planners ensured that a large amount of relatively dense development was built along the transit lines, making the city one of the world's premier examples of transit-oriented development.[26]

One common characteristic in cities in developing countries is the presence of large amounts of informal housing. Individuals or entire communities flood into the city from the countryside and construct housing for themselves with whatever materials are available on land that they do not own (typically public land or large parcels controlled by absentee owners). Such "invasions" are often coordinated by middlemen, who demand a price for their services. The result is that large shanty towns emerge literally overnight. The nature of these towns varies enormously. Some may be constructed of cardboard or plywood, but others may consist of multistory reinforced concrete apartment buildings. Without any regulation or oversight, informal settlements often feature winding, narrow streets in an "organic" urban form reminiscent of medieval cities, as well as very dense and mixed-use development. Over time, these communities become formalized. Buildings and services are improved. Local governments often cannot evict residents, since there is nowhere else for them to live and they may represent a powerful political force. Also, 50 percent or more of a city's residents may be housed this way. Instead, local politicians often extend city services to residents and eventually legalize the settlement. Decades after initial settlement the result may be an urban neighborhood that is highly compact, mixed-use, and tightly knit socially and economically. From a sustainability point of view these neighborhoods may be superior to the sterile, modernist social housing blocks that governments have often built for their citizens, or the lower-density, automobile-dependent suburbs created for the middle and upper classes by private developers.

Another distinguishing feature of many cities in developing countries has been the relative lack of motor vehicles and the presence of a wide range of creative transportation alternatives (though this situation is changing rapidly). People in such cities depend heavily on foot transportation or bicycles—both modes that planners in North America are desperately trying to reinvigorate. Also, these cities are served by informal systems of vans, jitneys, or taxis, in addition to public transportation. These privately owned services are often highly efficient and may be a better illustration of "free" markets at work than transportation in the United States, where cheap gasoline, limited public transportation options, and a failure to factor the externalities of automobile use into the price of driving have skewed the transportation system heavily towards automobile dependence.

Much of the suffering of cities in developing countries has arisen because of the lingering effects of centuries of colonization or other forms of exploitation. Frequently wealthy elites in developing countries have literally stolen the benefits of development for themselves through outright theft as government officials, or have gained them legally through close association with Western corporations, banks, and political leaders, who have reinforced each others' interests. Often colonization set up a class of Westernized elites, who garnered control of much of a country's land, resources, political machinery, and military, and have remained in power with varying groups of allies ever since. In those few cases in which populist revolutions have been successful, such as in Cuba and Nicaragua, governments have been undermined in almost every way possible by capitalist interests in the United States and elsewhere.

Despite their differences, cities in developed and developing countries face many similar problems. Planners in most countries must seek to preserve or create compact urban form, avoid sprawl, reduce automobile dependency, lower resource use, improve equity, promote locally oriented businesses and employment opportunities, and adopt green architecture and landscaping practices. These needs are nearly universal. Even the old, compact cities of Europe such as Amsterdam, Paris, or Florence are now surrounded by sprawling suburbs. The Japanese with their bullet trains face rising automobile use and traffic. Countries that are supposedly committed to equality face substantial problems in assimilating minority populations—Algerians in France, Turks in Germany, Koreans in Japan, and African Americans and Latinos in the United States—as well as growing disparities between rich and poor. Equity issues also concern the status of women everywhere. In developing countries, women can play a central role in bringing about many sustainable development practices—from the initiation of locally oriented micro-enterprise businesses to the use of less polluting stoves—and improving education and employment opportunities for women is seen as a central way to reduce population growth. In developed countries, the degree to which women feel comfortable and safe in public environments is one of the prime indicators of urban quality of life.[27]

Although their problems may be somewhat different, "developed" and "less developed" countries can learn from one another. The United States can certainly learn from the compact urban form, efficient transportation systems, and less materialistic lifestyles of other countries. Meanwhile, many other countries are copying elements of US environmental regulation and planning. A diffusion of sustainability planning ideas between both industrialized and developing countries—aided by international institutions and NGOs—can be a central part of the process of creating more sustainable communities.

20

NATIONAL PLANNING

Although their influence may be slowly waning as global and regional institutions gain in power as well as civil society, nation-states are still the dominant political force in the world, and a constituency whose leadership is essential for sustainability planning. Those of us in the US do not tend to think of our federal government as being engaged in planning activities, especially in the traditional sense of guiding local land development. The idea that the national government had a specific planning policy would be anathema to many Americans, and "planning" is not a word widely used on Capitol Hill. However, the federal government's actions do of course fundamentally shape physical, economic, and social landscapes around us. For example, national spending on highways, federal home loan guarantees, and tax deductions for home mortgage interest have promoted suburbanization,[1] while federal spending on military research and development has at times promoted economic development in the Sunbelt states at the expense of the "Rustbelt" states of the upper Midwest and Northeast.[2]

In many other countries national governments play a more overt and conscious planning role, establishing policy on land use, spatial planning, housing, public transportation, and energy and resource use that shape the physical, social, and economic environment of cities and towns. Britain, the Netherlands, and Sweden have had active programs to build new towns; South Korea and Britain have established large greenbelts around their capital cities; Germany and China have strongly promoted solar and wind energy; and France, Japan, Germany, China, and other countries have made long-term investments in high-speed rail networks. Many countries in the developing world have directed large infrastructure and industrial investments toward certain "growth pole" cities that were expected to leverage broader development in their surrounding regions. Although it has not had explicit planning policy in most of these areas, the US federal government has been a leader in other ways, for example in environmental and civil rights legislation that other countries have reproduced to varying extents.

A main challenge ahead is to more explicitly and consistently relate the actions of national governments to sustainability objectives. This might occur through programs implementing global treaties on global warming, environmental protection, human rights, and related topics; specific national initiatives related to subjects such as transportation, energy, housing, or green building; or through policies mandating and supporting state, regional, and local action in cases where these scales are the most appropriate level for planning.

WHO PLANS AT A NATIONAL SCALE?

The range of institutions—agencies, processes, legal frameworks, and organizations—that are involved in planning at a national scale is substantial. The process of setting the national agenda starts with constitutions, which may have sustainability-oriented principles written into them. National constitutions written in the post-Second World War period often provide a broader range of human and environmental rights than those documents created earlier, since these values had become part of the international political discourse by that time. Constitutions for countries such as Italy, Germany, and Japan, for example, were largely imposed by the victorious powers in that war, and may include more progressive principles than found in the US founding document itself.

Ironically, with the oldest constitution among the major industrial countries, the United States has a more limited and conservative set of guiding principles with which to work, as well as a deeply entrenched national reluctance to amend this document. Countries such as the United Kingdom that have no written national constitution represent another extreme, relying on, in the UK, for example, a large collection of statutes, precedents, and court decisions that serve much the same function, but which are more open to interpretation and amendment in some ways (for example land policy) while being bound by tradition in other ways (such as by perpetuating the apparently anachronistic institution of the House of Lords).

One of the ways constitutions affect urban planning is by helping to spell out the relation between citizens and land. Those frameworks that prioritize individual property rights and treat land as a commodity make the job of sustainability planning far more difficult than those that view land as having social value and intrinsic worth. The American Constitution, for example, places great weight on private property and under the Fifth Amendment asserts that "nor shall private property be taken for public use, without just compensation." This clause allows individuals to sue for "takings" by local governments if the development value of their land is diminished without clear justification. In contrast, the Swedish Constitution provides relatively weak protection for property rights and did not explicitly provide for compensation for property owners until 1994, when the country incorporated into Swedish law the European Convention on Human Rights, which requires that the right to compensation for state interference with private property be established.[3]

Legislative branches of government establish more specific policy in many areas, with legislation typically proposed and/or signed by executive branch leaders. Congresses or parliaments also approve budgets that determine the ability of public agencies to implement policies and programs. Meanwhile, national court systems play an extremely important role in interpreting and enforcing legislative or executive actions. At times, courts also establish policy themselves through their interpretation of constitutional language.

This role was particularly important historically in establishing civil rights in the US, but in the early twenty-first century has been used primarily by a conservative Supreme Court to undercut democracy, for example through the *Citizens United* decision allowing unrestricted amounts of private money to be spent on elections.

Other equity-related issues, such as a public right to equal educational facilities and fair housing opportunities within municipalities, have also been enforced by courts at various levels, though unevenly and sometimes without much success. In a famous set of *Mount Laurel* decisions in New Jersey, for example, the state Supreme Court ordered the town of Mount Laurel to "provide realistic opportunities for the construction of affordable housing constituting a fair share of regional need."[4] The National Association for the Advancement of Colored People had sued this community for its exclusionary zoning policies, which ensured that only relatively expensive single-family housing could be built within its borders. However, although the Court issued its initial decision in 1975 the municipality continued to resist change, in part by appealing the decision to the US Supreme Court. It took a second, even stronger decision by the state Supreme Court in 1983 and passage of a state Fair Housing Act in 1985 to begin increasing the amount of affordable housing in the jurisdiction. During the 1990s and 2000s cities throughout the state continued to drag their feet on affordable housing, and the court-mandated state agency to apportion such housing often proved sympathetic to these local governments. Only in 2008 after still more litigation did the state agency set strong affordable housing targets. Thus in the end the courts did change the planning process in favor of equity, but only after 30 years of struggle by affordable housing advocates.[5]

Agencies of the executive branch of government implement and enforce policy, and carry out substantial planning activities in their own right. Housing, urban development, and interior agencies often affect city planning directly; environmental and social welfare agencies are also major players. Staff in such agencies have the ability to shape public debate through regulation, incentives, or events that promote education and dialogue. For example, the US EPA under Clinton and Obama administrations has played a significant role in supporting nationwide Smart Growth efforts, joining with other nonprofit and governmental organizations to create a Smart Growth Network and underwriting educational efforts on this subject. Federal agencies have also made sustainable communities grants of various sorts available to local and regional agencies to support innovative planning efforts. The EPA, Department of Transportation, and Department of Housing and Urban Development (HUD) also help finance urban infrastructure and make many other forms of grant funding available to local government.

In Britain, the Department of Communities and Local Government and its predecessor the Department of Transport, Local Government and the Regions have set policy on many planning topics ranging from building regulations and urban regeneration to racial equality. These and other agencies formerly developed planning policy guidance documents for lower levels of governments, although after 2010 the newly elected Conservative government announced its intention to prepare only a single national policy planning framework as planning guidance for the UK.

Incentive grants and competitions sponsored by national agencies can be an effective way to stimulate local action. The US EPA's Sustainable Development Challenge Grant program helped stimulate local planning initiatives in the late 1990s, as did a multi-agency Partnership for Sustainable Communities during the Obama Administration in the early 2010s. "Best Practices" awards by the US HUD Department have also helped publicize and reward successful local initiatives. Since 2000 the Federation of Canadian Municipalities and the consulting firm of CH2M Hill have sponsored an annual Sustainable Community Award, likewise stimulating and rewarding local interest in the subject.[6]

Special national commissions and task forces represent an institutional mechanism to forge agreement on planning issues at the national level. British Commonwealth countries in particular sometimes use a royal commission or task force to develop policy or consensus on important topics. For example, the Rogers Commission in the UK produced an influential 1999 report *Towards an Urban Renaissance*.[7] This report emphasized the role of good urban design in bringing about the regeneration of cities and towns, and called both for a national urban design framework to disseminate key design principles and for demonstration projects to illustrate the nature of these techniques. The Rogers task force also called for 65 percent of transportation expenditures to be allocated to projects giving priority to walking or public transport. In the area of land use, it called for urban priority areas to be established with streamlined development permitting, expanded compulsory purchase powers by local government, and fiscal incentives for infill development. Such commission recommendations may be stronger and more unfettered than could be produced by a government body itself, and can expand the boundaries of a debate or focus public attention.

In 2006 a similarly influential British task force led by Sir Nicholas Stern developed a *Review Report on the Economics of Climate Change*, concluding that the costs of mitigating and adapting to global warming are significant but manageable.[8] This widely cited work helped stimulate climate change planning in the UK and elsewhere. In Canada, parliamentary legislation in 1994 established the National Round Table on the Environment and the Economy, which since then has commissioned extensive research, produced numerous reports, and involved many stakeholders in sustainable development-related discussions. A 2012 "Reality Check" report by the Round Table served to point out that Canada is not on track to meet its goal of reducing GHG emissions 17 percent by 2020 compared with 2005, and called for the government to enact additional measures.[9]

The US has been somewhat less successful with national commissions. However, the President's Council on Sustainable Development (PCSD) in the US did help promote dialogue on sustainability during the mid-1990s, and released a 1996 report, *Sustainable America*, that attempted to build national consensus on this topic and led to some more successful regional initiatives such as the Bay Area Alliance for Sustainable Development.[10]

Nongovernmental organizations are an additional type of institution that can very substantially influence national planning efforts. In Britain, the Town and Country Planning Association has been an extremely important force since its founding by Ebenezer Howard in 1899. Leading members such as Frederick Osborne authored a seminal 1944 White Paper on "The Control of Land Use," which led to the country's most sweeping piece of planning legislation, the 1947 Town and Country Planning Act. Groups such as the Civic Trust (which has the Prince of Wales as its patron), the Royal Town Planning Institute, and Friends of the Earth have also been highly influential.

In the United States, environmental legislation and land use policy (at least for park and wilderness areas) have been strongly influenced by NGOs such as the Sierra Club, the National Audubon Society, the National Wildlife Federation, the National Parks and Conservation Association, and the Wilderness Society since the early 1960s, when these groups following the lead of the Sierra Club under its then-director David Brower initiated large-scale national lobbying efforts. More recently, a broad coalition of organizations called the Surface Transportation Policy Partnership helped rewrite national transportation policy through the Intermodal Surface Transportation Efficiency Act of 1991, and has helped preserve these gains in subsequent reauthorizations of this bill.

In Australia, groups such as the Human Rights Law Centre have contributed to social aspects of sustainability by bringing court cases and critiquing government policy. In the early 2010s that organization won major decisions protecting the rights of non-nationals, prisoners, and disenfranchised voters.[11] The Wuppertal Institute (Germany), the Canadian Urban Institute (Canada), and Friends of the Earth (in many countries) have also been influential NGOs in the areas of urban and environmental planning.

Quasi-autonomous NGOs (QUANGOs) are a further institutional mechanism set up by governments to meet specific programmatic needs. These organizations receive authority and/or funding from the government, usually to provide services such as education, housing, or health care, but operate in a largely independent fashion similar to an NGO.

Although nongovernmental entities have been successful in promoting sustainability action at the national level in some areas, especially environmental policy and civil rights, their influence is limited in many other fields. They have as of yet had only modest success in reforming energy and resource policies in many countries, in creating stronger land use planning and growth management incentives, and in changing the basic inclination of many national governments to favor large corporations and wealthy investors in their economic policies. Civil society has grown in strength in recent decades, certainly, but is nowhere near strong enough to challenge the power of capital in shaping structural aspects of national policy.

NATIONAL SUSTAINABILITY FRAMEWORKS

Stimulated in part by the Brundtland Commission Report, Agenda 21, and other international activities, a number of countries have prepared national sustainability planning frameworks. Most of these efforts are still in the early stages and have yet to systematically affect national policy. Many countries have also focused their plans on traditionally "environmental" topics such as resource use and pollution rather than more fundamental questions of urban development, transportation, land use, population, and equity. But still, substantial progress is being made, and the idea of broad-based sustainability plans at the national level is an important one.

Comprehensive "green plans" have been adopted by such countries as the Netherlands, Sweden, Canada, and New Zealand.[12] The Dutch government first prepared an "urgent policy document" on the environment in 1972, and created several iterations of its National Environmental Policy Plan (NEPP) beginning in 1989. Intended to be a single, comprehensive, ecosystem-based policy, the NEPP integrated several hundred policy initiatives and requires each of the its 12 provinces to do a more detailed plan for its own jurisdiction. The fourth revision of the NEPP was approved in 2001 with a relatively long-term time horizon extending to 2030. As of the early 2010s the NEPP framework appears to have been superseded by more general national strategic environmental impact assessment documents coupled with national policy strategies for infrastructure and spatial planning.

Dutch sustainability planning has often stressed voluntary agreements or "covenants" between government and business, and in many cases seems to have achieved significant results. After such planning first went into effect industry reduced its waste production by 60 percent, recycling increased 70 percent, and sulfur dioxide emissions fell 70 percent.[13] In addition, Dutch spatial planning has been oriented around sustainability objectives, in particular the country's effort to preserve a "green heart" of agricultural land from urban development within a large circle of cities including Amsterdam, Leiden, The Hague, Rotterdam, Dordrecht, and Utrecht. The "green heart" concept has been criticized by some as a misleading metaphor when this central agricultural area is not homogeneous and contains some urban development,[14] but is supported by others who argue that the concept has helped a wide variety of Dutch efforts to plan compact new urban development and to protect the already intensively used countryside.[15]

Swedish land use and transportation planning has historically sought to meet sustainability objectives such as compact urban form and transit-oriented development. As the Danish government did with Copenhagen, the Swedish government planned the city of Stockholm since the late 1940s around a finger-like structure of commuter rail lines, initially with a strong focus on suburban new towns such as Vallingby and Farsta, but more recently with an emphasis on channeling urban growth into denser, more mixed-use infill neighborhoods such as Stockholm Sodra and Hammarby Sojstad. Sweden also has a long track record on many other environmental issues. It has decreased water consumption substantially since the 1960s, held energy consumption constant despite a growing economy, cut waste generation by 50 percent in many cities, and expanded use of renewable energy.[16] Virtually all local governments have initiated Local Agenda 21 planning processes, following a period of active encouragement from the national government. In the 2000s the Swedish government enacted environmental quality objectives legislation designed to set specific targets and deadlines in 16 areas of environmental policy, and allocated new funding. The country's Environmental Protection Agency monitors progress towards these objectives.[17]

Although Germany, like many countries, started adopting environmental legislation in the late 1960s, it began explicit sustainability planning efforts only in the early 1990s at the time of the Rio Earth Summit. Chancellor Helmut Kohl established the National Commission for Sustainable Development in 1991. Three themes that formed the basis of German sustainability policies were the precautionary principle (that environmental risks and damage should be avoided from the very outset), the polluter pays principle, and the principle of cooperation. The Federal Environmental Agency produced a sustainable development progress report in 1997, and a national green plan was drafted in the late 1990s that emphasized reducing global warming emissions, ecosystem protection, resource use, human health, and environmentally sound transportation.[18] A set of indicators measuring progress in these areas was also developed. One of the country's main successes has been in the reduction of solid waste, in part achieved through the Packaging Ordinance of 1991 that requires companies generating wastes (including packaging and used products) to take them back for recycling, and to integrate the costs of doing so into purchase prices. Even German automobile manufacturers have become used to recycling used automobiles. To reprocess waste in this way German corporations set up a private recycling system, the Duales System Deutschland, which recycles goods marked by a green dot for which the producer pays a fee. A number of German provinces (länder) also have strong sustainability-oriented programs.

British planning since the early 1990s has explicitly emphasized sustainable development, and Local Agenda 21 programs are underway in most British municipalities (the Labor government proposed a 100 percent LA21 involvement target in 1997). A UK Round Table on Sustainable Development formed after the 1991 Rio Earth Summit moved forward national discussions of this topic, and a 1990 White Paper on the Environment and a 1994 UK Strategy for Sustainable Development helped develop national policy. The UK government has set some ambitious targets for specific changes, such as for an 88 percent reduction in global warming emissions between 1990 and 2050.[19]

Unlike some other countries the United Kingdom has emphasized land use and transportation as key elements of sustainability planning, in addition to more traditional environmental topics such as resource use and pollution. In 1994 sustainable development became a featured goal within Planning Policy Guidance (PPG) 13 on planning and transport, while 1995's PPG 6 moved the government away from promoting exurban sprawl development, especially of retail stores.[20] In 1997 the Blair Administration called for all local governments to prepare Agenda 21 strategies, and three years later established the Sustainable Development Commission to further national action on this topic.[21] The Department of Transport, Local Government and the Regions (DTLR) prepared a 2001 Green Paper on reforming British planning, beginning with the premise: "We need good planning to deliver development that is sustainable and which creates better places in which people can live and work."[22] While simplifying many aspects of the planning framework—including eliminating most county planning—the Green Paper requires local governments to create community strategies oriented around the Three Es and a long-term vision:

> Local authorities have a new duty to prepare Community Strategies, which they develop in conjunction with other public, private and community sector organisations. Community Strategies should promote the economic, social and environmental well-being of their areas and contribute to the achievement of sustainable development. They must have four key components:
>
> - a long-term vision for the area which focuses on the outcomes that are to be achieved;
> - an action plan identifying shorter-term priorities and activities that will contribute to the achievement of long-term outcomes;
> - a shared commitment to implement the action plan and proposals for doing so; and
> - arrangements for monitoring the implementation of the action plan, for periodically reviewing the community strategy, and for reporting progress to local communities.[23]

Such local strategies were to be in accord with national goals and regional planning guidance developed by the eight regional agencies within the country. They were also to set out a formal "statement of community involvement" stating arrangements for involving the public. The Green Paper was criticized for eliminating existing county planning processes, but had the advantage of asserting a clearer set of national policies oriented around sustainability and improving public involvement.[24]

The Blair government also took steps such as establishing a national goal of reaching 60 percent infill development by 2008 in order to reduce suburban sprawl,[25] and adopted transportation planning policy emphasizing transportation–land use coordination and accessibility by public transport, walking, and cycling.[26] A far-reaching 2003 strategy entitled *Sustainable Communities: Building for the Future* called for increasing the quality and quantity of affordable housing, improving the design of cities and towns, reducing homelessness and social exclusion, protecting greenbelt areas, and designating four areas within the country for intensive growth (the Thames Gateway, Milton Keynes and the South Midlands, Ashford, and the London–Cambridge corridor).[27] Some £2.5 billion was designated for these purposes. A further 2005 *Securing the Future* strategic framework established national sustainability indicators. Unfortunately after 2010 the Conservative government of David Cameron had far less interest in sustainability planning. This government released a rather weak, seven-page "Mainstreaming Sustainable Development" report in 2011 which contained few specific proposals, and abolished the Sustainable Development Commission which for a decade had acted as a watchdog on government operations throughout the UK.

New Zealand has initiated some of the most comprehensive national environmental planning of any country, beginning with its 1991 Resource Management Act, which set a sweeping agenda for managing air, water, soils, ecosystems, and natural resources. In contrast to many other countries' approaches, no one is allowed to discharge pollutants unless specifically authorized by a permit. A follow-up 1994 Environment 2010 plan called for integrating economic, social, and environmental planning, which is to be achieved through decentralized, participatory decision-making. The national government prepared a Sustainable Development Programme of Action in 2003, which focused on practical applications in the fields of water, energy, urban planning, and youth development. It then announced six additional initiatives in 2007, including goals of moving public agencies toward being carbon neutral and zero waste. Business and trade groups have been relatively proactive in adopting standards conforming to these plans, and certain regions and cities in the country, such as Waitakere and Christchurch, have become models of sustainable community planning.[28]

Australia adopted a National Strategy for Ecologically Sustainable Development in 1992, in conjunction with the Earth Summit.[29] This strategy grew out of the deliberations of nine different working groups, combining industry, environmental, scientific, community, and government interests, over two years. It established policy in areas of reducing greenhouse gas emissions, protecting forests, managing waste, preserving biodiversity, and managing rangelands. All three tiers of the Australian government, Commonwealth national, state, and local, approved this strategy at a 1992 conference of heads of government. Other pieces of national legislation have sought to implement this strategy, such as the 1999

Environment Protection and Biodiversity Conservation Act. At that time Australia endorsed the idea of creating sustainability indicators, but due to changes in administration it was not until 2012 that a Measuring Sustainability program was finally approved. Under the leadership of Prime Minister Julia Gillard, the country took major leadership on climate change in the early 2010s, introducing a carbon tax that applies to about 500 large emitters and making major investments in clean energy technologies.

Japan's Basic Law on the Environment, adopted in 1993, instituted sustainable development as a central national priority and mandated preparation of local environmental plans.[30] Environmental strategies have also been integrated into the country's five-year economic plans and a "New Earth 21" plan launched to develop environmental technologies.[31] In keeping with national tradition, implementation of such environmental initiatives is primarily through voluntary compliance, loans to assist businesses with conversion, and consensus agreements mediated by various levels of government. In 2009 Japan adopted an Action Plan for Achieving a Low-Carbon Society, which allocates the equivalent of $30 billion toward developing renewable energy technologies.[32] Following the Fukishima nuclear disaster, Japan has accelerated these efforts, planning to take advantage of the country's rich wind and geothermal resources as well as solar energy while phasing out nuclear power.

Interestingly, there is a strong case for Japan to have been one of the world's most sustainable societies during the 300-year Edo period, from the early 1600s to the late 1800s.[33] During this time Japan was closed to the outside world, was self-sufficient, had a low rate of population growth, and recycled almost all of its materials. No foreigners were allowed to land and no Japanese to depart. This period came to an end with the 1853 arrival in Tokyo (Edo) Bay of an impressive US military force under Admiral Perry, sent with the explicit intent of negotiating a treaty opening the country to the outside world.

In the mid-1980s the work of the Brundtland Commission stimulated the Canadian government to establish a National Task Force on Environment and Economy, involving business, government, environmental groups, labor, academic experts, and representatives of indigenous peoples. This task force and the federal Environment Canada agency produced a national Green Plan in 1990. However, subsequent national administrations failed to back this plan and it has been in large part ignored. The federal government in Canada is relatively weak in any case, and much oversight of environment and development rests with the strong provincial governments. Particular Canadian cities have been leaders in sustainable development, especially Vancouver, one of the best examples of compact, high-amenity urban development anywhere, and Toronto, which has been known for progressive transportation, urban design, housing, and greenway planning.

In the United States the federal government has so far done relatively little to promote sustainable development. The most explicit national initiative has been the PCSD, active between 1992 and 1999. Appointed by President Bill Clinton and co-chaired by Jonathan Lash, president of the World Resources Institute, and David T. Buzzelli, vice president of Dow Chemical, this 29-member commission produced the landmark 1996 report *Sustainable America*. This document proposed a set of ten national goals for sustainability planning, including general principles such as stewardship, equity, economic prosperity, international responsibility, and civic engagement. The commission suggested sustainability indicators in each of these areas. Although the report mentioned the need for physical planning reforms to stop suburban sprawl, its recommendations on this front were limited to general suggestions for local government action. Instead the commission focused on resource use topics, recommending greater regulatory flexibility, use of market forces, pollution taxes, extended product responsibility by manufacturers, and collaborative planning processes.[34] Despite the vague nature of many of its recommendations, if the Clinton Administration or Congress had taken Sustainable America seriously it could have served as the basis for national sustainability planning. But unfortunately little implementation of the PCSD's ideas has taken place.

Another main federal initiative occurred in 2009 when the US EPA, the Department of Housing and Urban Development, and the Department of Transportation formed a Partnership for Sustainable Communities. This office attempts to coordinate federal activities on this subject and makes grants available to local and regional initiatives. However, its budget is relatively small. A wide range of NGOs in the United States is also working to promote sustainability-oriented planning policies at federal and lower levels of government. Such organizations include the Surface Transportation Policy Project, the Smart Growth Network, the Sustainable Communities Network, the Sierra Club, the Natural Resources Defense Council, and the US Green Building Council.

The passage of a wave of federal environmental legislation in the late 1960s and 1970s marked the active entry of the US federal government into environmental planning. The National Environmental Policy Act (1970), Clean Air Act amendments of 1971, Clean Water Act of 1972, Resource Conservation and Recovery Act of 1974, and Superfund Act of 1980 established the legal basis for regulating environmental quality in and around cities. At roughly the same time civil rights and housing initiatives helped address social and equity dimensions of urban sustainability, though measures such as the Fair Housing Act and the Non-Discrimination Act were just a beginning. This first wave of environmental planning established strong but piecemeal federal regulation in a number of key areas related to community sustainability.

The passage of the 1991 federal transportation bill—the Intermodal Surface Transportation Efficiency Act of 1991 (ISTEA)—marked another milestone in that many urban regions were given flexibility to reprogram federal transportation money for transit, bicycle and pedestrian planning, and even transit-supportive land use planning. This approach was continued under the bill's 1998, 2005, and 2012 successors. Under the Clinton Administration HUD adopted much improved policies relating to public housing, in particular converting many sterile, modernist housing projects to more human-scaled townhouse neighborhoods through its HOPE VI program. After eight years of inaction under George W. Bush, the Obama Administration emphasized development of renewable energy within its 2009 economic stimulus package, and took a lead in raising fuel efficiency standards for motor vehicles.

The US federal government has been very reluctant to develop policy related to many other dimensions of sustainable development, including land use, resource use, greenhouse gas emissions, or social equity (apart from civil rights legislation). Land use and spatial planning in particular are seen as local government concerns, with great deference given to private property rights. The country has no equivalent to Britain's 1947 Town and

Country Planning Act, which established the principle of public control over the develop-ment of land, required large-scale plan-making by counties, and subjected these plans to approval by a national minister. In the United States, planning tools such as zoning do not rest on any such national legislation, but rather on the vague concept of "police power," a residual legal ability of government to pass laws to protect public health, safety, and wel-fare.[35] The last significant bill attempting to establish national land use planning policy failed in the 1970s.

The situation in the United States, in fact, has been one of strong national policy that often undermines the cause of urban sustainability. US refusal to abide by the Kyoto Protocols on global warming emissions and its disregard of Agenda 21 and other United Nations initiatives are particular black marks on the US's sustainability planning record. Much more active leadership is needed from the national level if the country is to move towards developing more sustainable communities.

The starting point for a sustainability planning agenda at the national level, then, is to actively recognize global goals and treaties, and to make good-faith efforts to follow up on these through domestic policy. Anything less essentially amounts to anti-social behavior on a global scale. The development of explicit urban planning guidance and policy at the national level is important if for no other reason than to ensure that national expenditures on infrastructure, housing, transit, and other categories do not undercut local sustainabil-ity efforts. The British PPG statements and the national green plans of countries such as the Netherlands and New Zealand have been examples of such national policy frameworks.

But nationwide sustainability planning should ideally go far beyond this to provide active support of local sustainability planning through planning grants, targeted infra-structure funding, tax policy incentives, and environmental policy. Environmental and civil rights goals in particular are often necessary to establish at national level, since local gov-ernments may resist setting policies in such areas or at best will be inconsistent in doing so. Even in countries where planning policy-making is more decentralized, such as the United States, federal incentives for sustainability planning can be extremely useful. Historically funding such as the Section 701 planning grants, Community Development Block Grant (CDBG) funds, and Congestion Management and Air Quality (CMAQ) grants has been pivotal in bringing about new local planning and programs. As we will see, the state Smart Growth programs being developed in many parts of the country rely heavily on such incentives as well. Where no stronger policies or programs are politically possible, incentives for lower-level action represent the least a national government can do.

In recent decades national governments have had a tendency at times to devolve powers to more local levels of government. The trend towards devolution was particularly pro-nounced under Nixon, Reagan, and the two Bush Administrations in the United States, and under the Thatcher Administration in Britain. Green Party platforms also back devolution of power to promote grassroots democracy, calling for decisions to be made at the lowest feasible level. Devolution has been justified as reducing national-level bureaucracy, pro-moting the operation of markets, and enhancing local responsibility and democratic decision-making. However, its downside is that it tends to weaken national standards for civil rights, environmental protection, and a host of public welfare issues. It also under-mines the rationale for planning at the large scales at which it is often needed—for example to establish national land use planning goals, to coordinate national transportation spending with land use and environmental policies, to manage energy and resources effectively, and to establish affordable housing policies. The effect of devolution is often to weaken planning across the board, since it reduces or eliminates the framework of upper-level policies and incentives supporting planning. Therefore a balance must be struck between levels of government that maintains the active involvement of those higher institutional levels that can best take a big-picture view, and that are most immune from the parochialism and occasional corruption of local government.

NATIONAL PLANNING ISSUES

National sustainability planning issues of greatest urgency tend to be those in which lower levels of government do not have the perspective, resources, jurisdiction, or political will to effectively bring about change.

Addressing civil rights and other equity-oriented concerns at the national level is particularly important, since state and local governments tend to vary widely in their ability or willingness to develop such policies on their own. Without a national mandate, privileged communities have little incentive to accommodate less affluent citizens and minority groups, or to share resources with them. In the same way that international norms such as the Universal Declaration on Human Rights and the Geneva Conventions place moral, legal, or persuasive pressure on national governments to conform to higher human rights standards, national frameworks help raise standards regarding equity and civil rights among more local governments.

National steps to equalize access to financial resources across state and local jurisdictions also represent an important equity measure. Otherwise, jurisdictions with a low tax base and high need for public services, such as central cities and impoverished states, are at a disadvantage compared with more affluent jurisdictions (whose benefits in turn often rest on federal spending for infrastructure, research, or the military). This problem occurs both within states and nationally. Since the early 1970s various "revenue sharing" programs in the US have provided a modest amount of unprogrammed block grant funding from the federal government to cities across the country, but this program has diminished in size in recent decades. In Britain, the national government serves a stronger tax equalization function by collecting all business taxes at uniform rates and redistributing revenues according to population.[36] Such a strategy also helps eliminate pressure on state or local governments to give tax breaks to businesses in order to attract economic development, since such tax rates are set nationally.

Action on global issues such as climate change, ozone depletion, deterioration of marine environments, nonrenewable resource use, and endangered species protection virtually requires a national commitment as well. Lower-level governments alone are able to undertake only piecemeal action on such issues, though such small-scale efforts are certainly significant. Local and regional agencies have little incentive to think about global issues, since almost all political pressures on them are local. In the absence of a national mandate for action, such large-scale topics are unlikely to be dealt with. Other broad areas of environmental policy also typically have been the purview of national governments.

Broad environmental policy laws, endangered species legislation, air and water quality legislation, and other big-picture environmental topics have generally been addressed at this level. State governments frequently take action on these subjects as well, so what results is a landscape of overlapping policy frameworks from these higher levels of government. Typically the state actions expand on the national goals, providing more stringent regulations to meet state and regional needs. For example, California has adopted much stricter air quality regulations than the US federal government because of severe air quality problems in the Los Angeles basin and other parts of the state. Other states such as New York then adopted California standards with regard to motor vehicle emissions.

Policy on large-scale infrastructure such as transportation systems is also typically handled at the national level of government. Large-scale road or public transit systems require large amounts of funding, which is often not available at the local level, and their planning often crosses local, regional, or state lines as well. National commitment to high-speed rail by continental European countries has greatly boosted rail transit networks in recent years, for example. A national commitment to bicycling is bearing results in the Netherlands and Germany. In contrast, US national preference for funding motor vehicle-related infrastructure during the post-Second World War period, rather than passenger rail systems or bicycle and pedestrian facilities, has had enduring effects on US urban and regional development, providing the infrastructure for suburban sprawl and shifting the majority of freight transport onto public roads rather than railway lines.

A final area in which national policy is particularly vital concerns land use, even though this is seen in many places as a local prerogative. By establishing the basic legal relations regarding real property, national constitutions, legislation, and court decisions set limits on what local planners can do. If it is presumed that owners are entitled to a relatively unfettered ability to develop land, or else must be paid high levels of compensation by local governments, then public regulation of development is limited. Tools such as zoning and urban design regulation become suspect, and cautious politicians avoid using them proactively. Such is the situation in much of the US. If, on the other hand, endangered species, ecosystems, historic cultural landscapes, and traditional communities are acknowledged to have value in themselves, then the balance may tilt more towards a more conducive context for sustainability planning. To date some US initiatives such as the Endangered Species Act have begun to move in this direction. But such steps are limited, and rethinking national values regarding land development is likely to be a very long process.

21

STATE AND PROVINCIAL PLANNING

State and provincial governments form an intermediate level of authority that is present within most medium or large-sized countries such as the United States, Canada, Australia, Germany, Italy, Mexico, India, China, and Brazil. These countries either came together from a collection of smaller political entities or possess a sufficient scale of territory as to make a layer of governments below the national level desirable. Some have internal regions with their own cultures or quasi-autonomous institutions of government. The United Kingdom has England, Scotland, Wales, and Northern Ireland; Spain has Catalonia and the Basque country. Many of these large cultural territories might easily be countries in their own right, or once were. Most states and provinces, in contrast, have no such pretentions, and simply form a consistent tier of government between national and regional or local scales—a level that may prove best suited to some dimensions of sustainability planning.

States control a substantial amount of political power and in federal systems are the primary source of power, in that national governments derive their powers from a voluntary relinquishment of these by the states. An extreme model of state-centered politics is Switzerland, where the cantons are the primary locus of decision-making. Although not decentralized to nearly this extent, the United States and Germany also have systems in which the American states and the German *länder* both hold significant power. There is substantial popular fear of too much centralized authority in both countries, for different historical reasons.

For the purposes of sustainability planning, state or provincial governments often have key roles to play in overseeing land use planning, transportation systems, environmental protection, social equity, and social services. Land use and growth management planning in particular may be best handled at this level, as state or provincial governments are still relatively close to local territories and cultures, but far enough removed from local politics to be less affected by the parochialism that often afflicts city, town, and county government. Many states and provinces also have great power over the formation and character of municipal governments.

Certain relatively progressive states and provinces can at times go much further than national governments in developing sustainability policy and programs. California, for example, has led the United States in developing motor vehicle emissions regulations and climate change policy, while states such as Oregon and Vermont have been leaders in growth management legislation. British Columbia has been a leader in environmental planning within Canada, and subnational European jurisdictions such as North Rhine–Westphalia in Germany, Flanders in Belgium, and the Basque Country in Spain have taken a lead in developing sustainable development plans.[1] States can offer a smaller and more manageable political arena in which to bring about change than national government, and can benefit from local concern about particular sustainability-related issues. Such considerations argue for viewing state governments prominently as a locus for sustainability planning.

WHO PLANS AT THE STATE OR PROVINCIAL SCALE?

State governments in the United States are organized much like the national government, with two houses of a legislature, an executive (the governor), and a court system. Various agencies of the executive branch handle different program areas related to planning. Particularly important are transportation agencies, which are the dominant road-building organizations in the country and so help determine urban growth patterns as well (the federal Department of Transportation channels funds to the states, rather than building infrastructure itself). These state departments of transportation (DOTs) are often dominated by old-line engineering attitudes, and tend to see large new construction projects as a way to justify their own existence. It has frequently been a challenge to get them to prioritize instead alternative modes of transportation, and to establish funding programs for local cities and towns to carry these out. In the end these agencies may need to shrink and redefine their mission, as transportation planning moves away from its past agenda of road expansion.

State departments of housing and/or community development are perhaps the most directly focused on urban development issues. California's Department of Housing and Community Development, for example, administers more than 20 grant programs related to affordable housing, alleviating homelessness, public facilities, and infrastructure. Such programs can be used to leverage additional action at the local level. State environmental protection agencies and public health agencies also play leading roles in sustainability planning. Minnesota's Office of Environmental Assistance sponsors the state's Sustainable Communities Network, which conducts education and networking statewide. Officials from several agencies in that state have also developed the Minnesota sustainable building guidelines that will be applied to all new state buildings.[2]

Statewide NGOs are increasingly effective at shaping policy, especially in areas of growth management, environmental protection, and education. "1000 Friends" groups have proliferated in many US states, inspired by the success of 1000 Friends of Oregon. Chapters or field offices of groups such as the Sierra Club, the Audubon Society, the Surface Transportation Policy Project, and the Nature Conservancy are very active as well. Public interest research group (PIRG) organizations, originally inspired by Ralph Nader in the 1970s, are found in many states. Teachers' unions, good government groups such as Common Cause, associations of public health professionals, labor unions, associations of city and county governments, and state chapters of the American Planning Association or American Association of Landscape Architects are frequently influential at the state level as well. The building industry is usually a major player, often through a Homebuilders Association. Potentially such constituencies can be organized into political coalitions supporting sustainable development initiatives.

STATE OR PROVINCIAL PLANNING ISSUES

Land use and growth management

With the federal government in the United States essentially abdicating any substantial role in land use planning, sustainable resource policy, housing provision, or many other areas of sustainable development, some state governments have stepped in to fill the void. Vermont, Oregon, Florida, Hawaii, New Jersey, and Rhode Island pioneered the first wave of state growth management policy in the 1970s, sometimes referred to as the "quiet revolution" in state land use planning. Vermont, for example, established a state permitting process for large developments; Florida required both state and regional review of large projects and later mandated that local plans be consistent with regional and state planning goals, although local resistance undercut this objective.[3] Since then some states have developed second- and third-generation growth management frameworks.

Oregon's statewide growth management framework is particularly well known and has evolved substantially over more than 30 years (see Box 21.1). The state established statewide planning goals in 1968, required large metropolitan regions (i.e. Portland) to set up urban growth boundaries, and has provided both technical support and policy mandates to regional and local levels of government. The state is an example of the benefits of incrementally improving policy and institutions over time. In several works, planning historian Carl Abbott has documented ways that Oregon has developed a culture of planning that stresses continued innovation and improvement, something that he calls "the Oregon planning style."[4]

In the late 1990s Maryland became the national exemplar of Smart Growth planning with a number of initiatives under governor Parris Glendenning.[5] Under its 1997 Smart Growth Act the state enacted policy to restrict infrastructure funds available to cities that do not place new development within locally determined Priority Funding Areas. The state also set up a fund to acquire open space and forestland, and adopted policy to site state facilities within existing urban centers. Such a framework represents a stronger and broader form of statewide growth management in the US in that it not only attempts to protect land from sprawl, but provides incentives for good development and rethinks state investment priorities. Despite occasional political setbacks and a predictably slow process of implementation, Maryland's growth management framework continued to expand during the 2000s and early 2010s. It is also now perhaps the only American state to have an official Secretary of Planning.

New Jersey has also been a leader in statewide growth management planning, though on-the-ground results are mixed. Its legislature first passed a Statewide Planning Act in 1985 that established a State Planning Commission and Office of State Planning. This agency then created the first State Plan in 1992 and an updated plan in 2001, still in effect, with sustainable development as an explicit theme. Although local jurisdictions retain

Box 21.1 OREGON'S GROWTH MANAGEMENT PLANNING

The state of Oregon has a more than 40-year history as a leader in statewide planning in the United States. After an initial attempt to establish statewide planning
policy with a Land Use Planning Act, SB 10, in 1968, the state legislature passed a
much stronger bill, SB 100, in 1973 establishing a set of statewide planning goals.[6]
These included the requirement that cities establish urban growth boundaries, and
goals of protecting forest, farmland, and coastal resources. SB 100 also created a
statewide Land Conservation and Development Commission to oversee application
of these planning goals, and a Department of Land Conservation and Development
to provide state assistance and resources to local governments.

The state continued to expand its planning role in subsequent decades. A 1991
state transportation planning rule required cities and towns to plan to reduce automobile use. This regulation requires metropolitan areas to demonstrate that vehicle
miles traveled (VMT) per capita will be reduced by 10 percent within 20 years and
30 percent in 30 years, and directs local and regional governments to consider a
host of physical planning initiatives to reduce motor vehicle use. In the 1990s the
state also undertook to develop a set of Oregon benchmarks that would serve in
effect as sustainability indicators. In the 2000s the system weathered a series of
conservative challenges through ballot propositions and lawsuits, and the state
expanded its leadership in health care planning by including uninsured children in
2009 and coordinating physical, mental, and dental care in 2011. The result of these
and other initiatives has been a multi-tiered planning framework that helps lead the
state in many sustainability directions.

authority over land use, the New Jersey State Plan attempts to orchestrate agreement on
development principles that emphasize the revitalization of existing communities and the
preservation of open space.[7] State policy establishes a procedure known as "cross-acceptance"
through which local, regional, and state plans are made consistent with one another.
Sustainability planning in New Jersey has been advanced considerably owing to the efforts
of New Jersey Future, a statewide nonprofit organization that released a Sustainable New
Jersey plan in 1999 and has since emphasized a variety of Smart Growth issues.[8]

Minnesota has likewise taken the lead in a number of sustainability-oriented planning
measures since the 1960s, including authorizing establishment of regional government in
the Minneapolis–St Paul area and the adoption of revenue sharing to reduce tax base disparities between local jurisdictions in the Twin Cities region. Former Governor Arne
Carlson launched a Minnesota Sustainable Development Initiative in 1993 as a collaborative effort of business, government, and civic interests. This initiative sought to develop
consensus on more sustainable resource use, land use, energy consumption, and community development. The State Legislature approved 11 planning goals in its 1997 Community-
Based Planning Act, and in 2000 the state provided a model ordinance for local governments

to implement sustainability planning, focusing on sustainable land use (including use of UGBs and transfer-of-development-rights frameworks), urban design, energy-efficient buildings, water supply, sewage treatment, and economic development.[9] Though a lack of political leadership led to slower progress towards sustainability in the 2000s, the state did develop a Climate Action Plan in 2003, and a 2011 Executive Order required all state agencies to develop their own sustainability plans.

In other countries, states or provinces have also at times taken a lead in managing land and environment more sustainably. Canadian provinces, for example, have particularly strong authority compared with both federal and local government levels, and have at times taken great initiative. British Columbia initiated an Integrated Community Sustainability Planning initiative in 2005,[10] using gas tax monies to support local sustainability and climate change initiatives, and adopted a Water Sustainability Action Plan in 2004 and a BC Energy Plan in 2007. The province of Ontario approved a Greater Golden Horseshoe Plan in 2005 to manage growth and promote infill development so as to preserve a greenbelt around Toronto, and Quebec province has established a Department of Sustainable Development, Parks, and Environment, which produced a Government Sustainable Development Strategy for the 2008–2013 period.[11]

Environmental regulation

Most US states have duplicated many aspects of federal environmental regulation, passing their own environmental quality acts modeled on the National Environmental Policy Act as well as air and water quality legislation. California's Environmental Quality Act is substantially stronger than the federal version, and has been extended by the courts to regulate private development as well as publicly funded projects. That state has also seen its tough automobile emissions standards adopted by New York and other northeastern states.

Many states have taken the initiative on issues related to energy policy. California, for example, established Title 24 of its building code in 1978 to raise requirements for energy efficiency, and through this initiative now saves the energy equivalent of the production of 14 nuclear power plants.[12] However, nine US states still have no building energy codes, and many others only added them relatively recently when the Obama Administration conditioned receipt of federal stimulus dollars on their adoption.[13] States also regulate public utility companies that provide electricity and natural gas, and thus are in a position to require these companies to pursue energy conservation programs and renewable sources.

States have also often taken the lead on programs to promote reuse, recycling, and waste reduction. Back in the 1970s states such as Vermont, Oregon, New York, Michigan, Iowa, Maine, and California enacted legislation requiring deposits of $0.05 on beverage containers. Although this amount has not been raised in decades, it nonetheless helped promote recycling and resulted in substantially less roadside broken glass and litter. Actual reuse of glass bottles was the norm in the United States until the 1960s, and is still the norm in many European countries. Residents there bring soft drink bottles back to retailers to receive back a deposit, and the bottles are then washed and refilled at the regional bottler. It helps that bottle sizes are relatively standard, unlike the bewildering variety of plastic bottles and aluminum cans that replaced glass bottles in the US. Another main area of state initiative has been to establish requirements that local governments reduce the volume of solid waste produced by their residents, through recycling and other policies. In 1989 for example, the California legislature required the state's cities to reduce their solid waste by 50 percent by 2000. Actual reductions totaled 42 percent in that year.[14] After some years of delay, the state then established a new goal in 2011 of reaching 75 percent reductions by 2020.

Transportation

Although it is tempting for politicians to announce major road-building programs to provide jobs during a recession, or to equate progress with new freeway projects, state governments are gradually diversifying their transportation planning away from the traditional focus on road construction. However, most commuter rail and bus transit systems are developed regionally, and most bicycle and pedestrian programs are implemented locally, so direct state involvement in alternative transportation modes is often limited to a small handful of intercity rail projects. State agencies in the US pass most non-highway transportation money through to regional metropolitan planning organizations (MPOs), who then frequently make it available to cities, towns, and counties. But state agencies can provide small, targeted grant programs to encourage these other modes. "Safe routes to school" grant programs now offered in California and New Mexico illustrate such creative grant-making.

A far greater impetus for sustainable development would be to link state transportation investment with good local land use planning. Localities that did not plan for transit-oriented development, or that allowed motor vehicle-dependent sprawl to continue, might not receive funds for local road maintenance, for example. Such policies would help ensure that investments in new transportation systems were cost-effective by making sure land use patterns would support them. It makes little sense to invest in new rail transit systems in particular unless the local governments along each line will allow intensified development near stations—building ridership and ensuring that the line is well used. It also makes little sense to build or expand freeways if local governments allow sprawling, automobile-dependent development to flood them with additional commuters.

The threat of withholding state transportation funding—or, conversely, incentives for good local planning in terms of increased state transportation money—can be a powerful lever for change. Such conditionality is fiercely resisted by local governments, and has yet to be used widely. But as previously mentioned the state of Maryland has instituted some conditions on its infrastructure money, as part of its statewide Smart Growth program, making funding conditional on completion of local plans designating priority funding areas. Other state-level governments could follow suit.

Housing

State governments can play a major role in enforcing fair-share housing distribution among local governments, so that each city or town approves housing units affordable to those making less than 80 percent (for low income) or 50 percent (for very low income) of the area median income. Again, state infrastructure funding could be made conditional on local governments complying with fair share targets. Or incentive grants could be made available to those that exceed their requirements by a certain amount.

Some states currently offer low-interest loans or other assistance to nonprofit affordable housing providers, often in the form of tax credits through which wealthy investors are enticed to lend funds to build affordable housing. Other state actions to assist housing production can include brownfield cleanup assistance, legislation to reduce the risk of litigation by Nimbys against housing projects, and incentive grants to municipalities that exceed their previous performance in constructing affordable housing (as has been tried in California).

Equity

States can play a crucial role in promoting equity, and have already done so in many parts of the United States, albeit sometimes pushed by the courts. In addition to fair-share housing policies, programs to provide partial equalization of local school funds are common. Usually ordered by court decisions, these programs take tax money collected statewide and use it to subsidize school systems in the poorest communities, somewhat equalizing school funding across jurisdictions. Some states also have relatively progressive tax systems, with steeply graduated income taxes that help reduce disparities in wealth. Hawaii's tax schedule, for example, features marginal tax rates ranging from 3.2 percent at the bottom bracket to 11 percent at the top. A few other states such as Oregon have persistently avoided instituting a sales tax, widely viewed as the most regressive form of taxation in that it falls most heavily on the poor, for whom basic purchases of household goods and transportation are a higher percentage of income. Many states exempt food, shelter, clothing, and medical expenses from sales taxes for equity reasons.

In the future, states will offer an arena for addressing the serious inequities in tax resources that exist at the local level—and the intense pressure for "fiscal zoning" by local governments. Under the fiscal zoning scenario, cities and towns zone for as much commercial development as possible in an attempt to gain sales and property taxes, but in so doing merely steal tax base from their neighboring local governments and promote suburban sprawl. The benefits of this misguided local tax system typically go to relatively wealthy fringe suburban jurisdictions with ample vacant land and few existing urban problems, while older suburbs and central cities suffer from the flight of their taxable businesses. By reforming tax structures statewide to collect revenue through relatively equitable state income and property taxes and then distribute it to cities on the basis of population, state governments could vastly improve both equity and local land use decision-making.

Economic development

Most states or provinces have active economic development efforts, and routinely court large corporations or industries to locate in the state. Governors travel internationally seeking to attract new investors, and may offer substantial incentives to interested parties. States typically also promote tourism heavily, since it is a major source of revenue that can build on the area's natural advantages. The desire to compete successfully with their peers often leads states to reduce their tax rates, adopt anti-union legislation, lower workers' compensation requirements, or take other steps aimed at a "business-friendly" reputation. Needless to say, these efforts do not necessarily produce sustainable development. Too often they produce few new jobs, many of which may be low-wage and temporary (if footloose corporations relocate elsewhere again). If they are successful, traditional economic development strategies may produce overly rapid growth that strains public services and fuels suburban sprawl. These "business-friendly" policies may also reduce much-needed state revenues and deprive workers of decent wages and benefits, if state labor policies are compromised in a "race to the bottom" with other states.

Corresponding state efforts to promote green economic development are few and far between, but there are some likely possibilities for action. Many states have historically been dependent on particular natural resource industries, which in many cases are exhausting their resource base. Coal, oil, timber, mineral, and fisheries industries fall into this category. State governments can take proactive steps to plan for a transition to more sustainable harvesting methods in many of these cases, or else for a transition to alternative products. Legislators can also plan so that the state budget does not take an inordinate hit when taxes from such natural resource industries decline.

States can also invest in human capital rather than in corporate recruiting incentives, a strategy that is likely to produce a capable, educated workforce in the long run that can help a variety of local businesses be successful. Investments in state university systems often bear handsome dividends, especially as graduates start new businesses. Similar investments in K-12 education are perhaps even more important, and play a substantial role in improving human welfare as well.

Along with programs to develop human capital, quality-of-life related investments by state governments are likely to help produce a more diversified, sustainable economy in the long run, attracting creative professionals and high-wage companies. Economist Richard Florida, in particular, has famously argued that much economic development depends on nurturing a young "creative class" interested in social tolerance, music, arts, recreational opportunities, and quality of life.[15] Rather than creating a low-tax, low-amenity environment, the most sustainable economic development strategy may be to invest in education, parks, the arts, public health, and community facilities. This may not generate the quick jobs that a single large manufacturing plant might create, but such a strategy may instead lay the groundwork for both better-quality jobs and a better-quality living environment in the long run.

A leading example of a large-scale sustainability planning framework oriented around economic development and social issues as well as environmental topics is that of Scotland. This quasi-autonomous British region—a scale that might be considered a state or province in other countries—developed a "sustainable economic growth" strategy within its 2007 Economic Plan, updated in 2011. This plan tackles issues of equity, education, and a transition to a low-carbon economy as well as more traditional economic goals. Implementation strategies include a living wage requirement for lower-income workers, improvements to education from early childhood to the university level, and extensive support for renewable energy industries. Performance indicators for such policies include reducing GHG emissions, raising the proportion of income earned by the three lowest income deciles, improving health and mental well-being, raising the proportion of trips made by public transit, and increasing resident satisfaction with their neighborhoods.[16]

As with sustainability planning generally, examples of state or provincial action are still in the early stages in most places. Successes ebb and flow; progress under one political administration is often sidetracked under the next, with state offices and programs changing name or disappearing. Yet on the whole initiatives have been increasing both in the US and internationally.

In Spain, for example, the Autonomous Government of Catalonia set up an Advisory Council for Sustainable Development of Catalonia in 1998. This body then helped develop a Strategy for Sustainable Development of Catalonia for 2026, adopted in 2010, as well as a Green Economy Plan, a Green Schools Plan, and an Energy and Climate Change Plan 2012–2020.[17] As with plans in many other places, the overall Strategy emphasizes improving biodiversity, increasing energy efficiency, reducing reliance on fossil fuels for transportation, promoting green economic development, improving education, minimizing waste generation, and maximizing public involvement in decision-making.[18]

In India, the state of Kerala is well known for progressive policies relating to social welfare and environmental issues, resulting in highly efficient distribution of food, low consumption of resources, a male life expectancy 10 years longer than the Indian average, and a female life expectancy 15 years longer.[19] During the second half of the twentieth century, largely without foreign aid, communist governments of the state worked with popular reform movements to de-emphasize the caste system, improve the status of women, extend formal education to most of the population, reform land distribution, and ensure public access to food and health care. The state achieved 100 percent literacy by 1991. Large remittances from state residents working in Persian Gulf countries also helped reduce poverty. The state was still very poor, however, with economic development still languishing. As a result the government moved to a "new Kerala model" which devolved administration and decision-making to more local levels, with more explicit focus on fighting environmental degradation.[20] Unfortunately in the 2000s the state appears to have moved into a more conventional economic development phase under more neoliberal governments, with declining state spending on education and health care, and rising inequality.[21]

State and provincial governments, then, can occupy a crucial middle ground between national and local levels in terms of sustainability planning. The sustainability planning agenda at the state level often focuses on issues larger than individual cities, towns, or metropolitan areas can handle, particularly in the areas of urban growth policy, environmental protection, transportation, equity, education, and resource planning. But state planning frameworks are likely to be more specifically tailored for the area than national policies, and may be quite detailed and stronger than their national equivalents.

22
REGIONAL PLANNING

One of the paradoxes of planning is that many social and environmental problems are best approached at a regional scale, but this is usually the weakest level in terms of government institutions and public understanding. Issues such as air quality, water quality, transportation planning, suburban sprawl, and tax base inequities overlap municipal and county boundaries and virtually require a regional planning approach. In the postmodern landscape cities and suburbs have expanded so much that it makes sense to think in terms of "the regional city," as Peter Calthorpe and William Fulton have argued in their book by the same name.[1] Los Angeles, for example, often seen as a prototypical late twentieth-century urban region, stretches for nearly 100 miles in several directions and includes hundreds of cities and parts of four counties. The City of Los Angeles itself has become a relatively small part of the urban region.

Yet although many development problems are best dealt with at a regional scale, most politicians and members of the public do not think regionally. Daily life occurs at a more local scale—in neighborhoods and cities—while national and state or provincial governments are strong and established institutions that dominate public attention. Regional planning issues tend to fall through the cracks.

Still, many opportunities exist to promote sustainable development at this scale of planning. In recent years, a growing movement calling for regional solutions to sustainability related problems has emerged that might be termed a "New Regionalism."[2] This movement includes political scientists and sociologists concerned about equity within metropolitan regions,[3] environmentalists and urban designers concerned about growth and suburban sprawl,[4] and economic analysts who see urban regions as increasingly important economic actors in the global economy.[5] While few participants hold out much hope for strong new regional governments to tackle such topics, many instead call for flexible and sophisticated governance using existing state and regional agencies as well as ad hoc partnerships of regional stakeholders to bring about change.

TYPES OF REGION

The "region" has been defined in many different ways within the urban planning literature. In the early twentieth century several British thinkers took a broad and holistic approach to the study of regions. Scottish biologist and polymath Patrick Geddes, often viewed as the father of regional planning, called for a "synoptic vision"[6] of regions combining geographical, economic, social, and political dimensions. This vision was to be based on firsthand observation, quantitative data, extensive drawing and mapping, and historical study of the region's evolution. Criticizing the "artificial blindness"[7] of book-learning, Geddes proposed that planners survey the region from vantage points similar to his Outlook Tower in Edinburgh, walk through it, and prepare multi-media exhibitions illustrating regional evolution. Only after planners and the public had gained a thorough understanding of the region's character, he believed, should plans be developed.

Howard's Garden City proposals at the turn of the twentieth century were equally holistic, including physical planning and urban design concepts, economic mechanisms for local industry and cooperative ownership of land, and provisions to maintain fair rents and promote equity. According to Howard, "a town, like a flower, or a tree, or an animal, should, at each stage of its growth, possess unity, symmetry, [and] completeness, and the effect of growth should never be to destroy that unity, but to give it greater purpose."[8] Although his proposals were focused particularly at a town scale, Howard's vision integrated these towns with their greenbelts into a larger, carefully organized metropolitan region.

Following the lead of these British thinkers the Regional Planning Association of America (RPAA), especially through its members Lewis Mumford and Benton MacKaye, sought to develop an "ecological regionalism" in the United States in which city and countryside, industry and nature were viewed as a whole. Mumford and others developed a vision—unfortunately never extensively implemented—integrating economic development, management of natural resources, transportation, large-scale physical planning, and humanistic architecture and site design.[9] As with Howard, the unit of development was to be the garden city. The Great Depression curtailed plans to build a demonstration model, and only a small portion of it was actually built in Radburn, New Jersey. Rather than serving as a model for new regional development, Radburn laid the groundwork for a new form of suburban subdivision emphasizing cul-de-sacs and an interior green spaces network.

The objective of all these early regionalists was to decentralize population from the overcrowded, unhealthy nineteenth-century industrial city into a more balanced regional development pattern. Unfortunately, these regional thinkers wrote before the era of the automobile and did not foresee that cars would facilitate an extreme version of decentralization with devastating ecological and social impacts. They also did not foresee that the planning profession would become increasingly pragmatic and dominated by modernist science during the twentieth century. Their holistic and idealistic approach to the "region" was disparaged by later generations of professionals who focused on more narrow and technocratic specialties such as transportation planning, economic development, and the administration of zoning.

Another approach to regionalism arose in the 1930s and 1940s, based in large part at the University of North Carolina in Chapel Hill. Sociologists there such as Howard W. Odum and Harry Estill Moore analyzed cultural regions such as the American South and were known as "cultural regionalists."[10] A prime motivation for this school was to preserve the unique social values and traditions found in such regions. Such cultural regions were typically very large and difficult to define, since social groupings and traditions overlap considerably.

Since the late 1940s many academic regionalists have defined "regions" as fields of economic activity, and have focused on areas such as Southern California, Silicon Valley, the Tennessee Valley, the Emilia-Romagna area of northern Italy, or the even larger Sunbelt across the southern United States. Economic analysis and scientific method have been dominant within this group, rather than the physical design visions and normative values of earlier regionalism. The field of "regional science," founded by Walter Isard at the University of Pennsylvania in the 1940s, has concentrated almost exclusively on economic analysis. In their classic 1964 volume *Regional Development and Planning*, John Friedmann and William Alonso referred to the region as an "economic landscape,"[11] and Friedmann wrote that regional planning was concerned mainly with "problems of resources and economic development."[12] References to real places disappeared almost entirely within this analysis. Perhaps the ultimate extreme of modernist regionalism was Mel Webber's early-1960s concept of the "non-place urban realm."[13] In his view place-oriented urban community would disappear entirely within highly mobile, automobile-oriented regions, and individuals would somehow establish "community without propinquity." Although this vision did describe much twentieth-century development, it gave the shivers to increasing numbers of individuals in the latter part of that century who viewed such a landscape as environmentally, socially, and culturally disastrous.

In the 1960s and 1970s neo-Marxist regional economic geographers and sociologists such as David Harvey and Manuel Castells brought a new analysis of power to the study of regions, looking at the way economic capital, social movements, elite groups, and "growth machines" dominate urban and regional development. The region for these observers became a terrain of power—economic, political, and social. Simply developing technical analyses without an understanding of power dynamics, in this view, was pointless for planners. Instead what was important was understanding the structural changes needed in order for planning goals to be effectively reached.

More recently there has been a resurgence of interest in regional physical and spatial planning—addressing land use, infrastructure, urban design, ecosystem planning, and equity primarily within metropolitan regions. This reframing of the field has come about because of the rapid growth of urban regions worldwide, a new wave of environmental planning emphasizing bioregions and landscapes, and concern about growing income and wealth disparities between central cities and suburbs. To some extent this new regionalism represents a return to more holistic early twentieth-century definitions of the "region."

This form of regional planning concerns itself particularly with metropolitan regions, watersheds, bioregions, and other physical landscapes rather than more abstract economic or cultural space. The first of these categories, metropolitan regions, refers to collections of urbanized cities and counties in geographic proximity. Since the 1960s federal transportation and housing funding has required metropolitan planning organizations (MPOs) to be set up to coordinate efforts across these jurisdictions. There is sometimes argument over the exact extent of metropolitan regions—whether a region, for example, should include its entire commute shed, with some communities more than 100 miles away from the original core city. The Los Angeles region might or might not be seen as including the San Fernando Valley, Riverside, and other areas far removed from the core city of Los Angeles. (The Southern California Council of Governments does in fact include these areas.) Metropolitan regions have also expanded dramatically in recent decades, meaning that the borders of regional institutions have needed to evolve. In the mid-twentieth century the San Francisco Bay Area, for example, saw itself primarily as a 5-county region. Now the Association of Bay Area Governments (ABAG) consists of representatives from 9 counties. In another decade or two it may have to include as many as 14 counties as the region's development sprawls outwards into California's Central Valley.

Watersheds are ecological regions that may or may not overlap with urban areas. In contrast to the changing borders of metropolitan regions, a watershed is a naturally defined area rooted in the characteristics of particular drainage systems. However, since rivers and streams branch extensively there can be said to be watersheds within watersheds (fluvial geomorphologists speak of ten orders of waterways), and different scales of watershed may be useful for different sorts of planning. For example, Rock Creek within Washington, DC, forms a watershed draining dozens of square miles, with important issues of water quality, hydrology, habitat, and interface with recreational park uses. But Rock Creek empties into the Potomac River, whose much larger watershed encompasses hundreds of square miles in Maryland, Virginia, West Virginia, and the District of Columbia. The effort to clean up the Potomac was a major environmental campaign in the 1960s and 1970s. The Potomac in turn drains into the Chesapeake Bay, whose watershed covers some 64,000 square miles in parts of six states (adding Pennsylvania, New York, and Delaware). In 1983 a Chesapeake Watershed Partnership including most of these states plus the US EPA and the Chesapeake Bay Commission began one of the country's largest planning exercises, still ongoing, to protect and restore the entire ecosystem.[14] Over three decades the Partnership has developed a comprehensive set of goals and programs affecting land use, pollution control, wetlands, forests, recreation, urban development, transportation, and certain aspects of local economies—a wide range of sustainability-related issues—throughout the entire watershed region. This extremely large scale of watershed planning is vastly different from planning for more local watersheds in the same area.

Watershed programs have a long history that illustrates changing values within regional planning. Regionalists of the 1920s and 1930s promoted watershed planning to manage natural resources and to meet social objectives of reducing poverty and providing employment. The crowning achievement of this period in the US was the Tennessee Valley Authority, initially formed to provide electric power, employment, and social development for a large, impoverished area of the American South. Unfortunately social objectives were soon forgotten, and this project proved successful mainly at building large-scale infrastructure—dams, highways, power plants, and a nuclear power laboratory. As such it fell victim to modernist impulses to tame and manage the environment rather than to coexist sustainably with it.

Comprehensive river basin planning in the western US during the mid-twentieth century likewise created dams, reservoirs, and power plants at great environmental cost. (The drowning of Glen Canyon by Hoover Dam, viewed by former Sierra Club executive director David Brower as one of the environmental movement's greatest defeats, was just one of the more disastrous examples.) However, beginning in the 1960s and 1970s watershed planning shifted in a more ecological direction. People realized that large dams, levies, and artificially stabilized channels, combined with urban development in floodplains, often dramatically increased flood damage rather than reducing it. Other negative effects of dam-building worldwide became well known. Even the Army Corps of Engineers, responsible for much of the twentieth century's dam-building and river channelization, adopted a mission of sustainable development in the 1990s and undertook projects such as restoring the oxbows and meanders in the Kissamee River in central Florida, which had been channelized by a previous Corps project 30 years before.

Bioregional planning often overlaps with large-scale watershed planning and has been an attractive ideal to many environmentalists since the 1960s. However, there are as of yet few political institutions corresponding to this scale. Bioregions are probably best defined in terms of distinctive plant and animal communities that form a typical natural mosaic of ecosystems. Not coincidentally, they are also often characterized by particular human cultures with their roots in specific natural resource industries or forms of agriculture.

Yet, as with watersheds, bioregions are difficult to define. Is San Francisco's bioregion, for example, the area centered around San Francisco Bay itself, the watershed of the Sacramento and San Joaquin rivers that feed the Bay (which would include much of central California but not necessarily the coast), or all of northern California? Local bioregionalists have generally taken the latter approach, naming the area the "Shasta Bioregion" after Mount Shasta, a 14,000-foot extinct volcano in the northern part of the state. But this landmark is very remote from most areas within northern California, and ecosystems vary enormously throughout this area. The redwood and Douglas-fir forests of coastal northern California bear little resemblance to the grasslands of the Sierra foothills or the semi-arid chaparral of the many inland valleys. And to many residents of the northern California coast, San Francisco bears more similarities to Los Angeles than to their own communities. In such places clear boundaries of a bioregion are hard to come by, and so it is difficult to develop ideas about how to plan for such an elusive region.

Airshed planning is yet another form of regional planning whose borders are to some extent defined by the natural landscape. Airsheds are typically created by mountain ranges or hills that trap air and create local pollution, although long-distance transport of pollutants can add to local problems. A large metropolitan area in a hilly region may contain several discrete airsheds that may be grouped together for planning purposes. However, rather than focus on each air basin separately, air quality planning agencies tend to adopt metropolitan regional boundaries as determined by the politics of the region.

Finally, "landscape planning" is a growing field within the discipline of landscape architecture, pioneered by figures such as Ian McHarg and Richard Forman, that has a strong regional component. Landscape ecologists are concerned with distinctive natural landscapes that may cross the jurisdictional borders of cities, counties, and even states. During the last two decades of the twentieth century this field developed a sophisticated terminology with which to describe landscape structure. Landscape ecologists speak of "patches" of habitat, "corridors" between them, "edge" and "interior" ecosystems, a background "matrix," and an overall "mosaic" of these forms across a broad area. Landscape planning then becomes a task of managing these ecosystem elements to meet goals such as biodiversity.[15] Again, few institutions may exist corresponding to landscape regions, but landscape ecologists may be able to use many different levels of government to meet their planning objectives.

These, then, are many of the types of regions that are important for planning purposes. Despite definitional difficulties, the "region" for any given planning problem can usually be defined in practice. The main problem is that strong governmental institutions do not exist for many sorts of regions, or don't have significant power to affect land use, environmental planning, distribution of fiscal resources, or other areas of interest within planning.

Regional institutions have gradually been growing in strength—regional environmental planning agencies, park districts, and transportation commissions, for example, have expanded greatly in the past 50 years. But the process of institution-building is slow and incremental. Regional planners must therefore often make do with whatever institutional tools are available, while seeking to improve these in the long run.

WHO PLANS REGIONS?

A large number of often overlapping agencies exist to plan at many of these regional scales. Metropolitan planning organizations (MPOs) exist in every US metropolitan area as the agencies responsible for coordinating transportation decisions and dispersing federal transportation money to cities and counties. Since they handle large amounts of funding, these MPOs have considerable power over how the region develops. But they usually have no control over land use, which handicaps them in terms of managing regional growth. MPOs may or may not be synonymous with regional Councils of Government (COGs), which are voluntary associations of local governments in a metropolitan region generally set up in the 1960s or 1970s. These latter agencies collect information about the region, provide a forum for local cities and counties to coordinate strategies, and frequently provide technical support services to municipal governments. Their membership is entirely voluntary, and they do not have legal authority over land use or other important planning subjects.

Britain has seen a long and turbulent history of regional planning, in particular through the "county councils" that coordinated public housing and other planning needs beginning in the nineteenth century. These were abolished under the Thatcher Administration, which was generally opposed to such planning. The Blair government established a new set of eight Regional Development Agencies (RDAs) under a 1998 Act. Each agency was directed "to contribute to the achievement of Sustainable Development in the United Kingdom where it is relevant to its area to do so." However, this mandate was balanced with four other goals, and so the overall focus of these regional authorities was less than clear. In any case, the RDAs were themselves abolished by the Conservative-led coalition government of David Cameron that took power in 2010.

France, the Netherlands, Denmark, Sweden, and many other countries have a strong tradition of regional planning, often to coordinate development of large metropolitan areas such as Paris, the Randstad (the area including Amsterdam, The Hague, Rotterdam, and Utrecht), Copenhagen, and Stockholm. In many cases these agencies have established plans for the spatial development of regions. The "finger plan" designs of Copenhagen and Stockholm, in which development is channeled along rail transit corridors, are well known. Regional authorities directing development around Paris have created a series of new towns and development poles, including areas such as La Defense, Marne la Vallée, Val Maubuée, Evry, Senart, St Quentin en Yvelines, and Cergy Pontoise. Because of their large scale and modernist design, these projects have often been problematic in terms of social and environmental issues.

In China there are four different levels of "city" government, some of which can be considered regional in scale.[16] While the national government remains by far the strongest political authority in the country, these forms of city government have strong control over physical development, amplified by national reforms aimed at decentralizing responsibility to such regions. In a very rapidly developing country, such links between regional authorities and developers have unfortunately provided ample opportunity for corruption, and for what we might consider a Chinese version of the "growth machine" (the de facto alliance between political authorities, elites, and developers found in many countries). The result has been very rapid building of massive new cities and high-rise suburban districts. These projects often have the advantage of improved housing quality for residents and high density (which helps preserve agricultural land and build the market for public transit when compared with lower-density sprawl), but the disadvantage of being less pedestrian- and bicycle-friendly than past Chinese cities, less mixed-use, and relatively monotonous. Chinese regional planning is becoming increasingly concerned with livability issues and environmental quality, which has resulted in new attention to urban and regional parks as well as pollution regulation. However, environmental data collection, tracking, monitoring, and enforcement mechanisms are often weak.[17]

Actual metropolitan governments with statutory authority over land use planning are extremely rare in the United States. Portland, Oregon's, Metro Council is the best-known example. Since the 1970s this three-county regional government has gradually grown in power and respect, and is in charge of implementing Oregon's statewide land use planning goals in the Portland Area (see Box 22.1). A similar Metro Council in the Minneapolis–St Paul area, with members appointed by the governor, has somewhat less power over local land use, although it is able to review large regionally significant projects and has overseen a regional tax-sharing framework since 1972. Consolidated city–county governments have also been created in metropolitan areas such as Nashville, Indianapolis, Jacksonville, and Miami as a way of providing more effective regional coordination. However, such regional planning entities have generally been more concerned with providing efficient public services such as police, fire, sanitation, and public health to metropolitan areas than with rethinking patterns of development.

Even if broad-based regional governance doesn't exist, single-purpose regional agencies exist in most metropolitan areas and have authority under state or national law to develop policy and programs in fields such as air quality, water quality, parks, public transit, and public utility planning. In addition to its Association of Bay Area Governments, the San Francisco Bay Area, for example, has a Metropolitan Transportation Commission (the MPO), an Air Quality Management District, a Bay Conservation and Development District, a Regional Water Quality Control Board, and numerous water, sewer, park, and transit

Box 22.1 **REGIONAL PLANNING IN PORTLAND, OREGON**

Agencies in the Portland, Oregon, metropolitan area have been refining regional planning visions for more than 40 years. In 1966 the fledgling Metropolitan Planning Commission prepared three alternative scenarios under the heading "How Shall We Grow?" The options as planners saw them then were for the region to form a "lineal city" with development in a broad north–south corridor down the Willamette River Valley, "regional cities" with growth dispersed into twenty-five separate and relatively self-contained cities, and "radial corridors" in which development clustered around mass transit in several corridors radiating from the central city, with surrounding land preserved as open countryside.

As required by state growth management legislation, the strengthened Metro regional government approved an urban growth boundary in 1979 around the entire metropolitan area (except for that portion that had already sprawled across the Columbia River into the state of Washington). During the 1970s planning began for the MAX regional light rail system, and new parks, public spaces, and infill development projects revitalized downtown Portland.

In the 1990s the Metro Council developed a highly detailed Region 2040 plan emphasizing transit-oriented development, infill within existing urban areas, creation of a regional green-spaces system, and improved design of new neighborhoods. Regional planners then began a long process of working with local governments to change zoning codes and other standards to reflect the regional vision.

In the 2000s the region expanded on these initiatives further, adding new light rail lines, a central city trolley, and sustainable living programs for residents, as well as bringing a regional Willamette Greenway closer to reality.[18]

districts covering portions of the region. Many of those single-purpose agencies have come together at times within regional planning processes, such as Plan Bay Area following California's passage of SB 375, the Sustainable Communities and Climate Protection Act of 2008.[19]

Some single-purpose regional agencies are well funded and powerful (the transportation MPOs fit into this category). However, they do not necessarily have a mandate to coordinate with other agencies or local governments around the region, or to "think regionally" in any comprehensive sense. Rather, these agencies focus on a single, narrowly defined function. Regional branches of state or federal agencies (such the US EPA and HUD) likewise concentrate on carrying out particular policies of these higher levels of government within the region, but may not have a mandate to coordinate with other agencies on broad regional policy.

A final notable force in planning urban regions has been NGOs. These citizen groups have grown in number and influence in many areas and can play a significant role by developing plans themselves, influencing plans developed by official agencies, or lobbying for change at the regional level. The Regional Plan Association of New York and New Jersey is the oldest continually active US organization of this sort. Formed in the 1920s by civic leaders and the Russell Sage Foundation, this Association prepared regional plans for the New York metropolitan area in 1929, 1964, and 1996. Although they lacked governmental authority, these visions greatly influenced the work of cities and agencies throughout the region. An even earlier example of citizen-based regional planning was the Plan of Chicago prepared by Daniel Burnham in 1909. This pioneering document proposed regional park systems, transportation systems, and a network of grand boulevards and public spaces. Of these, the lakefront park system has been the most consistently implemented.

In the past two decades citizen organizations and networks have become a powerful force for regional growth management planning, environmental protection, and social equity.[20] For example, the Chesapeake Bay Foundation has led efforts to protect the Chesapeake watershed since 1967, and boasts more than 100,000 members and 200 staff.[21] The work of 1000 Friends of Oregon and allied groups has been crucial in supporting regional growth management and transit plans in the Portland area. Sister 1000 Friends groups now exist in more than a dozen states nationwide, including Iowa, Washington, Florida, Minnesota, Wisconsin, New Mexico, Maryland, and Hawaii. Starting with the efforts of three University of California faculty wives concerned about Army Corps of Engineers plans to fill much of San Francisco Bay, the Save San Francisco Bay Association succeeded in getting state legislation passed in the 1960s protecting the Bay from development and establishing a regional agency to monitor its health, the Bay Conservation and Development Commission. Both organizations are still highly active, with their mission in many cases shifting from protecting the Bay towards restoring wetlands and shoreline damaged by previous human uses such as for salt ponds, farming, military bases, and industry. In a more rural context, property owners and environmentalists in the Lake Tahoe area, led by the League to Save Lake Tahoe, have spearheaded the development of a regional plan to protect water quality and other environmental aspects of that large lake basin in the Sierra Nevada Mountains.[22]

Internationally, the London 21 Sustainability Network has produced an informative green map of that metropolitan region,[23] while other groups such as the Institute for Sustainability and Green London have worked on projects such as sustainable transport systems, low-carbon buildings, and eco-tours of ecological restoration sites. In Taiwan citizens organizations began in the late 1960s to protest pollution and land use, and by the 2010s more than 300 NGOs worked on regional or community issues in that densely populated island nation. Among other activities, NGOs frequently organized protests against the construction of golf courses, arguing that these inappropriately used land and water while undermining social equity. NGOs have also been integral to the emergence of democracy in that country, building international relationships, working extensively with legislators and endorsing "green" candidates.[24]

REGIONAL SUSTAINABILITY ISSUES

Those groups wishing to promote sustainability-oriented planning at a regional level typically focus on types of certain sorts of issues best coordinated at this scale, including

Box 22.2 OPEN SPACE PROTECTION IN THE VANCOUVER REGION

The Vancouver, British Columbia, area has taken a number of successful steps to protect its stunningly beautiful natural landscape from development (Tomalty, 2002). Many of these initiatives have been spearheaded by the Greater Vancouver Regional District (GVRD), established in 1967 and renamed Metro Vancouver in 2007. This agency approved its first Livable Region Plan in 1975, and a successor Livable Region Strategic Plan in 1996. Based on a compact region scenario, the latter plan included a green zone component protecting areas of great social or ecological value from development, and establishing a UGB for the urban area. The GVRD required local cities to designate lands to be protected under this framework, and more than 440,000 acres have been safeguarded in this green zone. Many of these lands are additionally protected through public ownership, park designation, or inclusion in the region's agricultural land reserve.

The Vancouver region is also trying to manage urban growth so as to limit the overall urban footprint, with mixed success. The core city of Vancouver has gained substantial population in recent years, but the largest proportion of growth is taking place in outlying municipalities such as Surrey and Richmond. Other goals of regional planning include to require that "complete communities" be built instead of single-use subdivisions, and to support alternative modes of transportation. For more information, see http://www.metrovancouver.org/Pages/default.aspx.

transportation planning, large-scale land use and growth management planning, watershed and environmental protection, air quality planning, regional equity planning, and regional economic development.

Transportation

Ask people in US urban regions about their greatest concerns, which pollsters regularly do, and, except during times of economic difficulty, "transportation" often tops the list. Traffic congestion is a fact of daily life for millions and has worsened over the years in most metropolitan areas. Developing strategies to reduce motor vehicle use, promote other modes of transportation, and reduce the amount that we all need to travel every day is a key challenge at the regional scale. Although all levels of government must be involved in so complicated a challenge, many of the most important actions are taken by regional agencies. Such transportation planning policies are also a crucial determinant of metropolitan land use and growth.

Regional agencies can play a role in all of the areas of transportation planning reform mentioned earlier: better alternative modes of travel, better land use, economic incentives for driving less, and less travel-intensive lifestyles. In the United States MPOs typically prepare regional transportation plans (RTPs) every two or three years to identify projects to be funded over a 20–25-year time frame. These plans help funnel federal and state money into regional transportation infrastructure and transit operations. Until the passage of ISTEA in 1991 such plans were almost exclusively focused on roads and large transit systems. However, this federal legislation gave regional agencies in areas suffering from air pollution increased flexibility to allocate monies formerly used for highways to transit, bicycle, and pedestrian planning and to programs adding amenities and urban design improvements in transit-supportive locations. Through such programs, regional agencies can fund measures to improve walking and bicycling provide incentives for better local land use planning, and initiate regional Smart Growth visions to better coordinate local planning.

American urban regions such as St Louis, Atlanta, Portland, Washington, DC, the San Francisco Bay Area, San Diego, Los Angeles, and Dallas have built major new rail transit systems in the past 30 years. Agencies in some of these areas have also begun offering incentives to local governments to plan for new development near transit lines, minimize sprawl, and emphasize pedestrian and bicycle planning. The Atlanta Regional Council's Livable Centers Initiative[25] and the Bay Area Metropolitan Transportation Commission's Transportation for Livable Communities program[26] are two examples of such incentive programs. Both give grants to local governments for the planning and construction of new amenities near transit, and the latter at times has offered grants of up to $2,000 per unit for new housing near transit.

As with state infrastructure funding, regional transportation funds can be used to leverage better land use by making grants to local government conditional on the adoption of good local land use plans that will reduce automobile use and sprawl. This step is more difficult politically and has yet to be taken by US regional agencies, although the US EPA has temporarily frozen transportation funding to metropolitan areas such as Atlanta and the Bay Area that failed to achieve compliance with clean air laws. But bills establishing this sort of linkage have been introduced into the California state legislature, and many observers believe that such "sticks" are needed as well as incentive "carrots" to induce local governments to change their ways.

Regional agencies can help coordinate local action in other areas of sustainable transportation policy as well, such as creating pedestrian- and bicycle-friendly streets, regulating parking charges, adopting transportation demand management programs, and ensuring transit- and pedestrian-friendly urban form. Transportation demand management programs are especially significant, and as discussed earlier, may consist of a wide variety of inducements for people to drive less. These incentives may include increased parking changes, EcoPass programs, preferential parking and rebates for carpools, new shuttle vans, bike facilities, "guaranteed ride home" programs in the event of family emergency for workers who carpool or take transit, and many other innovative services. Although no one TDM program is likely to greatly reduce driving by itself, many such initiatives combined with land use changes and better transportation alternatives can begin to reverse the unsustainable growth in automobile use.

Land use, growth management, and regional design

Stopping sprawl and revitalizing central urban areas through new development are major concerns in most metropolitan regions, even in relatively rural regions where low-density exurban development may be occurring around small towns and rural highways. Although they rarely have direct authority over land use, regional institutions can still play a major role in coordinating growth management and Smart Growth by bringing local governments together, facilitating consensus on better growth directions, providing incentive funding for local governments to implement regional growth goals, and potentially withholding funding from local governments who refuse to take regional needs into account.

As previously mentioned, transportation funding is a particularly important lever that regional planning agencies have to shape metropolitan growth. The extension of water and sewage systems also supports suburban sprawl. Regional decisions about where to build such infrastructure help determine the growth and form of a region. Building new suburban-serving freeways is likely to encourage further sprawl, as many metropolitan areas have found out. Conversely, regional decisions to forgo highway expansion and promote public transit serving existing urban areas may help limit sprawl.

Outside the US, regional agencies can potentially take a more overt and ambitious approach to shaping regional land use. During the past 100 years there have been several waves of efforts to design metropolitan regions in ways that manage growth; coordinate transportation, land use, and housing; and preserve open space. The garden city visions of Howard, Unwin and others, first conceptualized in the late nineteenth century, eventually bore fruit in the 1940s and 1950s in plans for many European metropolitan regions such as London, Copenhagen, Stockholm, and Paris. In South Korea, the country's government established a broad regional greenbelt about 10 kilometers wide around Seoul in the post-Second World War period, and undertook construction of several large new towns outside the greenbelt, following the British model.[27] In the early 1960s regional planners in the Washington, DC, area developed a "star" plan for development in that region, calling for future development to occur in radial corridors along transit lines. However, unlike the European examples, Washington had no strong agency to bring the vision about, and local governments neglected this plan.

Beginning in the 1990s many planners again focused attention on regional design.[28] New Urbanists such as Peter Calthorpe have been particularly active in this regard,[29] and have actively consulted with regional agencies in places such as Portland, Toronto, Salt Lake City, and Minneapolis. Many of these visions are strikingly reminiscent of earlier garden city concepts, in that they cluster transit-oriented communities along rail lines throughout a metropolitan area, often with lower-density areas or open space between these. This basic regional design concept has proven remarkably resilient and attractive to different generations of planners (see Figure 22.1). Portland, for example, has designated nodes of density along stations of its new MAX light rail system, and intensified corridors of development along heavily used Tri-Met bus lines. These transit lines radiate out from downtown Portland, creating a traditional star-shaped pattern.

A wide range of regional planning documents helps implement such visions. These plans help coordinate land use with transportation systems, create regional park and green-space systems, ensure economic development, and/or meet affordable housing needs across the region. Examples include the Region 2040 plan prepared by Portland's Metro, Vision 2040 prepared by the Puget Sound Regional Council in the Seattle–Takoma area, the

a.

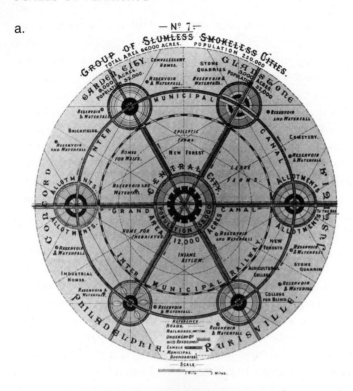

b.

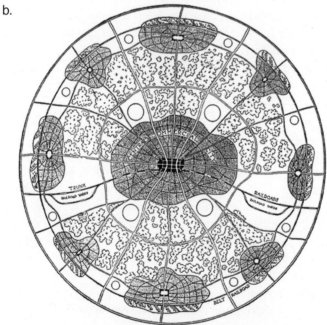

DIAGRAM OF SATELLITE TOWNS

Figure 22.1 Garden city regional design concepts. Similar regional design concepts have occurred time and again since Ebenezer Howard's famous "Garden City" vision of 1898, typically clustering development around transportation systems. Shown are diagrams from: a. Howard (1902 [1898]), b. Nolen (1916), c. Gruen (1964), and d. Calthorpe (1993)

c.

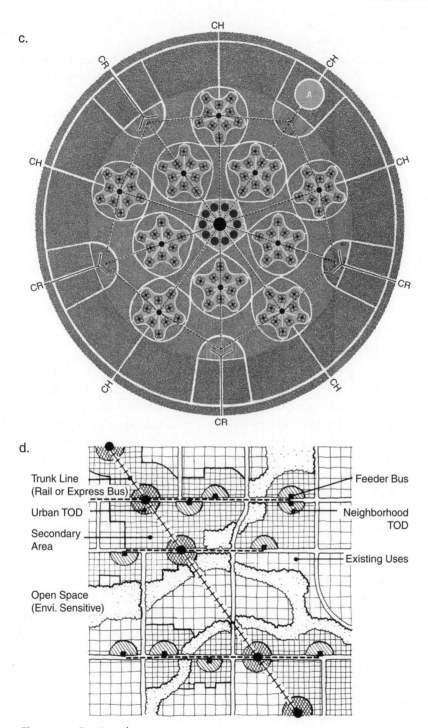

d.

Trunk Line
(Rail or Express Bus)

Urban TOD

Secondary
Area

Open Space
(Envi. Sensitive)

Feeder Bus

Neighborhood
TOD

Existing Uses

Figure 22.1 Continued

Livable Metropolis plan prepared by Toronto's Metro Council, and the New York Regional Plan prepared by the Regional Plan Association of New York.

A major question concerning regional design plans is how they are to be implemented. Such action requires a national, state, or regional agency with considerable leverage over local land use decisions. Generally such plans have been most successful in areas where strong higher-level government authority can help bring about regional and local land use changes, such as in the state of Oregon. National governments have played this role for metropolitan regions such as London, Copenhagen, Stockholm, the Randstad in the Netherlands, Paris, and Seoul. But in the absence of such higher-level leadership, extensive collaboration between local governments could potentially develop and implement regional design as well. Given the strong tradition of local government land use authority in the United States and resistance from federal and many state governments to overriding this, what will probably be necessary in the US is a combination of local and regional coordination with guidance and incentive from higher levels of government, perhaps focusing on greenhouse gas reduction goals as is happening in California. Over time such a decentralized institutional framework may be able to build consensus around new regional growth directions.

Open space, watershed, and environmental planning

It makes sense that plans for park systems, ecologically sensitive natural areas, and air and water quality should be prepared on a regional scale, since watersheds, airsheds, and ecosystems are regional in nature rather than limited to the bounds of any municipal jurisdiction.

Some metropolitan environmental planning was done as early as the nineteenth century, for example resulting in Frederick Law Olmsted's "Emerald Necklace" system of parks for Boston. Turn-of-the-century city beautiful planning also resulted in metropolitan open space plans that might be termed "regional." These helped to produce Chicago's waterfront parks, Washington, DC's system of parks, monuments, and public spaces, and the extensive set of parks in Buffalo, New York. In England, greenbelt planning helped create impressive rings of open space around metropolitan regions in the 1940s. London's Greenbelt, set forth in the 1944 Abercrombie Plan, remains one of the most significant regional open space achievements. But most systematic efforts at regional environmental planning— especially with habitat and ecosystem management goals in mind—are relatively recent, dating back at most to the 1960s.

A number of both rural and metropolitan regions have developed regional open space plans that seek to meet ecological objectives such as maintaining wildlife habitat, creating wildlife corridors between existing parks, and restoring watersheds, in addition to managing urban growth and providing recreational amenities. In rural areas in particular such planning is widespread as federal, state, or regional park districts—or consortiums of agencies—seek to enhance ecosystem and recreational values across large areas. The Sierra Nevada Conservancy, for example, is a regional agency overseen by both state and local officials that conducts programs in the Sierra Nevada mountains of California and Nevada. The agency both pursues environmental protection and restoration projects, for example allocating $54 million in state funds to conservation projects during the late 2000s, and works with a large collection of stakeholders to bring about sustainable development of human communities in the region.[30] In Florida, the South Florida Water Management District and the US Army Corps of Engineers jointly developed the Comprehensive Everglades Restoration Plan in 2000. One of the world's largest ecosystem restoration efforts, this effort directs $9.5 billion in projects through 16 counties and 18,000 square miles.[31] The focus is on improving natural water flow though this fragile habitat. Ironically, the Army Corps of Engineers is the very same agency that by channelizing and diverting rivers created many of the region's environmental problems in the first place.

In metropolitan areas coordinated open space planning is also on the rise. For example, in 1992 Portland's Metro Council adopted a Metro Greenspaces Master Plan that provided the blueprint for acquiring 14 key natural areas. The regional agency subsequently developed a Water Quality and Floodplain Protection Plan in 1996, crafted a Fish and Wildlife Habitat Protection Plan in 2002, and later in the 2000s developed an "Intertwine" consortium of local governments, businesses, nonprofits, and citizens devoted to developing the region's network of parks, trails, and natural areas. These plans fit together with the region's overall Region 2040 plan to help coordinate environmental planning and urban growth throughout the three-county, 1.8-million-resident region. The New York Regional Plan Association (RPA) proposed a "greensward" network of open spaces as a major element of its 1996 regional plan for the New York metropolitan area, and in 2012 expanded its vision of large-scale natural landscape conservation to the megaregion extending up the US eastern seaboard from Washington to Boston.[32] The former plan calls for the creation of 11 Regional Reserves to conserve waterways and working landscapes in the region, for increased investment in urban parks, and for a network of corridors and greenways providing recreational opportunities as well as benefits for wildlife. Although advisory in nature, the recommendations in this RPA plan, as in 1929 and 1968 versions, are likely to be taken very seriously by local governments in the region. In the UK, regional authorities in the vicinity of the Telford new town west of Birmingham, England, have established a "green network" covering nearly 49,000 acres. The former site of mining for coal, ironstone, clay, and limestone, this region is one of the world's oldest industrial landscapes and has been designated a World Heritage Site by UNESCO. Now officials are seeking to restore forests, meadows, and fens, in part through the planting of 6 million trees and 10 million shrubs, while creating 25 miles of footpaths and creating or preserving a mosaic of wildlife habitats.[33]

Regional air and water quality plans are usually developed in response to state and/or federal legislation by regional agencies specifically set up to handle those issues. In the area of air quality planning, many regional plans have been prepared to implement the 1972 federal Clean Air Act and various subsequent state legislation. Such air quality plans aim at meeting threshold criteria for ozone, nitrogen oxides, carbon monoxide, sulfur oxides, and particulate emissions established under this legislation, and can regulate motor vehicle use, smokestack emissions, dry cleaners, use of painting solvents, backyard barbecues, and many other sources of pollution. In the Los Angeles area, for example, the South Coast Air Quality Management District, established in 1977, prepared regional air quality plans in 1982, 1989, 1991,1997, 2003, 2007, and 2012 setting out hundreds of "control measures" to reduce emissions.[34] These measures included cleaner vehicle fuels, retrofits to bus engines, bans on burning garbage, a program of emissions trading among industries, and many other steps. Although still only partially successful, the agency's plans and other state regulation of motor vehicle emissions have resulted in a substantial decline in health alerts and emissions violations. If regions remain out of attainment with federal air quality standards because of motor vehicle use, the EPA has the power to freeze federal transportation funds that might be spent on highway projects. Environmentalists can also sue to force regional agencies to comply with federal or state air quality law.

Regional water quality initiatives attempt to implement standards set by the federal Clean Water Act and a variety of other legislation. Since the original 1972 Act these efforts have focused particularly on better sewage treatment facilities, since historically many North American municipalities have dumped poorly treated sewage into waterways, and such pollution still often occurs when wet weather causes systems to overflow in cities that do not have separate sewage and storm drain systems. Water quality efforts initially also sought to regulate "point sources" of pollution (pipes or contaminated sites discharging pollution into watersheds), and somewhat belatedly came to recognize the importance of "non-point sources," including pollutants that drain off roads and parking lots, that wash out of the air during rainstorms, that drain off agricultural fields, golf courses, or construction sites, or that likewise do not have a specific, discrete source.

Collaborative efforts in recent years have sought to develop consensus on some of the most large-scale and difficult water quality challenges in North America. These efforts include the Chesapeake 2000 plan for the Chesapeake Bay Watershed,[35] the 1994 Ecosystem Charter for the Great Lakes–St Lawrence Basin area (complemented by a St Lawrence River Plan for Sustainable Development in 2005),[36] and the CALFED and Delta Stewardship Council processes for the San Francisco Bay estuary system in the 1990s and 2000s.[37] In the Great Lakes region, for example, efforts to combat pollution—which had led to Lake Erie being declared virtually dead in the 1960s—began with the first Great Lakes Water Quality Agreement signed by agencies of the US and Canadian governments in 1972. These efforts continued in subsequent decades with work by many cities, states, provinces, regional branches of federal agencies, and collaborative forums such as a Great Lakes Commission consisting of officials from throughout the region.[38] The National Wildlife Federation and the National Parks and Conservation Association formed a Healthy Lakes coalition of more than 115 other environmental, civic, and business groups in 2005 to further advocate for Great Lakes restoration and related funding. This organization has been successful in securing billions of dollars for regional environmental protection, restoration, and education.[39]

The CALFED project was an enormous multi-year consensus-building effort between 25 state and federal agencies and many private stakeholders including farmers and environmentalists. Initiated in part because of the threat of unilateral action by the US EPA Region IX office, the process achieved a historic 2001 Record of Decision that set forth agreement on many basic principles, such as expanded water conservation programs, an additional 300,000 acre feet of water annually to improve ecosystem health, and policy that new water projects should be paid for by the users rather than government. The US Congress approved the agreement in 2004, and additional organizations such as the Delta Stewardship Council were set up to oversee these initiatives. Implementation during the late 2000s and early 2010s was a continuing struggle given California's fierce politics around water. Still, CALFED does represent one of the most far-reaching regional environmental plans prepared to date.

Stimulated by the rise of the environmental justice movement in the 1980s, the Smart Growth movement in the 1990s, and Occupy efforts of the early 2010s, activists have paid growing attention to questions of equity within the metropolitan region. Planners and public agencies have increasingly considered the degree to which lower-income neighborhoods and communities of color are exposed to toxic chemicals, pollution, and locally unwanted land uses (LULUs). Some organizations and political leaders have also begun working to address inequities in regional transportation planning, growing disparities in income and wealth between central cities and suburbs, and the perceived control of political institutions by the wealthy. However, equity initiatives to address all these problem areas are still in the early stages.

At the regional scale the widening gap in income, wealth, and tax base between suburbs and central cities is one of the most dramatic and rapidly growing forms of inequity. Within the suburbs themselves, a gap is widening between more affluent or rapidly growing jurisdictions and older, inner-ring communities that have a stable, working-class population and aging infrastructure. Essentially, lower-income and minority populations are penalized as businesses and upper-income residents move to more affluent or rapidly growing suburbs on the urban fringe. Central city groups in particular may no longer have access to many decent-paying jobs, and their city governments have fewer tax dollars with which to meet service and infrastructure needs. Authors such as Myron Orfield, David Rusk, and John Powell have argued that these disparities demand action.[40] One approach towards reducing such disparities, typically resisted by higher-income communities, is to ensure that each municipality zones and plans for a variety of job and housing opportunities suitable for all income groups. Another approach has been for municipalities to adopt living wage ordinances establishing a minimum wage appropriate to living costs in their region. Still another approach has been for states to partially equalize school funding across rich and poor school districts, reducing a main regional disparity of opportunity. Court decisions have often mandated such equalization; however, disparities usually continue and are made worse by the fact that parents in affluent neighborhoods are often able to contribute substantial time and money to improving the quality of their children's education both at home and in schools.

One more basic strategy to address resource inequities has been to advocate for regional tax-sharing, which helps even the tax base between rich and poor communities, and also has the benefit of reducing incentives for local governments to zone for high-tax generating (and sprawl-inducing) land uses such as regional malls and automobile dealerships. Under regional tax-sharing frameworks, local governments contribute all or some portion of sales or property tax receipts into a regional pool, which then reallocates revenues based on population or need. Currently the Minneapolis–St Paul area is the only metropolitan region in the US that does this on a large scale. Since 1972, 40 percent of new sales taxes generated by local development in the Twin Cities area has gone into a regional pool distributed by population. State legislator Myron Orfield credits this mechanism in its first 25 years with reducing disparities in the commercial and industrial tax base between rich and poor communities of substantial size from 18:1 to 5:1.[41]

Of course, affluent communities are likely to fight tax-sharing tooth-and-nail. Orfield recommends a "metropolitics" strategy in which central cities and declining older suburbs form a political coalition against wealthy new suburbs to pass regional tax-sharing in state legislatures. But, needless to say, this will not be easy. Another potentially more workable strategy would be for state or federal governments to step in and reform the tax system so that sales and perhaps property taxes are collected by higher-level governments and redistributed to localities through revenue-sharing formulas based on population. President Richard Nixon's federal revenue-sharing program begun in the early 1970s provides something of a model of how this might be done.

Regional equity has been sought in the realm of transportation planning by groups fighting to ensure that investment does not favor affluent suburban communities at the expense of central city residents. Often this debate occurs over whether funding is going preferentially to suburban-serving freeways and commuter rail extensions rather than to central city transit. In a 1993 lawsuit settlement, for example, the Bus Riders Union of Los Angeles won a promise from the Metropolitan Transportation Agency not to raise fares and to purchase 500 new clean-fuel buses. The agency had previously proposed to raise fares and cut service in part to pay for its expensive new Metrorail system. During the 2000s and early 2010s the Union continued to organize on behalf of low-income bus riders, winning the addition of more than one million annual hours of bus service, and redistribution of more than $2.5 billion in federal funds.[42] To take another example, in 1998 an environmental justice coalition in Atlanta sued the Atlanta Regional Commission over its policy of emphasizing suburban road-building, specifically exempting "grandfathered" suburban road projects that had supposedly been approved in the past from Clean Air Act restrictions on road-building. The coalition won a settlement eliminating 44 of 61 grandfathered road projects and freeing up money for transit projects.[43] Such transportation justice debates are just beginning in many metropolitan areas. (See Figure 22.2.)

Regional economic planning

A final area of regional planning concerns the field of economic development. Ironically, for decades regional planning within academia has focused on questions of economic development and economic geography, but real-life regional institutions and planning mechanisms for economic development are few and far between. Most economic development planning is handled at municipal or state levels instead. Moreover, scholarly regional economic research has not focused on sustainable development, but rather has tended to take for granted that conventional forms of economic growth are desirable and that the goal of regional planning should be to encourage these. Nevertheless, new directions in regional economic planning are possible if local governments coordinate their actions.

Important steps are likely to include encouraging greater regional self-sufficiency in agriculture and key industries, encouragement of small-scale local retailers within the region, developing regional programs for industrial recycling and sustainable materials use, coordinating investments in education so as to ensure a skilled regional workforce, and ensuring fair regional wages and hiring policies.

Regional economic planning is often a more pressing concern in developing countries than within industrialized countries. Their national governments often play a greater role in developing economic policy, channeling funding to particular regions, encouraging

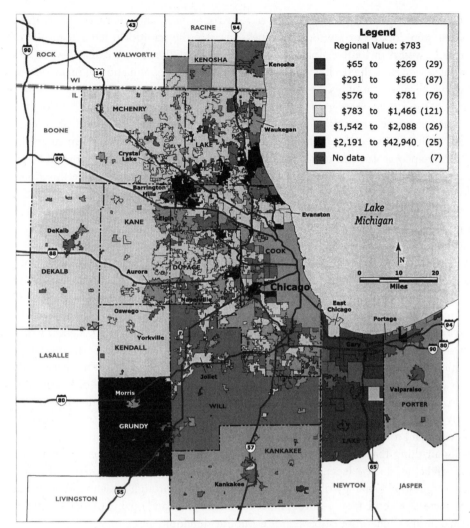

Legend

Regional Value: $783

	$65 to	$269	(29)
	$291 to	$565	(87)
	$576 to	$781	(76)
	$783 to	$1,466	(121)
	$1,542 to	$2,088	(26)
	$2,191 to	$42,940	(25)
	No data		(7)

Figure 22.2 Inequities in a metropolitan region. A map of tax capacity per household in the Chicago area shows wealthy areas concentrated to the north and impoverished areas south of the city

Source: Myron Orfield's *American Metropolitics*, published by The Brookings Institution.

businesses to locate there, and developing educational systems and infrastructure to promote regional development. In the United States, in contrast, the national government avoids specific spatial planning within economic development policy, with noted exceptions such as the establishment of the Tennessee Valley Authority in the 1930s and programs to reduce poverty in Appalachia in the 1960s.

Nevertheless, other federal decisions have had profound and largely unacknowledged regional implications. For example, decisions to expand the military budget and deregulate lending by savings-and-loans institutions during the 1980s promoted the growth of

Sunbelt states—especially southern California, Arizona, Texas, and Georgia—where these industries were located or saw opportunities. Likewise the traditional federal subsidy of water in the west contributed enormously to the boom of California, Arizona, Nevada, and other states, as well as the decline of agriculture, industry, and population in many eastern and Midwestern communities.

Many of the most controversial regional debates in recent decades, such as over logging in the Pacific Northwest, have focused on "jobs vs. the environment." As sustainable development with its "Three Es" philosophy became a more widespread movement in the 1990s, more people came to understand the falseness of this dichotomy. More and better jobs can be available in the long run through a sustainable economy. While traditional exploitative industries do suffer in the short term during the transition, opportunities emerge in new fields such as sustainable forestry, pollution abatement and control, energy efficiency, renewable resource use, recycling, public transportation, ecological manufacturing, and ecological education.[44]

Whether traditional concepts of economic growth will be part of a regional sustainability paradigm is doubtful. Urban regions that have experienced rapid economic expansion in recent decades have often paid a high price in regional side-effects such as population growth, suburban sprawl, traffic congestion, and unaffordable housing. Additional jobs may actually lower standards of living if the employers pay low wages or attract new workers to the region. A study of 100 metropolitan regions by Paul D. Gottlieb for the Brookings Institution found that employment growth leads to population growth but not necessarily to per capita income growth.[45] In times of rapid growth local governments often have difficulty keeping up with a flood of development applications, and have little time to plan proactively for smarter development. Conversely, during the "bust" phase of economic cycles cities and towns are often desperate for development of any kind, and will approve virtually anything that is proposed. A more stable middle ground is needed, in which regions and cities do not depend on continuous economic expansion for their well-being. Gottlieb argues for a strategy of "growth without growth" through which regions would seek to increase income levels—through a focus on the types and quality of jobs—without the usual business-chasing that passes for economic development.

Regions such as the Bay Area and Atlanta offer perhaps the most extreme examples of the negative effects of regional economic growth. Atlanta expanded very rapidly during the late twentieth century, but became the country's worst example of rapid suburban sprawl, surpassing even Phoenix and Los Angeles. Traffic congestion grew enormously, producing the longest average commutes in the country and causing air quality to decline, leading the US EPA to slap sanctions on the region. In the Bay Area, Silicon Valley generated extremely rapid growth in wealth and average incomes during the 1990s. However, this was accompanied by a dramatic increase in traffic jams, long-distance commuting, and housing prices. These trends were temporarily slowed by the recession that began in 2008, but heated up again in the early 2010s. The region has wound up with some of the most expensive housing in the country, and many workers are forced to commute 100 miles or more from new subdivisions in California's Central Valley. Yet growth was so deeply entrenched as a regional priority that concepts of reducing the extent to which cities zoned land for new jobs or requiring companies to also provide housing for their workers were rarely broached even among environmentalists.

Sustainability planning implies a sea change in economic development priorities. Instead of simply seeking to add jobs as fast as possible, regions will need to nurture industries that can provide quality jobs for existing residents. Instead of trying to become the next Silicon Valley, areas all over the world will need to prioritize growth in quality of life, community livability, ecological health, and human well-being. This more balanced set of goals is what sustainable regional development is all about.

At the regional scale, then, sustainability planning calls for much-improved coordination to strengthen weak institutions, and for action in areas such as metropolitan growth management, planning of large-scale transportation systems, air quality, water quality, and equity. For many of these areas, such as air quality and transportation planning, the region is the natural scale at which action can be taken. For others, such as growth management and equity, local actions are insufficient since neighboring municipalities can so easily undercut each others' initiatives. Although many debates exist over how regional planning agencies are best structured—multi-issue or single purpose, directly elected or appointed, with wide boundaries or relatively narrow—they can potentially play a powerful role in coordinating sustainable development regionally. Where new agencies cannot be developed, better coordination between existing institutions is needed instead.

23

LOCAL PLANNING

It is at the local level—usually consisting of city, town, or county government plus a wide variety of special-purpose districts—that most urban development planning is done in many countries. Local governments represent the front lines of planning in that they usually have primary control over land development, local streets and roads, bicycle and pedestrian facilities, recycling and waste collection, local parks and greenways, K-12 education, and many economic development, housing, and social welfare programs. Although higher levels of government establish the framework within which local planning takes place and may provide incentives, mandates, and funding, implementation of many important policies and programs occurs locally.

For decades local planners and government officials have thus played the lead role in approving suburban sprawl and other forms of unsustainable development. These local leaders face pressure from landowners, developers, and businesses with interests oriented around business-as-usual. Information related to more sustainable paths of development is often lacking, and short-term pressures related to the next election cycle work against a long-term planning perspective. Nimby attitudes of local residents also enforce parochial attitudes, as neighborhoods often seek to protect local property values and the status quo by keeping out affordable housing, community facilities, infill development, and even public transportation facilities. Though neighbors do at times have legitimate reasons for such protests, they frequently oppose any change even if it would benefit the city as a whole, the region, or the planet. An oppositional local politics has become a way for people to express their frustrations with modern life and their concern about larger-scale social and economic change. These frustrations are displaced onto local issues that residents feel they have more control over.

As a result of such forces there has been an enormous disconnect between local government decision-making and regional or global problems. The challenge is to put in place a different set of incentives and processes that can encourage more proactive local sustainability planning. Enlightened local officials and residents can and should be at the forefront of efforts to create more livable communities, preserve or restore elements of local ecosystems, promote green businesses, and improve social equity and human welfare.

There is already movement in this direction in many localities. In recent decades local governments have played a much more active role within international conferences, as a part of the broader growth of civil society, and have often networked with each other and with NGOs directly, bypassing national governments.[1] Cities and towns have increasingly adopted broad sustainability or climate change plans, as discussed further below. Examples of local action in some places have in turn inspired action in other cities and towns, regardless of national policy. In the more flexible, multi-level world of governance that lies ahead, there is certainly plenty of room for local officials to play a leading role in this way.

WHO PLANS AT THE LOCAL LEVEL?

As mentioned previously the planning profession originated in large part through local government efforts to implement zoning, build infrastructure, regulate housing, and ensure public health. However, the agencies doing local planning have proliferated over the years and are far more numerous and varied than is commonly realized.

Local government in the United States is made up of cities, counties, unincorporated towns or townships, and special districts created on an ad hoc basis to handle everything from schools to flood control to mosquito abatement. Cities are incorporated under state law to govern contiguous urbanized areas of land. City governments typically form when citizens in an urbanizing area (often led by local business or community groups) petition state legislatures or governance commissions for authority to govern themselves. The resulting municipalities may be of several types. "Home-rule" cities operate with their own charters that have been approved by state governments, whereas small towns and all counties operate under state law with a more limited set of powers. "Cities" are not necessarily large urban places; most suburban or rural towns are also incorporated in this fashion. A typical large metropolitan area now contains dozens or hundreds of city governments.

Over time, municipal boundaries often expand to include newly urbanizing land through a process of annexation. However, older cities may be hemmed in by other municipalities and cannot expand further. During the twentieth century most suburban communities in US metro regions incorporated as cities in part to prevent their takeover by older center city municipalities. The unfortunate result is a highly fragmented political landscape within each metropolitan area, with numerous suburban municipalities clustered around the original core, and each jurisdiction on the fringe rapidly gobbling up any remaining unincorporated county land. This fragmentation leads to competition between cities for economic development and tax base, and encourages each city or town to protect its own local interests rather than thinking regionally or globally.

Elected city councils, usually chaired by mayors, run municipal government. These councils vary in size, often from around 5 to 11 members. Development and planning decisions are typically handled by appointed planning commissions, zoning boards, design review commissions, and other city bodies, as well as by city planning staff who are permanent, paid municipal employees. Staff may in turn hire a variety of consultants to assist in planning processes, for example to examine environmental impacts of proposed development, to develop detailed designs for new public spaces, or to conduct public workshops. In many cities appointed city managers coordinate staff and run day-to-day operations of city government. All these individuals and groups have input into decisions that affect community sustainability, and can initiate actions that promote sustainability goals at a local level.

Counties are larger jurisdictions governing land use in non-incorporated, primarily rural areas, often providing services such as police and fire protection, parks, schools, and sanitation outside city boundaries. With control over rural or sparsely populated land at the urban fringe, counties play an important role in determining the character and pace of new development.

Typically county governments are more strongly pro-growth than incorporated cities, and have fewer planning and zoning mechanisms to limit or regulate development. They are therefore a key battleground in fights against suburban sprawl, for example in efforts to set up urban growth boundaries or to establish design standards for more compact and livable neighborhoods. Elected boards of supervisors, analogous to city councils, run county government, although these typically have less power since counties operate more closely under the guidance of state governments and do not have their own charters. Counties also maintain their own planning staffs and planning commissions.

Special districts are single-purpose entities set up to handle particular functions ranging from public schools to street lighting to parks across a geographic area. Special districts often overlap cities and may bear no relation at all to other jurisdictional boundaries. Large metropolitan areas may have thousands of special districts that make the governmental picture very confusing. The San Francisco Bay Area, for example, has 9 counties, 101 cities, and at least 721 special districts, not counting school districts. The New York metropolitan region covers parts of 3 states and 31 counties, and is estimated to contain more than 2,000 local governmental units.[2] This proliferation of special districts poses enormous problems of coordination, but also many opportunities, in that each can play a role in increasing urban sustainability. While local governments may be hemmed in by political or institutional limitations, autonomous or semi-autonomous special districts may be more free to take action. A park district may undertake an intensive ecosystem restoration program, a transit agency may work to improve bus service, or several cities together may establish a "joint powers authority" to purchase and protect open space.

Inside the US the landscape of local institutions varies somewhat from state to state. The state of Massachusetts, for example, has no county governments—cities or towns have jurisdiction over all the state's land. Some of these local governments are also still governed by the traditional town meeting format, one of the few forms of direct (as opposed to representative) democracy. In many western states, by contrast, cities are few and far between. Most land is controlled by counties, which are dominated by rural political interests and are often strongly pro-development. Some states have township governments that are less formal than cities and have less "home rule" power under state law. Louisiana uses the term "parishes" instead of counties. And so on.

Outside the United States, variation in the nature and extent of local government powers is even greater. In Canada provincial governments—analogous to US states—have almost total power over cities, and can dissolve municipalities or rearrange their boundaries at will. Provinces often review municipal land development decisions as well. In much of Europe and Asia, national governments play a more active role in overseeing local planning, especially for capital cities that are of great national importance. National governments in Britain, France, Denmark, and Sweden, for example, have historically made major physical planning decisions for the capital cities of London, Paris, Copenhagen, and Stockholm. The same pattern holds true for large Asian cities such as Bangkok, Tokyo, and Seoul. In China local governments have great power over the leasing of land for development, and often profit financially from such arrangements. Such an arrangement there, as elsewhere, lends itself to corruption. Indeed, the tendency of local governments officials to engage in corrupt dealings with landowners and developers is a powerful argument for regional, state, or national frameworks of land use planning.

LOCAL SUSTAINABILITY ISSUES

As with other scales of planning, some issues are particularly salient at the level of local government. The planning and regulation of land development is one of these. We have already discussed land use and urban design issues extensively in Chapters 10 and 11; suffice it to say that what sustainability planning requires is more proactive steps by local government to achieve desirable urban form values. Particular priorities include urban compactness, more highly connected road networks, more pedestrian-friendly streets and intersections, park and greenway systems, and a mix of land uses that balances jobs, housing, services, and community facilities across the urban landscape. City governments can use existing general plan and specific area plan processes to establish their visions for land development, as well as the many tools such as zoning, infrastructure provision, and tax policy mentioned earlier to bring visions into reality.

Along with control over land use goes control over the local housing stock. Housing affordability is often an important local sustainability issue that cities and counties can respond to. Unaffordable housing leads to excessive commuting by local workers, declining social welfare (for example due to overcrowding), social justice concerns (for example as low-income residents get displaced), and at times class and racial tensions. As we saw in Chapter 13, a variety of local government initiatives can help improve housing affordability and quality.

Bicycle and pedestrian planning is another area that local governments have great control over. Although funding for bike/ped improvements often comes from higher levels of government and can be obtained through grant applications, cities and counties design and maintain local roadways. Greening those and making them attractive to a wide range of human uses is one of the largest design challenges of future decades. Public transit is also generally a local or regional responsibility, and improving such systems gives residents additional incentive not to drive. The Latin American cities of Curitiba, Brazil and Bogata, Colombia are examples of local governments that planned their transit systems and land use very proactively to keep motor vehicle use low, in both cases using relatively new bus rapid transit technology (see Box 23.1).

Since local governments adapt and administer building codes, they have great influence over the sustainability of whatever structures are built in their jurisdiction. Building and zoning codes currently make impossible various forms of green development in a number of places, for example green roofs, gray-water systems, use of salvaged wood, and car-free housing. By changing local codes, municipalities can make these strategies possible.[3]

Box 23.1 CURITIBA'S TRANSPORTATION PLANNING

The Brazilian city of Curitiba, a metropolis of 1.8 million in the southern part of the country, has emerged as one of the world's leading examples of creative urban development.[4] The city's success began in the 1960s when planners first laid out a concept of growth concentrated along structural axes. Since the city could not afford a rail-based metro transit system, it opted for a low-cost but highly innovative bus network. Double-articulated buses speed along on their own rights-of-way, with feeder networks funneling passengers into the main routes. Raised "bus tubes," where passengers pay in advance, speed up boarding at each stop. The system cost $200,000 per kilometer, as opposed to estimates of $60 to $70 million per kilometer for a subway.

As planners hoped, very high-density development then grew up around the bus corridors, putting many of the city's residents within a short walk of public transit. Despite having one of Brazil's highest rates of automobile ownership, three-quarters of all commuters in Curitiba now take the bus. The city's transport system is an impressive illustration that intensive development along transit corridors, coupled with a highly efficient transit service, can dramatically reduce driving.

Beyond this, some cities are beginning to require private development to conform to LEED standards (generally only for commercial buildings above a certain size). Some such as in Austin, Texas, have also established their own green building standards that developers must conform to. Local agencies can also require that public buildings be examples of green development, and this move has helped catalyze green building in many places over the past decade.

LOCAL SUSTAINABILITY AND CLIMATE CHANGE PLANS

Since the early 1990s a growing number of cities in North America and elsewhere have explicitly oriented their general plans or other planning documents around the concept of sustainability. Other communities internationally have adopted Local Agenda 21 planning concepts or other broad sustainability planning frameworks. In many cases such efforts have represented more window-dressing than actual change. In the 2000 study mentioned previously (Chapter 6, Note 2), Berke and Conroy found that US general plans with specific language about sustainability were not significantly different from other high-quality general plans without such language.[5] Similar research in 2000 by Caroline Brown and Stefanie Dohr assessing national, regional, and local level plans in the United Kingdom found that although sustainability terminology has become widespread, these concepts haven't always made it into policy documents.[6] In another assessment of UK planning, Daniel Mittler found that although more than 80 percent of local authorities were expected to produce Local Agenda 21 plans, and this had substantially increased public involvement in many places, these efforts hadn't changed the basic power of planners to bring about more sustainable development.[7] Finally, researcher Kent E. Portnoy developed an "Index of Taking Sustainability Seriously" and has applied this to 24 large US cities.[8] He found these communities generally lacking in systematic implementation of sustainability initiatives in areas such as transportation, brownfields revitalization, and biodiversity. The top-scoring cities on Portnoy's ratings were all on or near the West Coast: Seattle, San Jose, Scottsdale, Boulder, Santa Monica, Portland, and San Francisco.

Such findings may simply indicate that sustainability planning is in its early stages, and that consensus or political backing has not yet emerged for the most meaningful changes. There may also be a tendency by mainstream planners and politicians to co-opt terms such as "sustainable development," using them to justify efforts that are essentially business-as-usual, even though some more committed jurisdictions may be embarking on dramatically new planning directions using the same concepts.

In any case, there is a large amount of variation in the nature and success of comprehensive municipal sustainability frameworks. Some very successful and progressive local governments, such as that of Portland, Oregon, used the term little until the 2000s, yet for decades have pursued planning that meets a wide range of goals that others might consider to be sustainable. Other cities such as San Francisco adopted broad sustainability platforms early on, but made little systematic progress for many years owing to political disinterest or opposition. What matters in the end is not the terminology but the results. Still, focusing directly on sustainability objectives within a general plan or other comprehensive planning document does provide a way to think through systematically a long-term approach to local planning, and to reconcile objectives of economy, environment, and equity. It may also be an opportunity to develop specific sustainability-related goals, and to use tools such as sustainability indicators to measure progress towards these goals.

Still, many local governments are now on their second or third generation sustainability plan, and are gradually making progress. Sustainability organizers have been active in London, for example, for many years. A mid-1990s effort sought to quantify materials flows through the metropolis, while then-mayor Ken Livingstone called for London to become "an exemplary sustainable world city."[9] An early 2000s London plan featured sustainability themes, although it was far from clear how the sustainability goal would be realized, and the assumption of rapid economic and population growth made this objective problematic.[10] Sustainability planning got another boost with the approach of the 2012 Olympics: 2007 and 2010 versions of a London 2012 sustainability plan called for a wide range of initiatives in the area of climate change, waste, biodiversity, inclusion, and healthy living.[11] The Olympics themselves helped illustrate green development, and other initiatives over the years have abetted the cause of sustainability, including the toll ring in 1998, charging a steep fee for motor vehicles to enter the center of the city, and ambitious bicycle planning since the early 2010s under mayor Boris Johnson.

Climate change plans are somewhat more recent than municipal sustainability plans, generally dating back only to the 2000s. They contain many similar initiatives, but focused on the goals of reducing greenhouse gas emissions and adapting to a changing climate. As mentioned in Chapter 7, these plans typically start with a package of actions to green the public sector—requiring that public buildings be LEED-certified at a certain level, that municipal vehicle fleets use alternative fuels, and that energy efficiency measures be taken in all facilities. Beyond this, local climate action plans can promote recycling, stimulate bicycle and pedestrian planning, improve public transit, cap landfills to reduce methane leakage, and change land use so as to discourage driving.

If cities do develop comprehensive sustainability or climate change planning frameworks, it seems essential to structure conditions to ensure that they have a decent chance of succeeding in the long run. A general sustainability vision must be connected in a meaningful way to specific policy and program changes, for example zoning revisions, changes in energy and recycling policy, and programs to redevelop certain areas or to ensure sufficient affordable housing. These steps must be monitored and evaluated over time, preferably by city or county staff rather than by nonprofit organizations, who may not always have time, money, or political backing to do the job. Policy changes must have political buy-in from key interest groups, politicians, and the general public. And they must be institutionalized so that implementing staff and watchdogs will exist long term to ensure that changes come about. All these conditions will come about slowly, through consistent effort by planners, politicians, and local residents.

With or without a comprehensive sustainability plan, planning at the local government scale is crucial to sustainable development, since this is the scale at which most day-to-day land use and economic development decisions are made. Efforts to balance the Three Es, to establish sustainability-oriented criteria for new development, and to monitor sustainability indicators are especially vital here. Planning for sustainability within local government is challenging because broader perspectives are easily lost at this scale and local politics may be very pro-development. However, opportunities for change exist in every local jurisdiction, even those that seem most conventional. If we can find ways to show residents how these changes will improve local quality of life as well as meet regional and global needs, much progress can be made.

24

NEIGHBORHOOD PLANNING

Neighborhoods are one of the basic building blocks of cities, modest-size physical units that make up the residential portion of the urban area and form the environment that we all inhabit every day. Planning and design at the neighborhood scale affect our daily lives, determining what services and amenities are available locally, how far we need to travel, the socioeconomic diversity around us, and much about our opportunities for interacting with our neighbors. This is also the scale at which individual development projects occur—whether they are large new subdivisions or smaller, more incremental changes to the urban fabric. Relatively small design and planning decisions such as the width and design of streets, the size of blocks, the mix of land uses, and the location and nature of parks and public spaces can have large implications in terms of urban livability and sustainability.

"Neighborhood" is a subjective term. For some, it refers only to a few blocks around the home, or even just a few neighboring houses or buildings. For others it may include a square mile or more, a large area containing hundreds of blocks and tens of thousands of residents or workers. The "hood" may also denote a particular cultural or social grouping of people living in proximity to one another, with little relation to any physical attributes of blocks or streets. All of these definitions can be useful at different times. But typically the term has been applied to an area that a resident can easily traverse on foot—that is, with dimensions of one mile or less. This area should also possess some unifying social, architectural, economic, historical, or physical characteristics so that it can be distinguished from surrounding neighborhoods.

In practice, neighborhoods often have their own historical self-definition, and neighborhood associations that have defined their own boundaries based on local tradition may exist. The original land developers may have also established defining features such as village centers, gateways, and unifying street design or architecture, and may have given neighborhoods distinctive names as a part of marketing schemes. Thus, appellations such as "Forest Hills," "Glen Ellen," and "Elmhurst" are common. A frequent joke is that newly built neighborhoods are named for the natural elements the developer has destroyed in the process of construction, for example "Emerald Glen" or "Woodland Estates."

Main transportation routes such as freeways, railroad tracks, or major arterial streets often serve as de facto neighborhood boundaries. Many neighborhoods have also traditionally been built around schools and parks, particularly by developers following the classic "Neighborhood Unit" model first proposed by Clarence Perry in the 1920s.[1] This influential model established the principal of an inwardly focused residential community centered on a school and park, with most traffic and stores relegated to large peripheral roads with shopping centers at their intersections. Partly in an attempt to insulate neighborhoods from automobile traffic many twentieth-century developers followed this model, although often without creating the interconnecting street network, parks, shopping, and community facilities that Perry envisioned.

Each neighborhood tends to be characterized by a particular type of urban form. It typically possesses a certain sort of street fabric—a particular type of grid, or else looping streets or cul-de-sacs—used by the developers who first platted the area. This street fabric determines to a large extent how pedestrian-, bicycle-, and transit-friendly the neighborhood will be, and how well the neighborhood is connected to the surrounding urban area. Neighborhoods may also have consistent street design (street widths, sidewalks, street trees, and the like), a particular type or era of housing, typical building forms and setbacks, and a particular style of park or greenway system. In addition, demographics, income levels, culture, and housing prices may be similar through the neighborhood.

The neighborhood scale is particularly important because of a widely perceived need to reinvigorate a sense of community in postindustrial society. For at least 50 years sociologists have lamented the weakening of ties between people and the growth of individualistic attitudes instead. David Riesman's 1953 classic *The Lonely Crowd* described this process within conformity-oriented postwar society.[2] More than three decades later Robert Bellah and his colleagues chronicled the often-desperate search for community within 1980s America in *Habits of the Heart*,[3] and Harvard political scientist Robert Putnam documented the long-term decline of individual ties to social organizations in his 2000 book *Bowling Alone*.[4] The reasons for the decline of community ties are varied, according to Putnam, but include the physical design of neighborhoods and cities in addition to other factors such as the growth of electronic media, economic pressures, the loosening of traditional family structure, increased mobility, and generational change.

Neighborhood design is also responsible for a large number of environmental problems, including loss of open space, destruction of wildlife habitat, and excessive resource consumption. Because they have historically been in charge of regulating land development and subdivision, planners have a great deal of control over the character and form of neighborhoods, and can potentially help bring about more sustainable types of neighborhood design. Legal and institutional mechanisms, including zoning ordinances, subdivision controls, design review standards, and the processes of development approval, are relatively well developed for action on this scale. Neighborhood-scale planning is therefore one of the more promising areas for sustainable urban development.

WHO DOES NEIGHBORHOOD PLANNING?

Historically, private developers have played the most active role in planning neighborhoods, since they actually plat and/or construct them. Within the large-scale subdivision model, these entrepreneurs obtain large chunks of land, lay out streets, subdivide property into small parcels, sometimes add parks and other amenities, and either construct homes and infrastructure themselves or sell parcels to other builders who do so. Even if buildings are not constructed immediately but are added decades later, the street and subdivision plan decided on initially determine much about the eventual character of the neighborhood. Although most large-scale developers are for-profit corporations, nonprofit builders also have opportunities to engage in neighborhood planning, usually on a more incremental scale around specific building projects.

Having developers as the primary designers of neighborhood layout, typically working within a framework of relatively weak public sector regulation at the urban fringe, has been a major problem whether in China, South Africa, or the United States. These builders are not necessarily motivated to produce sustainable communities in the long run, just buildings and lots that will sell for a substantial profit within a few years. Developers also have little incentive to examine how their own projects relate to other neighborhoods or the city as a whole. The result is a fragmented, chaotic urban landscape in which different neighborhood-scale developments—including residential subdivisions, office parks, and shopping malls—do not connect well to one another or meet broader urban or regional objectives. Streets may not meet one another, sidewalks may not be present, buildings and site design may be unattractive and unecological, and no public places may exist that are comfortable for pedestrians or children. Leftover land between large construction projects develops haphazardly and adds to the confusion. Some suburban areas may be quite dense, as in many Asian countries currently, but lack an overall neighborhood design that is livable and sustainable.[5]

Local officials, including planners and zoning boards, also exert substantial control over designs for new neighborhoods and improvements to existing ones. These authorities can approve or deny project applications and grant requested zoning variances. However, while planning staff often engage in extended negotiations with developers on project details they rarely exercise the authority they might to revise subdivision layout. The current development approval process often simply codifies conventional wisdom, creating low-density, single-family-home subdivisions with excessively wide streets, no shops or neighborhood centers, little variety of housing types, and little connection to surrounding areas. The public sector role in bringing about more sustainable neighborhood form remains weak.

To be sure, some cities are revising zoning codes and adopting design guidelines to produce more livable places, often under the influence of the New Urbanism. Changes to allow mixed land uses (homes, shops, and jobs near one another), mixed housing types, more pedestrian-friendly streets, and appropriate forms of infill development are particularly common. Some municipalities are also proactively preparing area plans for new or existing neighborhoods, ensuring for example that street networks connect well to serve bicyclists and pedestrians, and that greenway corridors are included. But such actions are still in the early stages and often lack the political backing to truly meet regional and municipal needs. A basic conflict exists between sustainability needs and capitalist land development processes,[6] so that even carefully thought-out sustainable neighborhood master plans are often weakened by dozens of amendments during the development process, as happened with the Clayton neighborhood in Surrey, British Columbia.[7] In that case, developers persuaded local authorities to substantially weaken plans for a compact, mixed-use neighborhood by allowing conventional single-family houses rather than a mixture of housing types and commercial development in a more pedestrian-friendly format.

A third main set of groups doing neighborhood planning—and often leading the charge for more sustainable communities—are advocacy organizations, community development corporations (CDCs), business groups, and neighborhood associations. These groups have greater freedom to develop a strong vision for a neighborhood than city planners hemmed in by political concerns. However, advocacy organizations must work with city planners or developers to get their visions implemented; they have little power to undertake development or fund improvements themselves. Consequently these plans usually work best if done in consultation with other actors who can provide resources and implement agreed-upon recommendations.

NEIGHBORHOOD SUSTAINABILITY ISSUES

Certain sustainability planning issues are particularly important at this scale. Achieving compact and relatively mixed-use neighborhoods is often seen as a prime goal, since these more traditionally urban land use characteristics reduce suburban sprawl and the distances people need to travel to go to shops, schools, workplaces, and recreational facilities. The mix of jobs and housing can be balanced at a neighborhood scale, as well as at city and regional scales. Integration of high-quality public transit into neighborhoods is vital. Doing this usually means zoning for higher densities around transit stops, so as to put as many residents as possible within walking distance of public transportation.

Traffic-calming is often best planned on a neighborhood scale, with a carefully thought-out package of street design features throughout the neighborhood. Preservation and restoration of creeks, wetlands, slopes, wildlife habitat, and other ecological features is a frequent concern at the neighborhood level, along with creation of local parks, community gardens, greenways, and other public spaces. Other urban design and planning initiatives at this scale can help create a "sense of place" by preserving historical structures, celebrating local culture and tradition, adding art and whimsy, and linking the neighborhood to the natural landscape.

Neighborhood design

Many urban form elements contribute to the feel and function of a neighborhood. The nature of the street fabric is one of the most basic and determines many other elements of neighborhood form. Grids, curvilinear streets, cul-de-sacs, and other street forms all produce radically different neighborhood characteristics, even with the same overall density and building form. Some street patterns connect neighborhoods to the surrounding city much better than others; some are best adapted for exclusive, insulated communities that want little to do with the outside world. The size of blocks and lots, the design of streets, the arrangement and design of parks, and the presence of shops or community facilities also greatly affect neighborhood livability and sustainability. As Jane Jacobs argued in 1961, small blocks make for a more pedestrian-friendly environment, giving those traveling by foot or bicycle far more potential routes between two points and avoiding long, monotonous streetscapes with large-scale buildings. Sidewalks and safe intersection crossings are two other neighborhood design features that greatly improve pedestrian friendliness; both are featured in the "Pedestrian Quality Index" developed by planners in Portland. Park and greenway design, discussed further later, are also key factors in neighborhood livability.

The evolution of neighborhood street fabrics has gone through a number of stages, influenced by evolving transportation technologies, urban design philosophies, construction techniques, planning regulation, and development economics. Most North American development in the early to mid-nineteenth century used a grid with relatively small, square blocks. A look at the map of most cities or towns will show this square-block grid still existing at the historic urban center, though now perhaps streets are lined by office buildings or stores instead of homes. In the late nineteenth century as streetcar lines proliferated (the first electric streetcar was introduced in Richmond in 1879), a new grid form emerged with longer, rectangular blocks. Streetcar company owners frequently bought land and platted large new neighborhoods along their routes. This "streetcar suburb" grid can also be identified on maps of most North American cities. Within these nineteenth-century grids, blocks filled in slowly over decades with buildings built individually by hand, typically using post-and-beam construction as opposed to the more rapid "balloon frame" method that came later. The gridded streets of each new neighborhood did not always link up to the existing grid perfectly but generally formed a relatively well-connected overall street network.

In the late nineteenth and early twentieth centuries a new neighborhood design philosophy arose—the "garden suburb" based on ideals of English country living. This vision of green, leafy, picturesque suburbs was propagated in the United States through the writings and drawings of Andrew Downing, a popular arbiter of taste in the 1840s and early 1850s as editor of the *Ladies' Home Journal*.[8] Ironically, picturesque design principles were initially applied in the United States to cemetery design, beginning with Mount Auburn cemetery in Cambridge in 1831.[9] These facilities pioneered naturalistic scenery and curvilinear streets to provide a semi-rural atmosphere in contrast to the gridded regularity of the nineteenth-century city. A few pioneering mid-nineteenth-century developments, such as Frederick Law Olmsted's 1869 design for the Chicago suburb of Riverside, then applied this philosophy to residential development in North America. With its curving roads, wide lots, and large setbacks, this subdivision on a commuter rail stop helped establish a new ideal of the low-density suburban environment.

But, except for a few experiments such as Riverside, the garden suburb model would wait for more widespread implementation until the advent of the automobile in the early twentieth century made such decentralized development more feasible. In his turn-of-the-century garden city concept Ebenezer Howard presented a different and more radical suburban vision—a constellation of relatively self-sufficient new towns circling a large older city, connected by rail and separated by greenbelts with agricultural land and public institutions. In Howard's vision a grid of streets, albeit molded into a circle, would still form the framework for each community, but other qualities would change, especially the regional form, the placement of employment and shopping along central avenues, and ownership patterns, which would become essentially cooperative rather than private. But with the partial exception of the British and Swedish New Towns built after the Second World War and the three American Greenbelt communities built in the 1930s, this vision of community form remained unexplored. Garden suburbs, with their winding streets, lower densities, and exclusively residential land use, triumphed over garden cities.

Modernist philosophies of neighborhood design, promoted by the Congrès International d'Architecture Moderne (CIAM) in the 1930s, advanced another alternative. This neigh-borhood model emphasized sleek, functional buildings, large public spaces, the creation of much larger block sizes or "superblocks" with internal pedestrian circulation, and (as implemented in Europe) good public transit and extensive public services. "Towers in a park" described many modernists' vision of the neighborhood. This model was mainly embraced by public authorities in socialist or social democratic countries and in North American public housing projects. But unfortunately these developers gave too little thought to how these environments would actually fit the needs of residents or natural ecosystems. The result was frequently disastrous. Bleak, sterile public spaces and buildings dwarfed residents while frequently proving just as automobile-dependent and wasteful of land and materials as suburban tracts.

Meanwhile, the private market embraced the garden suburb approach to neighborhood construction and expanded on this during the twentieth century. As automobiles prolifer-ated designers realized the advantages of residential streets with no traffic, and began to add loop roads and dead-ends to their new neighborhoods. In 1916 American developer Edward Henry Bouton praised this model of the non-connecting residential street although the French term "cul-de-sac" had yet to be widely applied to it:

> In many places where topographic or other conditions make it difficult or undesirable to extend a street to its intersection with another, such streets may be designated with "dead-ends," or returned upon themselves, forming "places," to which great charm is attached by the sense of privacy and seclusion which they impart.[10]

Developments such as Roland Park in Baltimore, built in the 1910s, added a bulbed end to short dead-end streets, producing the distinctive cul-de-sac form that would dominate later suburban neighborhoods. Cul-de-sacs proliferated steadily throughout the twentieth century, their numbers reaching a peak in 1980s and 1990s development. In Canada devel-opers took a somewhat different approach, favoring loop roads over cul-de-sacs, but the end result was the same: to create neighborhood environments completely insulated from the rest of the metropolis.

Towards the middle and end of the twentieth century a growing chorus protested both modernist and garden suburb philosophies of neighborhood design. Mumford had been a strong critic of mainstream development throughout his career,[11] but was joined by writers such as William H. Whyte, who argued for clustering suburban development to save open space,[12] and Jacobs, who described the advantages of traditional urban neigh-borhoods with highly connected street patterns.[13] The discipline of environmental design arose in the 1970s in part to look at how people actually use neighborhood environments, employing means such as post-occupancy evaluations to see how residents liked the build-ings and exterior spaces that had been created for them. The results were often shocking, and pointed to the need for more human-scaled public spaces, "eyes on the street" to reduce crime, better places for children to play, and a variety of semi-private outdoor spaces that apartment residents could personalize and have control over.[14]

In the 1990s the New Urbanism proposed a rethink of neighborhood design by returning to many features of the old streetcar suburbs—in particular highly connected street patterns, pedestrian-oriented street design, mid-block alleys, and architectural features such as front porches and garages behind the house. Referred to initially by labels such as "traditional neighborhood development," this movement began with a few pioneering projects such as Seaside, Florida (designed by the husband-and-wife team of Andres Duany and Elizabeth Plater-Zyberk in the early 1980s), and The Crossings in Mountain View, California (designed by Peter Calthorpe in the early 1990s). The movement was institutionalized through annual Congress for the New Urbanism conferences beginning in 1993, and quickly spread into the mainstream of planning thought if not into actual neighborhood development. Movements for Smart Growth, livable communities, and sustainable development dovetailed with many aspects of the New Urbanism.

The result of this re-evaluation of urban form was a much more sophisticated understanding of how neighborhoods could be better designed to meet human and environmental needs.[15] In fact, the New Urbanism does not simply repeat the streetcar suburb model of neighborhood design, but creates a range of highly connected street forms that include many small parks and public spaces. Rather than rigid grids, New Urbanist communities often employ more organic, slightly curving street forms, but the end result is the same: a highly walkable urban fabric (see Figures 24.1 to 24.4).

Figure 24.1 The beginnings of the garden suburb. British designers pioneered picturesque garden suburb design in the early 1800s. In a late 1820s perspective showing homes similar to today's "McMansions," Decimus Burton envisions villa-residences at Hove

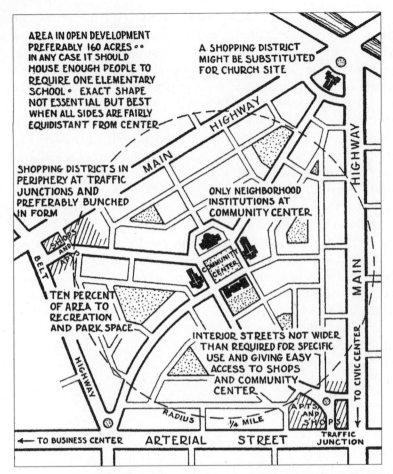

AREA IN OPEN DEVELOPMENT PREFERABLY 160 ACRES °° IN ANY CASE IT SHOULD HOUSE ENOUGH PEOPLE TO REQUIRE ONE ELEMENTARY SCHOOL ° EXACT SHAPE NOT ESSENTIAL BUT BEST WHEN ALL SIDES ARE FAIRLY EQUIDISTANT FROM CENTER

A SHOPPING DISTRICT MIGHT BE SUBSTITUTED FOR CHURCH SITE

SHOPPING DISTRICTS IN PERIPHERY AT TRAFFIC JUNCTIONS AND PREFERABLY BUNCHED IN FORM

ONLY NEIGHBORHOOD INSTITUTIONS AT COMMUNITY CENTER

TEN PERCENT OF AREA TO RECREATION AND PARK SPACE

INTERIOR STREETS NOT WIDER THAN REQUIRED FOR SPECIFIC USE AND GIVING EASY ACCESS TO SHOPS AND COMMUNITY CENTER

HIGHWAY   MAIN   HIGHWAY   MAIN

COMMUNITY CENTER   TO CIVIC CENTER

SHOPS AND APTS   BELT   HIGHWAY

RADIUS   ¼ MILE   APTS AND SHOPS

◄— TO BUSINESS CENTER   ARTERIAL   STREET   TRAFFIC JUNCTION

Figure 24.2 The neighborhood unit. Clarence Perry's ideal of the "neighborhood unit," first presented in the 1929 Regional Plan for New York and New Jersey, proposed an inwardly oriented neighborhood with traffic kept to arterial streets on the edge. In practice the development industry has taken this model to an extreme, isolating the neighborhood from the surrounding context. Parks, shops, and civic buildings are often missing, leaving neighborhoods with no social center

Density

For many North Americans, "density" is a four-letter word. They associate the term with large, impersonal apartment buildings, public housing projects, or physical environments like Manhattan. In other parts of the world as well, many wealthy individuals and ordinary citizens seek a decentralized, low-density living environment on the fringe of urban areas. Yet higher-density living environments have many advantages if they can be made safe, green, and well designed. Adding residents, jobs, and businesses to a neighborhood can improve safety; increase the viability of local businesses, cafes, and restaurants; provide sufficient ridership for transit; enhance community interaction; and save farmland and open space.

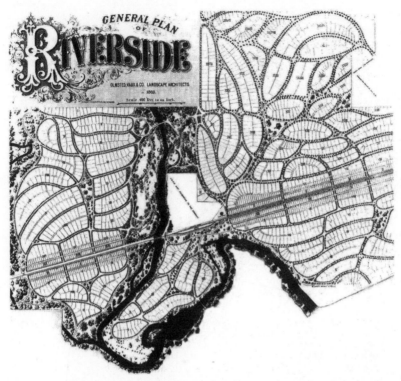

Figure 24.3 An early garden suburb. Frederick Law Olmsted's 1869 development of Riverside, outside Chicago, proved an influential model of the low-density, affluent, exclusively residential garden suburb

Rather than use the "d-word," planners and elected officials often talk instead about "compact development," "Smart Growth," or "walkable neighborhoods." Former Oakland mayor Jerry Brown used the phrase "elegant density." One approach pioneered by Rutgers professor Anton Nelessen to help reduce local objections to denser development has been to conduct a "visual preference survey" of local residents. Researchers show people images of typical low-density suburban development and other types of higher-density communities such as turn-of-the-century streetcar suburbs and well-designed urban infill projects. Most residents find they prefer the higher-density alternatives because these include more attractive streetscapes, local shops and restaurants, and a greater diversity of housing choices. Over 25 years Nelessen and his colleagues have administered this survey to approximately 50,000 people nationwide, with fairly unanimous results in all geographic regions.[16] Public workshops and design charettes are also useful tools to help citizens see that increasing neighborhood densities can be desirable. Again, when asked to choose among many housing and land use patterns, residents often select traditional town forms with higher densities and mixtures of land uses rather than typical suburban sprawl.

Recent US suburban densities have been relatively low, often 4 to 8 dwelling units per net acre before local roads and public facilities are factored in (gross densities are even lower). Average densities were lowest after the Second World War, when residential lots of

Figure 24.4 The Radburn model. At Radburn, built in suburban New Jersey in the late 1920s and early 1930s, Clarence Stein and Henry Wright created a new version of the garden suburb: the superblock with houses located on cul-de-sacs and an internal network of green spaces. Only the darkened portion was built due to the arrival of the Great Depression. Mainstream developers copied the cul-de-sac and superblock ideas without the green spaces

a quarter-acre or more were common, and have risen in recent decades as land prices have escalated, though much development is still below 8 units per acre. By contrast, densities in many older neighborhoods built around the turn of the century—including the street-car suburbs that often score best in visual preference surveys—are often 10 to 16 units per acre. Densities for apartment buildings in downtown locations can range above 200 units per acre for attractive five-story buildings that fit well along existing streets. (A five-story, 50-unit apartment building on a quarter-acre, 100 by 100 foot lot represents a density of 200 units per acre, and still can have an attractive courtyard, entry plaza, and rooftop deck.)

Traditional British suburban densities are somewhat higher than in the US. The garden suburb designs pioneered by Raymond Unwin and Barry Parker in the early twentieth century were around 12 units per net acre, while areas of London such as Bloomsbury, Regent's Park, and Bedford Park achieve densities of up to 40 units per acre (100 units per hectare) with many single-family homes and substantial amounts of private open space. Friends of the Earth in the UK proposes 28 units per acre (69 units per hectare) as a sustainable urban density;[17] some North American Smart Growth advocates call for similar intensities of land use. The minimum density usually seen as necessary to support frequent public transit service is 12 units per acre, so this average density, if combined with a highly connecting street fabric and good street design, should make for a walkable and transit-oriented neighborhood environment.

As sociologist Amos Rapoport pointed out in 1975, perceived density can be wildly different from actual density in a neighborhood, and is a function of traffic, noise, safety, greenery, reduced open space and many other factors rather than the number of dwelling units or people per acre.[18] An urban neighborhood that is green, quiet, attractive, and composed of modest-sized buildings will strike observers as far less dense than it actually is. Conversely, an environment that is dirty, noisy, and full of traffic is likely to be perceived as more urban and dense, even if it takes an extremely low-density suburban form. Perceptions of density are also highly related to culture; in many parts of the world having a large number of people in a neighborhood, such as in the fashionable *arrondissements* of Paris with their five-story buildings, is considered quite civilized and desirable. Planners and developers can make density livable—and lessen perceived density—by combining relatively intensive residential development with attractive streetscapes, public spaces, and amenities such as parks, shops, restaurants, and child care centers (see Figures 24.7).

Infill development

According to British writers David Rudlin and Nicholas Falk, "the most fundamental feature of the Sustainable Urban Neighborhood is its location—the fact that it is located within existing towns and cities."[19] Such development is the main alternative to continued suburban sprawl on greenfield sites. If done well infill can create not just attractive new buildings and housing units in existing urban areas, but entire neighborhoods that are more pedestrian-oriented, vibrant, diverse, and ecological than our present communities.

Four main types of infill opportunities exist within our cities—neglected downtowns, underutilized arterial strips, existing single-family-home districts where property owners might add additional units, and large sites formerly occupied by factories, malls, office parks, or military bases. If contaminated the latter are often known as "brownfield" sites.

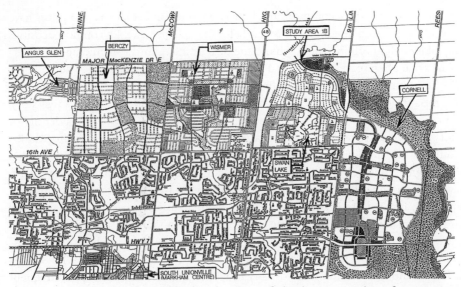

Figure 24.5 New Urbanist neighborhoods. One of the best examples of integrating New Urbanist neighborhoods into a broader citywide fabric is the Toronto suburb of Markham. The Cornell community designed by the firm of Duany Plater-Zyberk and others is at lower right. 1970s and 1980s subdivisions are at the bottom

Taken together, these areas can accommodate a great deal of development. Arguably, 100 percent of new development in many countries should be infill, since such development has a host of sustainability benefits such as producing far lower greenhouse gas emissions.[20]

Large-scale infill development opportunities are available in many downtowns that deteriorated enormously during the second half of the twentieth century. These centers of small towns as well as large cities are now characterized by vacant lots, surface parking lots, shabby older buildings in need of rehabilitation or replacement, and one-story fast-food restaurants and other low-intensity land uses. All these properties could be redeveloped to create new urban neighborhoods. Three- to five-story apartment buildings or residential towers with ground floor shops and restaurants can help make older downtowns into exciting 24-hour communities rather than depopulated wastelands after office workers go home for the day. Americans have an inordinate fear of high-rise buildings, perhaps associating them with failed public housing projects of the 1950s and 1960s, but cities such as Vancouver, British Columbia, show that high-rises can help create attractive urban neighborhoods. The key to Vancouver's use of high-rises is to keep the buildings slender and to set them back from the street, so that they don't block light and they allow the street to be a green, human-scaled environment. Two- to three-story townhouses—attached homes with private front and back yards—are also a traditional, highly livable urban housing type that could be reintroduced in many urban neighborhoods. Washington, DC, Baltimore, Boston, and many other older American cities are full of attractive townhouse neighborhoods built almost 100 years ago.

Such infill may need to be coordinated by city redevelopment agencies in many cases, since these entities have the power to assemble parcels of land, improve infrastructure, and

HOUSING TYPES AND DENSITIES

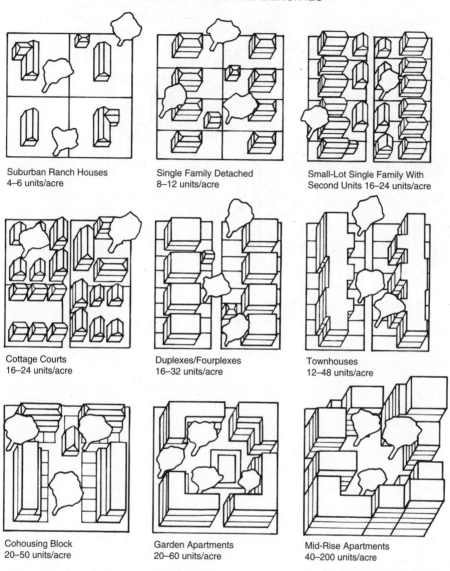

Suburban Ranch Houses
4–6 units/acre

Single Family Detached
8–12 units/acre

Small-Lot Single Family With
Second Units 16–24 units/acre

Cottage Courts
16–24 units/acre

Duplexes/Fourplexes
16–32 units/acre

Townhouses
12–48 units/acre

Cohousing Block
20–50 units/acre

Garden Apartments
20–60 units/acre

Mid-Rise Apartments
40–200 units/acre

Source: Stephen M. Wheeler

Figure 24.6 Housing types and densities

add parks and streetscape improvements to support infill. Care should be taken to respect historic buildings and existing residents. Rather than repeating the bulldozer-driven urban renewal projects of the twentieth century, a much more sophisticated, contextual, and incremental process of rebuilding urban neighborhoods is needed. Generally existing housing should not be redeveloped; instead, sites such as parking lots, failed shopping centers, and old industrial areas offer infill potential without risk of destroying historic properties or dislocating residents. Since they are such a ubiquitous feature of the North

BUILDING GUIDELINES

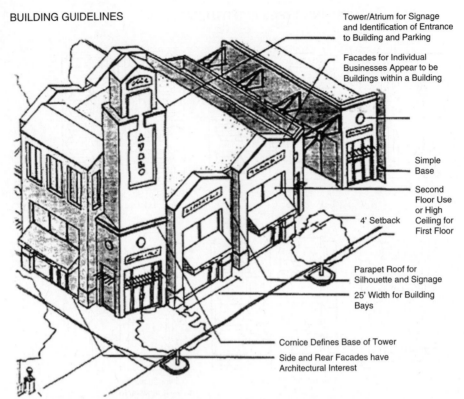

Tower/Atrium for Signage
and Identification of Entrance
to Building and Parking

Facades for Individual
Businesses Appear to be
Buildings within a Building

Simple
Base

Second
Floor Use
or High
4' Setback   Ceiling for
First Floor

Parapet Roof for
Silhouette and Signage

25' Width for Building
Bays

Cornice Defines Base of Tower

Side and Rear Facades have
Architectural Interest

Figure 24.7 Design guidelines. By adopting urban design guidelines for particular neighborhoods, cities can speed up the development review process, improve results, and
create greater certainty about what is expected

American landscape, arterial strips also represent enormous opportunities to create new
infill neighborhoods. However, steps must be taken to make these wide, often heavily traf-
ficked streets more attractive and pedestrian-oriented, for example, by adding sidewalks and
street trees, by reducing or narrowing lanes when traffic volumes are not too high, or by
creating plazas, mini-parks, or courtyards off the main streets (see Figures 24.7 to 24.10).

Infill can take place very unobtrusively in existing neighborhoods through adding
second units to existing single-family homes. In many older neighborhoods these have
already been added, legally or illegally, by converting basements, attics, or garages into
small apartments. These units can house students, elderly parents, and a variety of other
residents who do not need large amounts of space, while providing supplementary income
to homeowners. But city zoning codes often prohibit such accessory units or require
extensive procedures for use permits. Making second units legal would be a great step
towards adding additional housing and raising neighborhood densities to levels that better
support public transportation and local shops. Allowing duplexes or townhouses on vacant
lots within single-family-home districts would be a further step towards accommodating
additional housing without significantly changing the form or character of existing low-
density neighborhoods.

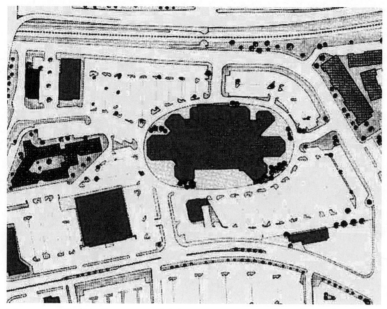

Figure 24.8 Redeveloping an old shopping center as a new neighborhood (before)

Large reuse sites—once occupied by factories, railyards, shopping malls, office parks, airports, or military bases—represent a final type of infill challenge. These locations offer the possibility of creating entire new neighborhoods from scratch. But often these brownfield sites must be cleaned up, in the worst case by removing many feet of soil and transporting it to a landfill. Local governments may need to work with property owners to

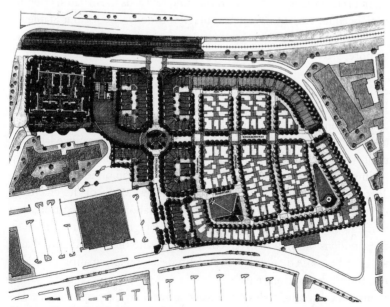

Figure 24.9 Redeveloping an old shopping center as a new neighborhood (after)

Figure 24.10 The Crossings. Figures 24.8 to 24.10 show The Crossings in Mountain View, California, which resulted from the redevelopment of a defunct 1960s shopping center with 359 homes on 18 acres next to a new CalTrain station. The new neighborhood includes townhouses, apartments, cottages, and single-family detached homes. The city has helped make such transit-oriented infill development happen by preparing "precise plans" for the sites

develop an area plan governing redevelopment of these sites, and political obstacles may need to be overcome, but the rewards can be enormous. Old railyards in Portland, Oregon, and San Francisco have become vibrant new urban neighborhoods. The former Stapleton Airport in Denver has been redeveloped as a New Urbanist infill community. And former port facilities in Baltimore, New York, and London have become significant new additions to those cities.

Infill development is often vigorously resisted by residents and local politicians, and Nimby opponents have killed many promising projects. This resistance may be due to fears that property values will decline (many studies have shown that they do not), desires not to have less affluent residents or people of color living nearby, or a generalized fear of change. Yet infill can provide enormous advantages for existing residents of a community, for example by providing new restaurants, cafes, parks, transportation options, and public spaces. It can also increase rather than diminish property values. Much of the challenge will be to win over opponents through communication, collaboration, education, and specific responses to their concerns. Cities can help overcome neighborhood resistance by conducting a neighborhood visioning process in conjunction with the preparation of area plans. This task is not easy, but skilled facilitators can help develop participation processes that are constructive rather than oppositional.

As discussed previously, local zoning codes often work against new development in existing urban areas. It is literally impossible in most places to recreate the thriving downtowns of a century ago because they would be illegal under current zoning. These codes often limit building heights, prevent mixed-use buildings, require setbacks from the street, and mandate large amounts of parking. The solution is for local governments to review zoning codes line by line, and make sure that they allow desirable forms of infill. For example, a town might raise its downtown height limit to five stories to accommodate housing above stores, while establishing a minimum height of at least three stories. Municipalities might also require buildings along main streets to establish a solid street frontage, and to include retail or restaurant spaces along the sidewalk (see Box 24.1).[21]

Compact, mixed-use development

Before the advent of single-use zoning for broad areas of cities, neighborhoods typically contained offices, small shops, grocery stores, and restaurants in addition to housing. Zoning for most post-Second World War subdivisions prohibits these non-residential land uses. One of the main tenets of the New Urbanism and sustainability-oriented design in general is to include this variety of land uses within communities once again, typically within neighborhood centers or along main streets. If jobs, housing, shops, and recreational facilities are closer together, the theory goes, then people will need to drive less and neighborhoods will be more vibrant and livable.[22]

Although mixed-use development does provide these benefits, it has proven somewhat difficult to bring about in practice. In most actual New Urbanist developments, such as Seaside, Florida, Kentlands, in the Maryland suburbs of Washington, DC, and Celebration,

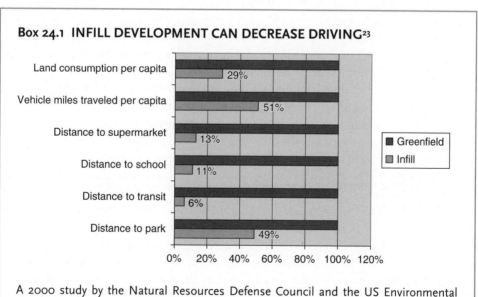

Box 24.1 INFILL DEVELOPMENT CAN DECREASE DRIVING[23]

A 2000 study by the Natural Resources Defense Council and the US Environmental Protection Agency of an infill subdivision in Sacramento vs. a greenfield counterpart found that the infill neighborhood substantially reduced driving and travel distances.[24]

in Denver, the "mix" includes a few workplaces, home offices, and only a very small number of retail storefronts in a village center. Developers have had to subsidize many of these shops or allow the storefronts to remain empty, since particularly in the early stages of a new development there is not sufficient population density to support them. Local shops also face competition from a wide range of big-box retailers, chain stores, and shopping centers in the surrounding area.

Although a number of good mixed-use developments have been built in North American cities, this more integrated mix of land uses remains more an ideal than a reality, especially in suburban locations. Token amounts of retail within large residential projects, New Urbanist or otherwise, on the edges of metropolitan regions are unlikely to change existing patterns of motor vehicle use or long-distance commuting. Much more fundamental changes in land use are needed instead.

What sustainability planning might envision is a more radical mix of land uses throughout urban and suburban neighborhoods, coupled with strong restrictions on single-use and large-scale developments, and requirements for 100 percent infill development. Real neighborhoods need to include grocery stores, hardware stores, drug stores, cleaners, child care centers, places of worship, medical or dental offices, fitness centers, and much more—all within close proximity. Office buildings and much light industry should also be located near where people live. This, after all, is the model of the traditional town before the age of the automobile.

To help truly mixed-use neighborhoods come about, cities will need to end current zoning for large expanses of homogeneous land uses such as office parks, shopping centers, and residential tracts. Cities may well also need to amend their zoning codes to prohibit stores of greater than 30,000–50,000 square feet, since these big-box stores tend to kill smaller, neighborhood-oriented retail businesses and cause a great deal of long-distance travel. (Certain types of big-box stores are in fact known as "category killers," in that they drive every other merchant in the same category out of business.) Exact size limits will depend on the type of retail; grocery stores may need the higher amount to offer a wide range of products; hardware stores and most other types of retail business do not. Once an environment is created that does not actively undercut neighborhood-scale, mixed-use development, then more locally oriented businesses can emerge.

Placing a range of jobs near residential neighborhoods is essential. In an age in which most forms of economic activity take place in office buildings or other nonpolluting forms of workplace, there is no reason for businesses to be widely separated from homes. Instead, offices can be located along arterial streets and transit lines, in transit villages at commuter rail stops, and within city, town, and neighborhood centers (where they can be joined by new housing). Planners and community development groups should encourage types of workplaces that are appropriate to the income level of residents in a given neighborhood. Jobs and housing should be balanced, in other words, not just regionally or at a county scale (as progressive planners often seek to do currently), but within each city, town, and neighborhood. The result will be greatly reduced pressures for long-distance commuting and improved quality of life for residents who can walk or bike to work.

A growing number of new neighborhoods illustrate principles of relatively compact, mixed-use design. In the United States, new communities such as Orenco Station outside

Portland, Oregon, and Playa Vista in Los Angeles (though controversial for environmental reasons) provide such examples. In Britain, Poundbury, a 400-acre extension added to Dorchester, is a good illustration.[25] This new neighborhood, designed by Leon Krier under the aegis of the Prince's Foundation, is almost two-thirds mixed-use buildings, and provides 20 percent affordable housing along with 21 commercial and 7 retail businesses. To be inhabited by 5,000 people, Poundbury replicates local architectural styles and uses local and recycled building materials.

Streetscape design

One of the biggest challenges at a neighborhood level is making arterial streets more pedestrian-friendly and livable. These streets are often congested with traffic and lined by strip businesses—fast-food joints, gas stations, auto repair shops, and other one-story, drive-in businesses. Writers such as J.B. Jackson have suggested that they are an essential and underappreciated part of the American cultural landscape, and perhaps in some ways they are. Yet to many of us these arterial strips are unpleasant and stressful places to be whether on foot or in a car.

These arterial corridors within almost any city or town offer extensive opportunities for infill development. Luckily, there are well-established traditions of large streets in many countries that both carry substantial volumes of vehicle traffic and are green, pedestrian-friendly places to be. In particular the "multiple roadway boulevard" model places fast-moving vehicles on center lanes and separates this traffic from slow-moving side streets using landscaped medians.[26] Designers such as Frederick Law Olmsted originally developed this strategy in the nineteenth century. Examples are provided by the Champs Élysées in Paris, the Paseo in Barcelona, Atlantic Parkway in Queens (designed by Olmsted), and the Esplanade in Chico, California. Similar designs can be used today to retrofit some of today's least attractive arterials.

Converting existing arterial streets into pedestrian-oriented boulevards will take proactive planning by local government. Planners can rezone land along these streets for denser, mixed-use development, redesign streets to include wider sidewalks and landscaped medians, and, where possible, create fully fledged boulevards with separate fast and slow travel lanes. The traffic volume on these routes can be reduced by taking steps to move jobs and housing closer together, improve public transit, and provide economic incentives for people to use other modes of transportation. Gradually such actions can help humanize one of the most unpleasant and problematic elements of many neighborhoods currently (see Figures 24.11 and 24.12).

Traffic-calming

As automobiles multiplied rapidly in industrialized countries in the early to mid-twentieth century, many observers realized that they were degrading neighborhood quality. Proliferating motor vehicles were a central element of the inhuman "technopolis" that Mumford warned against. Jane Jacobs believed that too much traffic leads to an "erosion of cities" and that there should be an "attrition of automobiles" through widening sidewalks, narrowing streets, bottlenecking traffic lanes, and other steps that would make driving

Figures 24.11 and 24.12 Redesigning a suburban arterial as a walkable boulevard (before and after). This visual simulation shows how a typical suburban arterial in Pleasant Hill, California might be transformed into a pedestrian-oriented street.

Source: produced by Steve Price/Urban Advantage (http://www.urban-advantage.com/)

less convenient.[27] Later in the 1960s establishing "streets for people" was part of the vision of humanist Bernard Rudofsky, who provided an erudite, illustrated history of American streets and examples worldwide of how streets have historically served a glorious profusion of public uses other than conveying motor vehicles.[28] This line of thinking has expanded in more recent decades through the work of scholars such as Donald Appleyard, whose book *Livable Streets* was a milestone of research into neighborhood traffic management,[29] and activists such as Australian David Engwicht, who successfully rallied opposition to a local expressway by chalking out its route through a neighborhood park.[30]

The modern traffic-calming movement began in Europe in the 1960s in response to accidents in which children had been killed or injured by speeding drivers. Cities in Germany and the Netherlands adopted particularly active programs to reduce vehicle speeds; the English term "traffic-calming" is in fact a translation of the German *Verkehrsberuhigung*. In cities such as Cologne citizens began organizing in the early 1970s to close or calm residential streets where speeding drivers had endangered children. In the Netherlands, advocates developed *woonerf* ("living yard") designs through which local streets were made into pedestrian-priority areas. *Woonerven* generally include measures such as trees or planters within the road area, lack of differentiation between pedestrian and automobile space, and special paving treatments to designate certain street areas as part of the living area associated with houses. The traffic-calming concept has spread steadily in North America since the late 1970s.

Traffic-calming mechanisms fall into two main categories: those that seek primarily to reduce vehicle speeds, and those that focus on lowering traffic volumes. Speed humps—relatively broad "vertical deflection devices" six feet or more across—are perhaps the most common form of traffic-calming and have appeared in countless cities throughout the world. They are very effective at reducing vehicle speeds to around 15 miles per hour without placing any restrictions on traffic volume, and have the additional advantage of being relatively cheap. Speed humps have been probably the most widespread method of traffic-calming in the United States. In contrast, sharper speed bumps are used extensively on private access roads such as around shopping mall parking lots. Cities in Mexico and other parts of the world use them on public streets as well, where they are very effective at reducing speeds to around 5 mph, but are seen as a nuisance by many drivers. Both speed humps and bumps are viewed as a health hazard by some disabled activists who dislike being jarred by paratransit vans traveling over them.

Other devices to reduce traffic speeds include speed tables (raised sections of pavement much broader than speed humps), chicanes (offset planters or sidewalk extensions forcing vehicles to travel a zig-zag path down local streets), bulb-outs (curb extensions into the street at intersections to facilitate pedestrian crossing and slow cars entering the street), chokers (parallel intrusions into the street mid-block), traffic circles at neighborhood intersections, and various forms of colored and/or textured pavement (which can remind drivers they are in a pedestrian priority zone).

The other main approach to traffic-calming—focusing on traffic volume rather than speed—uses means such as diverters (preventing traffic from entering streets), semidiverters, diagonal diverters, and other forms of street closure. Berkeley was one of the US pioneers with such devices in the late 1970s, using rows of concrete barriers called bollards to divert traffic from neighborhood streets south and west of the University of California campus. This method greatly improved quality of life on the calmed streets. But as Berkeley and other cities have discovered, these devices have the disadvantage of diverting traffic onto other streets, which may then experience even worse problems. In effect they make a gridded urban neighborhood into a system of cul-de-sacs, with through-traffic concentrated on arterial streets. Such cities now approach diverters with great caution.

The current wave of traffic-calming devices in North America appears to encompass circles, humps, and bulb-outs, which slow traffic without such diversion. Circles in particular are used extensively by Portland, Seattle, and Vancouver. Planting trees along streets, adding mid-block chokers, and narrowing street or lane widths can also help reduce vehicle speeds.

Many European cities have gone much further by creating pedestrian-only districts in central cities. Almost every major city on the continent has such an area where motor vehicles are only allowed in the early morning for deliveries, if at all. Frequently these districts are the most dynamic, interesting, and economically vibrant parts of the city. Copenhagen's Stroget, a 2-mile-long pedestrian shopping street closed to traffic in the 1960s, was the first major example. There have been several waves of efforts to create pedestrian streets or malls in the United States since the 1960s, but these have not worked well in situations in which a critical mass of residents or tourists did not exist to support local shops. On the other hand, examples such as Santa Monica's Third Street Promenade have been successful (in its post-1990 incarnation). Closure of certain streets in urban parks such as Rock Creek Park in Washington, DC, and Golden Gate Park in San Francisco, especially on weekends, is also on the increase.

As far as motorists are concerned, the ideal (given that unlimited high-speed travel is not possible) may be residential streets that have relatively slow speeds (20–30 mph) but smooth and continuous travel. This approach minimizes the aggravation of stopping and starting, while creating a street environment that is relatively safe and pleasant for other users. Practical implications of this approach include minimizing use of stop signs in residential neighborhoods (relying on traffic circles and other methods to keep traffic slow and intersections safe) and synchronizing lights on arterial streets to serve traffic moving at relatively slow speeds. This approach also has significant benefits in terms of fuel economy, pollution, vehicle wear, and noise (see Figures 24.13 to 24.19).

Parks, gardens, and open space

Since a better connection between human and natural environments is a central challenge of sustainable development, neighborhood planning should seek to create a variety of open spaces and natural areas. Historically many neighborhoods were built with little regard for natural or public space. Schoolyards, with their asphalt playgrounds and standardized athletic fields, form the most common type of open space in many neighborhoods. However, school facilities do not serve the recreational needs of many residents and with their bare expanses of hardtop or grassy playing fields hardly represent a local connection with nature. New forms of neighborhood open space are essential—neighborhood parks,

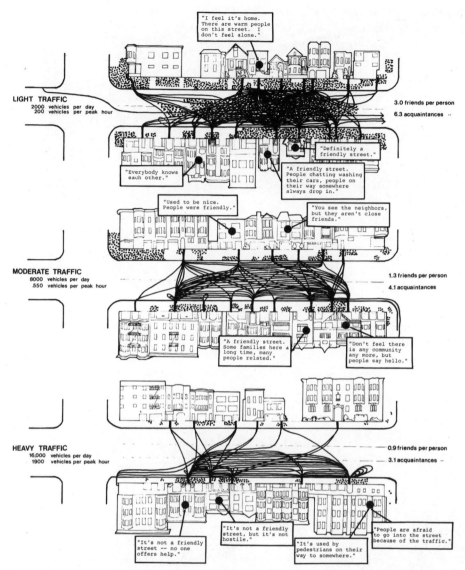

Figure 24.13 Traffic volume vs. neighboring. In this classic study by Donald Appleyard of three streets in San Francisco, the amount that people knew their neighbors varied in inverse proportion to traffic volume

trails, gardens, greenways, and outdoor recreational facilities—as urban areas become more intensively developed.

Whereas nineteenth-century park designers often sought to create picturesque urban parks as picnic grounds for working-class families, today's urban green spaces need to serve a far greater range of functions. Different types of space are needed for gardeners, young children, older children, teens, elderly residents, team sports players, runners, bicyclists, hikers, and many other groups. The traditional model of the one-square-block neighborhood park doesn't meet the needs of many of these groups. As previously discussed,

Figure 24.14 Traffic-calming devices: a speed table. Speed tables, as here on a residential street in Utrecht, create a broad, raised area of paving

Figure 24.15 Traffic-calming devices: chicanes. Chicanes, as here on a street in Basel, alternate parking and landscaping from one side of a street to another to slow traffic

Figure 24.16 Traffic-calming devices: bulb-outs. Bulb-outs, as here in Vancouver, push out the curb at an intersection to slow traffic

Figure 24.17 Traffic-calming devices: traffic circles. Traffic circles, as here in Arcata, California, force traffic to move more slowly at intersections. Neighborhood traffic circles are smaller than large roundabouts used on arterial and collector streets and focus on slowing rather than distributing traffic

Figure 24.18 A pedestrian district. Most European cities and towns have pedestrianized their downtowns, as here in Heidelberg. Deliveries are often allowed in the early morning. These pedestrian areas succeed because of their relatively high density, mix of land uses, and excellent access by public transit

Figure 24.19 A pedestrian street. Third Street Promenade in Santa Monica, California offers an example of a US city creating a new public space by closing a street to motor vehicle traffic

neighborhood open spaces should also be thought of as part of a regional green infrastructure network, providing habitat and wildlife corridors throughout the urban region. Restored creek corridors, wetlands, shorelines, and riverbanks are particularly important in this regard, and can also provide long, interconnected jogging or biking trails for active recreational needs.

Urban gardening is on the increase in many urban areas, although gardeners have often had to fight for scraps of land to serve as community gardens. New York City urban gardeners had to struggle especially hard in the 1990s against Mayor Rudolph Guiliani, who wanted to sell off city parcels that were being used as community gardens for development. Luckily, most of these areas were saved through a public campaign featuring celebrities such as Bette Midler. Rather than fitting community gardens in haphazardly on leftover scraps of land, municipalities should look on their gardens as an essential strategy for making dense neighborhoods livable. Planners can identify surplus city land or acquire strategically located parcels for this purpose or can require developers to add garden space within large development projects.

With an increasingly elderly population in North America, Europe, and Japan as baby boomers age, cities also need to focus on providing neighborhood open spaces for seniors. The elderly (especially those who can no longer drive) benefit from attractive public spaces that are within walking distance of homes and that are accessible to individuals with wheelchairs or other disabilities. Vast playing fields are not particularly useful for this population compared with smaller, attractively landscaped parks or "healing gardens" with multiple walkways, benches, and a variety of plantings. Community centers with programmed daily activities are also desirable for senior citizens.

For children, the standard playground with its swings, slide, and climbing equipment offers a very limited range of outdoor activities. Less structured play areas, including fields, woods, wetland areas, and areas with movable objects that can be actively rearranged, are also valuable. Children often benefit from having some "magical spaces" or "areas of mystery" that are not neatly created for them by adults. Vacant lots, patches of woods, and areas of overgrown landscaping have often served this purpose historically. Of course such areas must be safe. Ensuring their safety can be accomplished both through site design and by having an entire city or town that is safer. Means to achieve this include building design that puts "eyes on the street," relatively compact and walkable neighborhoods that put more people on the street, and, at a much broader level, steps to increase equity, reduce disparities between the rich and poor, and provide decent services, education, and employment for the less well-off.

Creating a range of attractive open spaces needs to become a much more integral part of neighborhood planning. For municipalities this may mean adding or strengthening open space requirements within zoning codes and subdivision regulations, developing specific design guidelines for such spaces, and actively acquiring land or easements for open space. State and federal governments can assist them through grant funding and modifications to subdivision law requiring additional open space. In many cases city governments will also need to actively schedule and manage events in these public spaces, as is already done by cities such as Portland in its Pioneer Courthouse Square and Tom McCall Waterfront Park. It is not enough simply to create great public spaces—a range of scheduled activities in them ranging from farmers' markets to concerts and cultural fairs can ensure that they are extensively used.

Ecological restoration

As discussed in Chapter 9, ecological restoration can add greatly to neighborhood livability and sustainability. Restored ecological features can improve habitat for wildlife, create recreational amenities, and help bring urban residents closer to the natural landscape. Natural areas within a neighborhood are particularly important to kids who may explore them in far more intimate detail than adults, learning a great deal about the non-human world by doing so.

Virtually every neighborhood of any size has a creek, waterway, or wetland associated with it, although some of these may have been destroyed or covered up by development. Traditionally developers or other landowners have channelized, culverted, straightened, or otherwise extensively altered creeks. Riparian vegetation (species adapted to growing along waterways) has been completely or partially removed. Invasive species have frequently taken over from native plants that are usually best suited to provide habitat for birds, amphibians, reptiles, and mammals. The water flow itself has often been radically altered by large amounts of pavement in the neighborhood or upstream, typically creating strong, "flashy" runoff after storms that can quickly erode creek banks and cause floods.

Impermeable paving in the neighborhood—roads, parking lots, driveways, school grounds, and even some backyards—also allows less rainfall to enter the groundwater table, lessening creek flows and water for vegetation during dry periods. Creek restoration offers a prime opportunity to remedy these problems at a neighborhood scale. It can help bring neighborhoods together and create valuable scenic and recreational amenities for nearby residents and property owners. Relatively easy creek restoration activities include removing trash and debris from stream channels, improving signs and trails, and digging out invasive, non-native plants and replanting native species.

More difficult or long-range activities include removing concrete rip-rap, stabilizing banks with log lattices and willow plantings instead, and unearthing and reconstructing culverted creek channels. The banks and floodplains of larger rivers can be similarly restored, in some cases by removing existing development that was inappropriately allowed to encroach on the waterway and its floodplain. Beginning in the 1980s a citizen movement in North America has sought to carry out creek restoration. Early Californian examples include Wildcat Creek in Richmond, Strawberry Creek in Berkeley, the Guadalupe River in San Jose, and St Luis Creek in the center of downtown San Luis Obispo.[31] In the eastern United States, community groups such as the Anacostia Watershed Council in Washington, DC, have taken on responsibility for cleaning up and restoring their local waterways, in this case Washington's "other river," the Anacostia.

Habitat restoration activities—often in association with creek restoration—gained momentum at about the same time. Within such programs, a particular strategy is worked out to restore or create wildlife habitat on a given site, taking into account the history of the land, the local microclimate, soils, and hydrology, and any problems resulting from urban use or contamination. Invasive plant species are typically removed and replaced with natives, and native fish or animal species are restocked. Attention must be paid to creating both "edge" and "interior" habitat spaces, since different species may inhabit each area. The size of the patch of restored habitat determines much about what species can be introduced and how, and corridors linking to other habitat patches in the vicinity must be considered. The ideal is to create a series of habitat patches—including some backyards—throughout urban neighborhoods, and to link many of these into a broader mosaic using greenway corridors, often along creeks or rivers. (See Figures 24.20 to 24.22 and Box 24.2).

Improving neighborhood equity

Equity concerns are present at the neighborhood level, as at others. Potential equity-related questions include: neighborhood accessible to people of all ages, races, and economic groups, both as residents and visitors? Does it contain a variety of affordable housing options and transportation options for different groups? Does it expose certain groups of residents to toxic contamination problems in soil or groundwater, or to air pollution? Does it saddle some groups with the externalities of other urban activity, such as the noise and pollution of freeways or contamination from oil refineries? Do concentrations of poverty exist because of the reluctance of other neighborhoods to plan for or accept affordable housing and social services?

Historically, the neighborhood scale is one at which segregation has been most vividly expressed in the United States, especially through exclusion of African Americans. Sustainability at the neighborhood level implies making every neighborhood accessible

Figure 24.20 Creek restoration. Volunteers plant seedlings of native shrubs along a recreated stream channel for Sausal Creek in Oakland

Figure 24.21 A constructed wetland sewage treatment marsh. Sewage from Arcata, California, receives primary treatment in the structures at lower right center, then secondary treatment in oxidation ponds at lower right, and tertiary treatment in constructed wetlands at upper center, which double as a public park

Figure 24.22 Tertiary treatment ponds in the Arcata marsh

**Box 24.2 ARCATA, CALIFORNIA, ECOLOGICAL WASTEWATER
TREATMENT MARSH**

For nearly 30 years the city of Arcata, California, has processed its sewage through
a series of constructed wetlands that double as a wildlife sanctuary, educational
facility, and popular recreational site. The idea for the project first came about in the
1970s, when the city was faced with the prospect of spending millions of dollars to
help construct a traditional wastewater treatment plant. The proposed plant was
energy-intensive, required much new infrastructure, and raised problems related to
ocean disposal of effluent. Instead, citizens worked with the city to explore environmentally sound alternatives. The current system was approved by the state in 1983
and began operation in 1986.

Wastes first pass through a settling tank from which solids are removed and
composted on site. Next the effluent passes through oxidation ponds heavily
planted with bulrushes, where microbes help break down wastes. A final stage of
treatment consists of larger wetlands in the Wildlife Sanctuary, which are accessible
to the public and ringed by pedestrian trails. The treated water eventually passes
into Humboldt Bay.

The 100-acre system of freshwater and saltwater marshes now attracts more
than 150,000 visitors annually, and is one of the best birding sites on the West
Coast, with more than 200 species sighted. The project has also proven reliable and
extremely cost-effective.

to all, particularly less affluent individuals and families and persons of color. The objective of sustainable development is not green enclaves in upper-middle-class areas, but well-
rounded neighborhoods that are diverse and equitable as well as ecological and livable.

Taking steps to ensure a range of housing types and sizes and a sizable number of
affordable housing units in each neighborhood is essential. Some cities seek the latter
through inclusionary zoning requirements. But this housing does not reach the very poor.
Cities need to target very low-income households earning 50 percent or less of area median
incomes as well, typically by providing rental housing or "supportive housing" which
offers services such as employment counseling, substance abuse programs, and childcare
for working parents. Whereas currently local governments often only require low-income
housing to remain affordable for 20 or 30 years, the affordability of these units should be
locked in permanently through deed restrictions and binding language in city permits. The
objective of a diverse range of housing types should also be sought through revisions
to zoning codes requiring a mix of unit sizes in all new developments. Other strategies
historically have included limited-equity cooperative housing and public housing (cities
building this now tend to opt for scattered-site housing that is identical in appearance to
market-rate housing, rather than the large-scale public housing projects of the past). Strict
state, federal, and local enforcement of regulations prohibiting discriminatory lending and
real estate practices is also essential.

Along with affordable housing, neighborhoods should feature a range of jobs and services available to all residents, the less affluent and less educated as well as the wealthy and well trained. Municipal governments can seek to ensure this through economic development strategies emphasizing a diverse employment base, and through requirements that businesses give preference to applicants living locally, and through living wage ordinances.

Beyond these steps, equity can be improved through efforts to ensure that decent education, recreational facilities, and other public services are available to all neighborhood residents. Neighborhoods that have suffered historically from entrenched poverty and social problems will need special attention not just to improve the housing stock and physical form of the neighborhood, but to address the historical factors that have led to its marginalization, including poor education, economic isolation within the region, and a lack of protections for renters and families. Cities may also want to map and identify particular hazardous materials threats, mitigating these where necessary to ensure that exposure to toxic pollution does not affect some segments of the population more than others.

Addressing equity at any scale is a difficult challenge, but it is especially so in neighborhoods where residents these days are so likely to oppose any change at all, let alone initiatives that might introduce poor people or other groups into an existing residential environment. There is no easy way to get around this problem. Nevertheless, a number of incremental changes, including mandates for fair lending and real estate practices, and a great deal of public education and political leadership can help more inclusive, equitable neighborhoods come about.

If done through inclusive and participatory processes, preparation of neighborhood or community plans can be an important mechanism to enhance local democracy and equitable decision-making. It can give residents a greater degree of control over their everyday environments, and provide them with an opportunity to get to know one another and to work together around common concerns. It can even be a path towards fuller involvement in citywide politics—many neighborhood activists have traditionally gone on to seek elected roles within city government. However, public participation at the neighborhood scale is often problematic. Although local opposition to heavy-handed planning schemes is needed at times, too frequently public involvement becomes channeled in negative directions, essentially opposing any change, even if it is to include affordable housing, calm traffic, add community facilities, or improve public transit.

The challenge is to promote decision-making that involves every constituency in the community, that is constructive, proactive, and broad-minded, and that keeps regional and global needs in mind. There is no magic way to do this, but early public involvement, full notification of interested groups, careful advance work with community leaders, and context-sensitive proposals and project designs can help.

Economic issues

Neighborhood-scale economic development planning frequently dovetails with equity goals. The question is how to ensure an appropriate mix of jobs and services for local residents, preferably building on local culture, skills, and resources. This goal is often best met through a diverse range of locally oriented businesses—a situation that is radically undercut by the large-scale manufacturing and retail economy that is ascendant in most industrial countries. Limiting the invasion of big-box stores and chain retailers is not easy. As previously mentioned, setting maximum building footprint limits within zoning codes can help. Zoning to prohibit particular types of businesses has a long history as well, and can be used to restrict fast-food outlets, drive-through businesses, or other types of unwanted enterprises. Zoning restrictions or urban design standards prohibiting large parking lots or one-story buildings—frequently desired by strip-type developers—can also be useful.

Meanwhile, a variety of proactive economic development strategies can help a more local economy emerge. Business incubators are one frequently used technique. Basically, a municipality or a local nonprofit corporation working with local officials sets up a building in which a number of entrepreneurs can rent cheap office or manufacturing space and take advantage of shared office equipment and support services. The city's economic development staff may offer these newly started businesses technical support and training in terms of analyzing markets, developing business plans, accessing capital, or establishing accounting procedures. Micro-enterprise loan funds are another technique used to support new businesses. Essentially these funds bring together a number of local entrepreneurs who take turns borrowing small sums for business start-up and support one another in these efforts. The Grameen Bank in Sri Lanka is the best-known example in a developing world context.

The globalization of manufacturing—and the consequent flight of local manufacturing jobs from communities within industrialized countries to low-wage labor locations in the developing world—is a fundamental force undercutting local employment. Although local economic development activities can help nurture other forms of business less likely to be affected by globalization, attention to larger-scale policy is necessary as well. Action at national and international scales to moderate or reverse global free trade policies is likely to be essential in the long run to address this problem.

Another economic issue relevant to neighborhood development concerns the scale of local development. In the nineteenth and early twentieth centuries most homes were built individually or in groups of two or three by local builders. During the second half of the twentieth century that situation changed, so that increasingly large-scale "production homebuilders" became responsible for creating vast tracts of housing at once. This scale of development fuels suburban sprawl—such builders vastly prefer greenfield land at the urban fringe where it is possible to build large numbers of units at once—and works against the emergence of a more stable local construction industry in the long term. The challenge here may be to promote a smaller-scale economy of local development, in which incentives favor small builders who hire locally and can take the local context into account when developing projects.

Public health

Many public health issues are particularly pressing at the neighborhood level. One of the foremost, increasingly recognized these days, is the degree to which the neighborhood supports walking and other outdoor activity. Such activity can help reduce obesity, improve cardiovascular health, promote community interaction, and enhance a sense of personal connection to the local landscape and place. For elderly people particularly, a daily walk is one of the best ways to maintain health and well-being. Many specific steps to improve neighborhood walkability have been mentioned previously. Particularly vital are the development of a highly connected street fabric; good sidewalks and pedestrian-friendly intersections; the addition of local destinations such as stores, parks, and community centers; and traffic-calming to make streets safer and more pleasant for pedestrians.

Provision of parks and open space at the neighborhood scale is important to public health for similar reasons. In many places this will require municipal governments instituting or strengthening requirements that developers include parks and greenways within new neighborhoods. Municipal investment may also be required to add parks to existing neighborhoods. Making small grants available to neighborhood-based groups for ecological restoration projects or the creation of new mini-parks can be a particularly effective way for cities to improve parks and public spaces within neighborhoods.

Reducing toxic hazards and moving towards use of safer products and techniques for local businesses, landscaping, lawns, gardens, and other neighborhood spaces are further public-health-related neighborhood goals. A variety of persistent organic pollutants and other toxins may be present in neighborhoods from past dumping practices, local gas stations with leaking underground storage tanks, past businesses, and utility transformers and facilities. The use of pesticides, herbicides, and fungicides on residential landscaping, road medians, or nearby farms is another potential threat. A good starting place for concerned neighborhood associations is simply to identify the potential threats and establish a dialogue with local officials on how best to address them.

Other local air quality and water quality problems may be further threats to public health. Differences in topography, hydrology, and prevailing wind patterns may mean that pollution problems are substantially greater in one neighborhood than in another a relatively short distance away. Ideally cities will collect and analyze information on pollution problems, warn neighbors, and take steps to reduce the extent of these problems in advance. But needless to say this does not always happen, and proactive investigations by neighborhood groups may be required to bring matters to public attention.

Undertaking sustainability planning at the neighborhood scale, then, is an opportunity to address very specific, on-the-ground issues related to the integration of the built and natural environments, the creation of walkable communities, and the existence of equitable housing options, employment, and educational opportunities. One of the main problems is that the basic components and physical structure of neighborhoods are frequently left up to the private sector builders who first develop communities. To ensure more sustainable neighborhood form and character, more proactive public planning will be necessary. City and county officials will need to establish stronger guidelines for new neighborhoods and for infill of existing neighborhoods, enforceable through zoning codes and development approvals processes. Nonprofit community development corporations, affordable housing builders, environmental watchdog groups, and other neighborhood advocacy organizations can help enormously in this process of making neighborhoods both more livable and more sustainable.

25

SITE PLANNING

The buildings we inhabit and the lots these structures sit on represent the smallest scale of development affected by planning, a domain that overlaps with the professions of architecture and landscape architecture. Issues at this level include how development fits into and affects the natural landscape, how it uses energy and materials, how it influences the neighborhood and social interaction, how accessible and useful facilities are for different groups of people, and how they affect the health and daily lives of residents or workers.

We experience the results of site and building design decisions every day. Are the structures we live and work in comfortable, safe, and adaptable to our needs? Do they relate well to the street and sidewalk, creating a pedestrian-friendly neighborhood? Is the landscaping attractive, climate-appropriate, and respectful of local ecosystems? Is the flow of energy and materials into the site or building kept as low as possible, ideally zero net energy? Are waste materials reused or recycled? In general, does this building or site represent a model that could be replicated widely in a sustainable society?

Such questions are often overlooked in architectural education and practice, where the emphasis instead is often on the aesthetic merits of different building forms or the desire to establish a distinctive look. Many current buildings are also built only for short-term utility, and their nature is dictated largely by economic rather than environmental or equity concerns. Even if well-built, structures are often bland and colorless, using generic architecture that makes every place look like every other place and acculturates us into a homogenized global culture run by multinational corporations. The result is a world of generic office buildings, chain stores, and tract housing that looks much the same in Jakarta as Tucson. We can do much better. Site and building design that enhances local ecologies, cultures, and communities is one of the core challenges of sustainability planning.

WHO PLANS AT THE SITE AND BUILDING SCALE?

The number of individuals and institutions who influence site and building development is much greater than one might think. Historically of course the process was simpler, and many people built houses and workplaces for themselves with little regulation or oversight. Self-built housing was relatively common on the outskirts of North American cities and towns until the middle of the twentieth century,[1] and is still the norm in much of the developing world. But in the process has grown much more complex. Key actors now include individual property owners, for-profit developers of varying sizes, nonprofit developers, architects, landscape architects, urban planners, engineers, environmental consultants, bankers, insurance companies, construction contractors, neighborhood residents, and city, county, and state government.

Developers—the individuals or companies that purchase or lease land and coordinate subdivision and building—are the best-known and most often vilified players within this scale of activity. However, many different types of developers exist, some doing much more sustainable types of development than others. The mix varies considerably from time to time and place to place. One of the big stories of the twentieth century in terms of urban development was the rise of production homebuilding corporations that built subdivisions on a grand scale, using mass production techniques first pioneered at developments such as Levittown, on Long Island, in the late 1940s.[2] (The earlier advent of "balloon-frame" construction, using relatively cheap, small-dimension lumber, helped pave the way for the mass production of housing.) Though their efforts provided cheap housing for millions in the decades after the Second World War, these companies institutionalized the motor vehicle-dependent, single-use, suburban tract style of development. Their ability to provide cheap, mass-produced housing also drove many small and medium-sized development companies out of business, fundamentally reshaping the development industry.

Other types of builder have specialized in office or commercial development, and likewise grew in scale and standardization during the past century. But some small-scale developers persisted, and new niches have appeared in recent decades. With people rediscovering the virtues of city living, some developers now specialize in urban lofts, townhouses, infill sites, or rehabilitation of existing buildings. Others focus on multifamily buildings containing apartments or condominiums. As affordable housing has vanished in many regions, public-spirited individuals have founded nonprofit development companies to meet the needs of the lower end of the market neglected by the for-profit development sector. Some of these nonprofit housing organizations have become very large—San Francisco-based BRIDGE Housing, for example, builds as many as 2000 units of affordable housing each year. Habitat for Humanity, founded in Atlanta in 1976 by Millard and Linda Fuller, has built more than 100,000 homes around the world largely through volunteer labor. States and municipalities often assist the efforts of these affordable housing entrepreneurs by making available loans, tax credits, and other forms of financial or technical assistance. Nonprofit builders are typically more willing to work on small, infill sites and to consider the surrounding neighborhood context.[3]

Further diversification of the development industry is essential if it is to become better at meeting the needs of sustainable development. Since many large development companies are wedded to particular "products" that are not particularly sustainable, smaller and more flexible developers may need to take the lead. Planners may want to ensure that a variety of incentives is in place to nurture alternative developers who are interested in creating green buildings, affordable housing, rehabilitated historic structures, and infill neighborhoods. In the long run the public sector and professional associations can perhaps entice mainstream developers as well towards more sustainable methods of construction and design through regulation, incentives, education, and infiltration.

In addition to the developers themselves, city planners, engineers, and inspectors employed by local government represent a main influence on site and building development. As should be clear by now, government action establishes the framework within which development takes place. Zoning codes, subdivision regulations, and parking requirements often drastically limit development options. Municipal inspectors also enforce building codes that regulate certain design elements, building materials, and construction methods. As discussed in Chapter 14, organizations such as the International Code Council promulgate standard codes initially, intended to promote human health and safety (and recently energy conservation as well). National and state governments then adopt versions of this model building code, and cities in turn apply versions of the state code with relatively minor modifications to meet local conditions. In the past overly rigid building codes have often prevented use of alternative building materials such as rammed earth and straw bale, and have prevented exploration of alternative building techniques in general. As interest in alternative materials has grown, some municipalities have added special sections of building code to accommodate them. Some cities have also changed their codes to allow or require gray-water systems, passive solar design, and energy efficiency.[4]

Architects wishing to incorporate sustainability principles must also design buildings that conform to these codes as well as meet client needs. This task is far from easy. Building designers may be constrained by the client's budget, choice of site, and lack of interest in alternative methods. In some cases there is leeway to suggest ways to save energy, minimize nonrenewable resources, avoid potentially toxic materials, and relate the building to neighborhood, city, regional, and global contexts. However, in other cases clients are inflexible. Some ecological architects turn down potential projects rather than work with a site and building program that they do not feel is appropriate—for example oversized houses on the urban fringe. In addition, many large developers these days have dispensed with architects altogether, relying on generic designs within a mass-production framework. As a result, many architects are in the often frustrating situation of wanting to design better projects but feeling they have little leverage with developers or individual landowners.

Landscape architects, often crucial to the livability and sustainability of particular development projects but not necessarily employed at all by developers trying to cut corners, represent an even more underappreciated profession. Key landscape strategies for sustainability include preserving or restoring natural drainage on the site, specifying native or climate-appropriate plantings, using trees and shrubs to maximize natural heating and cooling for buildings, planting to provide habitat for local species, and creating attractive outdoor spaces to meet the needs of all building users. Traditionally many builders and their landscape consultants have simply stuck in a lawn and a minimum of shrubbery to hide a building's foundation, or have created a more elaborate landscape planting plan exclusively for aesthetic purposes. Ecological considerations and the needs of building residents were often not considered. But this situation is slowly changing. In particular, more detailed knowledge is being developed of how to design sites to meet the needs of children, the elderly, the disabled, and particular cultural groups. Since 1968 the

Environmental Design Research Association has held conferences and published studies in part to share information on such strategies.

Like architects, landscape architects serve at the will of their clients and face a challenge of educating developers about green building and landscaping practices. However, the number of clients willing to experiment with native species, drought-tolerant landscaping, permeable paving, and other alternative techniques is growing these days, in part as overall public consciousness about the natural environment rises. Interest in creek restoration, innovative park design, and habitat protection, all areas in which landscape architects may be involved, has also grown.

Lenders are a hidden player influencing the character of many development projects, since developers (especially nonprofits) typically must borrow large sums for site acquisition and construction. Traditionally banks and other financial institutions have favored conservatively designed projects similar to others that have proven track records of repaying loans. Financiers have been skeptical about alternative building materials and mixed-use buildings in either greenfield or infill locations, often ignoring the retail component of a mixed-use project entirely in their calculations. Another major problem for developers wanting to build infill projects has been lack of the "comparables" required by lenders—past projects of the same sort of development in the same neighborhood. What results is a chicken-and-egg problem: a desirable project cannot be financed if comparables don't already exist, but those comparables can't be created if no one will lend for them. This situation is changing as more infill projects get built, some with government assistance, and more progressive lenders emerge who are willing to commit to the revitalization of older neighborhoods. However, this obstacle is still substantial.

Other groups also influence the character of particular development projects. Neighbors and/or future residents may enter into extended negotiations with developers, planners, and architects to express their particular needs and desires. Engineers and building inspectors frequently determine what technologies can be used for the building or site landscaping, and can nix innovative practices. Hydrologists may provide information about groundwater, soils, and waterways, determining, for example, what kind of drainage or sewage system is needed on the site. On large development projects, other environmental consultants may be called in as well.

Each of these participants has a role to play in improving the sustainability of a given development. One major problem in the past has been that the various parties involved in putting together a development project have often worked in isolation from one another, each with a different set of assumptions and expectations. Ecological design consultants often did not have a chance to talk to all parties and often were not involved early enough in the process. Getting them involved at the beginning to discuss opportunities and share knowledge with other team members can yield creative solutions and avoid unnecessary roadblocks later on. This is one of the most essential strategies for sustainable design—to begin with environmental and social goals at heart.

SUSTAINABLE SITE DESIGN ISSUES

At the site level how does one "design with nature," to use Ian McHarg's phrase?[5] More properly, from a sustainability point of view, how does one "design with nature and community and equity and a sustainable economy?" Many authors have developed lists of ecological design principles as a way of addressing this question.[6] Systems such as LEED and BREEAM also set forth such principles. The discussion that follows emphasizes factors not covered in Chapter 14.

Lot location

In many ways the most significant sustainability consideration is the location of the lot. Is it contiguous with existing urban development? Is it near transit, shops, and existing neighborhood centers? If the site is to be used for housing, is it near jobs, schools, shops, and parks? Would development on this site avoid disrupting open space or habitat? If the answer is "no" on one or more of these counts, then perhaps this parcel should not be developed at all. Problems arise particularly with sites at the urban fringe or in rural areas, where available cheap land may not be appropriate to build on. If local governments have done their job they will have protected such areas by zoning them for agricultural uses, purchasing them for parks, placing them outside urban growth boundaries, or protecting them through other means. But even if the land is not protected, development on it should still be avoided. A basic mandate in sustainable site planning, then, is to develop in the right places.

This guideline raises difficult questions for design professionals, developers, and landowners. Obviously, one solution is not to acquire such sites or accept work related to them in the first place. Alternatively, planners and designers can sometimes convince clients to preserve much of the site as open space, or can arrange for open space agencies to buy the land or purchase conservation easements preventing development while compensating the landowners. Questions of professional ethics may be involved if planners and designers go forward with conventional projects on these sites. The American Institute of Certified Planners' Code of Ethics prohibits planners from working on projects that may cause significant environmental harm (though there is little enforcement of this provision). It can also be considered unethical to work on projects that will impose large costs on society in terms of traffic, resource consumption, or social inequality. Consequently, it is best both for planners, designers, and developers to avoid questionable projects from the start and to focus instead on sites and forms of development with clear benefits.

Appropriate use

A related consideration is whether the proposed development is making appropriate use of the particular site. Is the proposed use (single-family homes, apartments, offices, stores, or other types of enterprises) already over-represented in the surrounding area? Should other uses be sought instead? Should the site contain mixed uses? Are the proposed densities of housing or office buildings appropriate? Is the development creating housing types that meet the needs of the local community, especially less wealthy individuals and families? Again, planners and designers may be at odds with landowners about the appropriate use of the site and may need to try to persuade clients to consider better alternatives.

If location and use reflect sustainable development needs, then the focus can be turned onto the site itself. "Site planning" organizes the development of a single parcel of land by locating building(s) or other facilities in particular places, arranging for roads, sewers, water, electricity, and other infrastructure, and developing plans for grading, drainage, landscaping, lighting, and other site improvements. Developers must do site planning at the outset of a development project, and typically pay architects, landscape architects, or planning consultants to assist with this task. Site planning may be a relatively simple proposition for a single house on a small lot, essentially treated as part of the building design. Or it may be a very complex process for a large development covering many acres, done by a design and planning team entirely separate from the building architects.

One main site-planning question is whether particular ecological features should be preserved or restored. Creeks, wetlands, and wildlife habitat on the site are of special importance and may be regulated or protected by local ordinances or state or federal law. Section 404 of the federal Clean Water Act in the US, for example, requires permits for the discharge of dredged or fill material into wetlands or waterways. Planners or developers wanting to fill such areas on a site must therefore receive a permit from the Army Corps of Engineers. But the far better course of action is usually to preserve waterways and to establish buffers around them, for example, setting development back from a creek to leave a riparian corridor with both wildlife and recreational value. Hillsides and wetlands should also be left alone for the practical reasons of avoiding flooding, landslides, or other forms of ground movement. The presence of such landscape features on the site may lead to a site plan that clusters development at one side to avoid these areas, or that maintains greenways and open spaces throughout the site.

When it comes to positioning buildings on the site, additional questions come into play. How should the building(s) be positioned to create an attractive, walkable street frontage? To maximize yard or garden space for inhabitants? To maximize solar access? To respect neighbors? Such considerations must be addressed both within the building siting and within architectural design. They require a detailed study of the site and an ability to understand the many different issues and needs that converge on any particular development project (see Figure 25.1).

Landscape design

Sustainable landscape design strategies may emphasize drought-tolerant species in arid or semi-arid areas, trees that provide shade and evaporative cooling in hot climates, and bushes or trees that furnish protection from the elements on windy, cold, or exposed sites. Planting a variety of native species can also often provide habitat and food for local bird and animal species. Edible landscapes—bearing fruit or nuts that can be consumed by humans—can provide food and interesting educational opportunities for adults and children alike. Including garden space within the site design provides opportunities for urban residents to grow some of their own food and develop their connections with the earth. Appropriate forms of irrigation are important for sites where additional water is necessary, for example, drip irrigation systems that minimize water loss in dry climates.

Good site planning can help restore ecological health on most sites. Invasive, non-native plant species can be removed, and native shrubs, trees, flowers, and groundcovers planted instead. Soil contamination from past uses of the site can be cleaned up. Impermeable paving can be removed and replaced with gravel, porous asphalt, or stone, concrete, or

Figure 25.1 Fitting buildings to the landscape. Buildings at Sea Ranch, on the northern California coast, are designed to blend into the landscape. Fences and decorative landscaping are not allowed

brick pavers that let water enter the ground to recharge the aquifer. Natural drainage swales planted with reeds, rushes, sedges, grasses, and other riparian species can also help storm water runoff recharge the local water table (see Figure 25.2). Existing fences can be removed to recreate wildlife corridors and increase habitat. Additional buildings can be

Figure 25.2 On-site drainage. Planted drainage swales in this parking lot in Portland, Oregon help filter runoff and allow it to drain into the soil

constructed at underused sites within existing urban areas, making more efficient use of them, adding needed housing, stores, or office space, and preserving open space elsewhere.

The nonprofit organization TreePeople and the US EPA held a series of design charettes to develop ideas for retrofitting existing urban sites in the southwestern US. Participants looked at five examples throughout Los Angeles: a single-family home site, a multifamily housing site, a public high school, a light industrial site, and a Jiffy Lube and mini-mall. Recommended sustainability strategies for these sites included adding systems to collect roof water and parking lot runoff in cisterns for later landscaping irrigation, shade trees to reduce air-conditioning needs, "parking orchards" with trees and edible landscaping to improve surface parking lots, vegetated swales in parking lots to filter runoff, porous paving for some surfaces, gray-water systems to reuse shower water for landscaping, and general reductions in paved surfaces.[7] Although the focus of this exercise was on saving water in a hot climate, a similar approach could be taken with other sustainable site planning goals such as saving energy and restoring elements of the natural landscape (see Figures 25.2 and 25.3).

Site planning is typically also concerned with connecting a given parcel of land to road, water, sewer, gas, and electrical distribution systems. Sustainability questions arise in each of these areas. Road connections are important to connect a large site to existing

Figure 25.3 Native plants. Use of native species in residential landscape design can provide food and habitat for wildlife and reduce needs for watering, fertilizer, or pesticides

neighborhoods all around, and site designers can seek to integrate their development into existing grids or street networks, thus improving the ease of biking or walking through the whole neighborhood. It is also important to minimize paved surfaces, especially if impermeable paving is used. A sustainable development project is likely to have relatively narrow roads or driveways and as few surface parking areas as possible, both to use the land efficiently and to avoid runoff and excessive use of asphalt.

ECONOMICS AND SUSTAINABLE BUILDING

Part of the reason green building isn't more widespread is that a whole set of economic incentives favors unsustainable building practices. Not only are social and environmental externalities not factored into the cost of building materials, but many architects, engineers, and contractors have little familiarity with alternative techniques, and the development industry itself has proven surprisingly resistant to new practices. This may be in part because the industry has become dominated by large companies that have become entrenched in particular ways of doing business. National production homebuilders construct tens of thousands of units each year, and their profits depend in part on cheap, mass-production techniques.

This situation is the opposite of the ways that homes were built prior to the twentieth century. As late as the 1940s most homes were built individually by relatively small-scale, locally owned companies, or in small groups of a few units at once. In older urban downtowns and many nineteenth- or early-twentieth-century neighborhoods, the buildings that appear distinctive and interesting today were added gradually over many years. This slower, more incremental style of building, epitomized by Europe's medieval and Renaissance cities, is at the core of the process proposed by Christopher Alexander and his coauthors in their classic urban design manifesto *A Pattern Language*. To create a beautiful and livable city, Alexander argued, it is necessary to add buildings one at a time and to design them according to time-tested principles that have been used in cultures the world over. Each building is designed in such a way as to add to the urban context around it and to help "heal the whole."[8] Development in this slower, more thoughtful manner may also provide greater opportunities to consider green building techniques than today's mass-production construction practices.

We certainly don't need to return to nineteenth-century styles of homebuilding to have greener building or more sustainable development. But a move towards building practices that are smaller-scale, more locally oriented, and more sensitive to particular contexts does seem likely to produce significant social and environmental benefits. Ways of producing such housing so that it is still relatively affordable—or to increase the incomes of lower-wage workers so that they can afford housing produced in this way—must also be found.

Further, true-cost pricing of building materials could greatly increase use of more sustainable components. The price of wood products, for example, has traditionally not reflected the costs of deforestation and the subsidies provided to the logging industry by the US government in the form of underpriced timber and road construction on federal lands. Life-cycle pricing of materials can likewise help "level the playing field" economically between green and conventional buildings, and ensure that building construction takes place with the lowest level of impacts. State or national governments can begin to establish true-cost pricing of building components by establishing green taxes on materials that are nonrenewable or have major environmental costs. They might also consider extra charges for development that doesn't recycle construction debris (burdening municipal landfills), that exacerbates urban heat island effects (through excessive paved surfaces and poor landscaping), or that creates excessive amounts of runoff from surface parking lots (burdening municipal storm drain systems, exacerbating flooding, and reducing natural recharge of the water table). A related step to ease and lower the cost of green building would be to revise building codes to allow and encourage use of alternative materials, meaning that developers would no longer face the added uncertainty of getting new methods approved and the expense of hiring engineers to justify "alternative methods requests" under current building codes. Concerted action of these sorts by public officials can begin to make green building practices more economically viable.

Sustainability planning at the site and building scale therefore dovetails with architecture and landscape architecture to offer opportunities to apply sustainability principles to our everyday environments. The design and construction of buildings and landscaping affect us very personally, since these processes create the settings in which we live. As with any other scale, there is a great deal of momentum behind existing, unsustainable ways of doing things. Yet at this most immediate level we have the opportunity to do many things ourselves, even if these simply involve choosing locations and buildings to purchase, taking simple energy conservation steps, or installing native or climate-appropriate landscaping. These daily decisions can make a big difference, and set an example of personal action for a more sustainable world.

Part Four

Conclusion

26

HOW DO WE GET THERE FROM HERE?

Although the challenges of sustainable development seem overwhelming at times, the point of this book is that it is indeed possible to plan for a better future. Much positive change is happening every day, and through our actions we can bring about even more.

Taking a long-term perspective on social evolution helps support this claim. Our species is slowly learning many things, such as to respect basic human freedoms and rights, and to develop institutions protecting these. It has learned much about human psychology and communication, and has developed great spiritual wisdom in many different traditions. And it is coming to understand and respect the Earth's ecosystems, though this too is very much a work in progress.

Nevertheless the situation is often daunting. Even where sustainability-oriented planning initiatives have been adopted, we often face what Owens and Cowell call an "implementation deficit."[1] Too often promising efforts succumb to entrenched social values, gridlocked institutions, or economic and political forces that promote self-interest, short-sightedness, and the phenomenon that historian Barbara Tuchman used to call "wooden-headedness"—the persistent following of strategies that are obviously counterproductive even when clear evidence exists that these are not working.[2] Americans particularly, isolated in our nation that epitomizes many of the excesses of capitalism and materialism, often seem unable to "connect the dots"—to recognize the hypocrisy of many of our leaders, to acknowledge the ways that power and money corrupt our ideals, and to see that we are interdependent with others in the world. Other countries share many of these problems. Many structural forces promote this blindness, including television, advertising, social norms, the power of economic institutions, and the physical environment we have created around us. At least some of these structural conditions will need to change for social and political values to change. Or perhaps, if we are lucky, both inner and outer changes will happen at the same time, in response to stimuli that we can as yet only dimly see.

DEVELOPING AN AWARENESS OF OPPORTUNITIES

The foundation for change at either a personal or a collective level is awareness of problems and the ability to see opportunities for new ways of doing things. As a society we are
in an enormous amount of denial about many of the problems around us—the ways that
we have degraded the landscape and impoverished certain groups, for example, and the
ways that lifestyles in affluent countries are related to suffering elsewhere in the world. The
flurry of daily activities, economic pressures, myopic news media, advertising, and many
other cultural forces constantly distract us from this awareness.

In terms of planning, we can all become more aware of the physical, social, and economic landscapes around us, better at understanding problems, and more proactive in
identifying opportunities and priorities for positive change. In particular we can cultivate
skills of "looking at cities," as former San Francisco planning director Allan Jacobs recommends in his book of that name.[3] We can observe the environments around us carefully as
we go about our daily lives, note how people interact and use places, experience what
different settings feel like, and practice identifying opportunities for ecological restoration,
neighborhood revitalization, a more vibrant local economy, and improved equity.

This sort of experiential learning often goes by the academic title of "phenomenology,"
and has been the focus of a small but growing movement.[4] One of its proponents, David
Seamon, defines phenomenology as "an interdisciplinary field that explores and describes
the ways that living things, living forms, people, events, situations and worlds come
together environmentally."[5] Whatever the label, these skills of seeing are particularly
needed right now. Mainstream culture and even graduate planning programs often discourage such direct observation, asking us instead to experience the world through computers, television, other electronic media, or highly abstract forms of analysis. Granted,
theory and quantitative analysis are very important. But gaining better understandings of
the on-the-ground nature, history, and potential of particular places is an essential starting
point for sustainability planning. From that awareness, opportunities for action emerge.

Understanding opportunities for change also depends on an appreciation of the forces
and dynamics that affect how societies evolve. This is where theory and analysis particularly come in. Gradually we all need to become more aware of these forces and ways to
change them. Ecological economist Richard Norgaard has proposed one useful model of
"coevolutionary" process through which society evolves.[6] In Norgaard's view, human
values, knowledge, and organization constantly evolve in conjunction with the natural
environment and technology. Each of these factors influences the others, and all are important. Figure 26.1 shows how he diagrams his model.

The exact labels here are not as important as understanding how such forces interact to
bring about social evolution. For "organization" the word " institutions" might also be
used, referring to the whole range of government agencies, NGOs, private corporations,
public–private partnerships, economic systems, and laws and regulations that structure the
environment within which any form of action occurs. Changes in any of these institutions,
such as the continued rise of NGOs promoting an environmentally and socially oriented
agenda, can help society evolve towards greater sustainability. Along with the word "knowledge," other terms such as "cognition" might be associated, referring not just to the body
of information available to people, but to the ways they process experience and the mental
frameworks or paradigms they use for evaluating what goes on in the world.

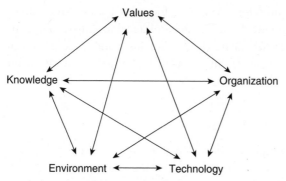

Figure 26.1 Coevolutionary forces

For "values" we might substitute terms such as "culture" or "society," that is, any set of social constructs that helps us decide what is important and what isn't. These priorities can evolve substantially over time. Political scientist Ronald Inglehart, for example, has tracked values within industrialized countries over several decades, and finds a pronounced shift towards valuing quality of life and the natural environment over the past several decades.[7] This is a hopeful trend for those of us interested in creating more sustainable communities.

A theoretical framework such as Norgaard's could go by many names—"coevolutionary," "ecological," "process-oriented," or "a systems perspective." What matters is our growing understanding of how different factors interact to bring about social change, and of how by influencing one element or another we can help society evolve. Helping people see the world differently (changing cognition), through teaching, writing, art, or even architecture, can thus have a very great effect in the long run. Changing institutional structures by the creation of new agencies and laws or the reform of existing ones can also have a large impact. Helping values change, through spiritual practice, teaching, personal example, work with children, or other means, can likewise help lay the groundwork for social evolution. Each of these strategies begins at one point on Norgaard's diagram and eventually affects all the rest.

One implication of such a coevolutionary or ecological perspective on the world is that contexts can be structured to maximize the chances that good outcomes will happen. In the case of sustainability planning, the structuring of institutions is particularly important. For example, the existence of federal and state Environmental Protection Agencies in the United States since the 1970s has made possible far more systematic approaches to environmental regulation and protection than before. At a local level, the establishment of design review commissions and zoning reforms can help set up an environment in which better-designed and more appropriate forms of development will occur. Giddens and others have written extensively on "structuration"—the processes through which we structure social systems through rules and institutions—and other institutionalist researchers focus on this as well. By structuring institutions well, we can create many opportunities for positive social change in the long run, and a healthier social ecology overall.[8]

All the tools mentioned in previous chapters can be used to nudge urban planning in the direction of sustainability, structuring a context in which the ecological or social health of communities can improve. Rational comprehensive planning, communicative planning, advocacy planning, grassroots organizing, visioning, development of best-practice examples, sustainability indicators, ecological footprint analysis, GIS systems and mapping, and last but not least frameworks of intergovernmental incentives and mandates can all help. Other more general factors such as leadership, inspiration, and insightful analysis are also important. Which strategies are most worth focusing on will vary from time to time and place to place.

TAKING A STRATEGIC PERSPECTIVE

What is essential, then, is a strategic perspective in which all of us as planners or citizens look for ways to bring about more sustainable development by incremental improvements in institutions, knowledge, values, society, or the physical environment. Mutually reinforcing initiatives in all these areas can help bring about change. Hence sustainability planning's emphasis on coordinated action at different scales. International agreements such as Local Agenda 21 or climate change treaties can help national policies change; state policy, incentives, and leadership can help regional and local action occur; and so forth. No single level of initiative is enough by itself—a mutually reinforcing set of actions at all levels is necessary.

A strategic perspective on sustainability planning must be a long-term one, because change often does not happen rapidly and it may be most useful to develop contexts in one generation that can support positive action many decades in the future. Setbacks occur, and unfortunately there will be a great deal of permanent loss along the way (of species, ecosystems, cultures, and individuals). None of this should be countenanced without a fight. But the long-term sustainability, evolution, and health of life on this planet are the ultimate criteria of our success.

It is crucial to be aware of how small, incremental steps can lead to larger long-term goals. Each new, appropriately designed building in a city, each small piece of a natural ecosystem restored to health, each human being empowered to contribute to society in a meaningful way represents a major step forward. Over even 5 or 10 years, such incremental steps can make a big difference in a city or town. Over 20 or 30 years, they can be revolutionary. It is up to us to figure out how best to contribute to the small as well as the large changes taking place.

The pace of change in urban environments and human values since the beginning of the Industrial Revolution, barely two hundred years ago, has been phenomenally rapid and is still accelerating. Widespread acceptance of goals of civil rights and environmental protection has occurred only in the last generation or two. Modern environmentalism dawned in the 1960s, and sustainable development first emerged as a theme in the 1970s. Environmental justice concerns came on the scene in the 1980s. The New Urbanism, Smart Growth, and livable communities became strong movements in the 1990s. Urban food systems and healthy living became central concerns in the 2000s. What has been happening, in other words, is a rapid growth in awareness of the problems related to creation of a global industrial society and in consciousness of alternative ways of designing and managing communities. Though events around us are often discouraging, the emergence of movements such as those above should give us optimism for the future.

DEALING WITH POWER AND IMPROVING DEMOCRACY

Action towards more sustainable communities will depend in large part on having a political system that can respond to current situations and lead sustainable development activities. Democracy is widely accepted as the political ideal of our age and appears to offer many sustainability benefits itself in that it can be resilient, adaptable, equitable, and inclusive. However, democracy is often subverted by power, especially economic power. What we have in the United States and many other countries is more plutocracy—government by wealth and powerful economic constituencies—than democracy.[9] One of the Three Es, economy, has dominated the other two and has profoundly shaped the institutions and value structure of the culture. With such a flawed system, only imperfect and partial progress towards sustainable development is possible. What seems needed is a democratic system in which social and environmental values really do balance economic ones, and in which the public sector and civil society can both play a more substantial role in asserting these values. The tradition of social democracy offers one such path, as Giddens and others have argued.[10]

As discussed earlier a truly functional democracy would rest on a foundation of three components: a clean political system, an enlightened electorate, and real choices within elections. Sadly, none of these elements is strongly present in the United States currently. Our political system is notoriously open to the influence of money and large corporations; this wealth allows certain interests to distort the political agenda and sets up a merry-go-round in which officials shuttle back and forth between corporate boardrooms and public office. The need for constant fundraising discourages many good people from running for office, from city council seats to the presidency. Americans often lack the most basic knowledge about the world and about history, and are more influenced by television and commercial culture than any other source. For its part the media fails to provide information about many of the most pressing issues of the day—including many sustainability planning topics—preferring to focus instead on crime, disasters, personalities, and scandals. Finally, we lack the full spectrum of political views that could give us real choice in elections. The two-party system, the enormous financial requirements for serious campaigns, and the "winner take all" structure of our representative democracy prevent third parties and alternative points of view from emerging. In contrast, within European parliamentary democracies (which are not perfect either) Green parties and social democratic viewpoints have been able to gain some degree of power.

Progress on all three of these fronts will be required if countries are to progress toward the ideal of democracy and create an environment in which sustainability planning can most effectively take place. Many authors have suggested ways in which a healthier politics can be created. The communitarian philosophy promoted by Amitai Etzioni,[11] the discursive democracy of John Dryzek,[12] and the "politics of meaning" promoted by Michael Lerner[13] represent several of these approaches. A variety of nuts-and-bolts political organizing, coalition-building, and educational strategies will be needed as well.

If equity and environment are to have equal standing with economic goals in our political system, we will need to reduce the influence of economic interests. We will need to place limits on the power of wealth and large corporations to capture government institutions to serve their own ends. This will mean taking steps such as reforming election financing, increasing participation in elections, adopting a more progressive tax system, and instituting extensive programs designed to assist lower-income individuals and small businesses (that is, those with relatively little power). Doing these things will depend on a fundamental growth in public consciousness about the nature of power in our society which may seem far-fetched in the current environment. Yet such a change is likely to come sooner or later, and in future eras the present time may be seen as an age of excessive inequality, materialism, and corporate greed. Giddens also urges a redefinition of rights and responsibilities, away from the present condition in which individuals and corporations are far more interested in receiving benefits from social institutions than in contributing to the common good. As he puts it, "One might suggest as a prime motto for the new politics, *no rights without responsibilities*" (emphasis in the original).[14] Or as John F. Kennedy urged Americans, "ask what you can do for your country." Perhaps future leaders can reawaken this civic spirit.

Whatever our country, we are often reluctant to acknowledge the realities of power within our society. But we must. Sustainability planning will necessarily include efforts to build alternative forms of political power through grassroots organizing, coalitions, urban social movements, lobbying, occasional litigation, and political leadership. Hard-headed political organizing has almost always been necessary to bring about progressive change, whether in civil rights, environmental, women's, or peace movements. A whole set of strategies devoted to nonviolent social change has arisen to assist such movements, spearheaded historically by figures such as Ghandi and Martin Luther King Jr. Current organizing against economic globalization represents a recent phase of such activity. Such political organizing includes founding groups (in particular nonprofit advocacy organizations), planning demonstrations and conferences, building coalitions, lobbying elected officials, and contributing to such efforts through donations, labor, and professional creativity.

One question that often arises is how sustainability-oriented improvements are to be financed. Federal, state, and local governments complain that they are strapped for cash. Anti-tax crusaders seem constantly to be launching ballot referenda or legislative initiatives to lower taxes further. Public interest groups and even private philanthropic foundations have relatively limited resources. Each of us individually often feels financially strapped and unable to afford green products or additional charity. For one thing, it should be clear by now that much sustainability planning does not require expensive new investment. It simply involves rethinking existing spending and regulation, and making small strategic amounts of funding available to catalyze new programs. Changing zoning codes, for example, requires no new expense. Designing new communities well does not take much more money than designing them badly, and may actually save money. Increasing equity across an entire metropolitan region may be a question of distribution, not of new spending. On a personal level buying organic produce or adding green features to our houses does not cost much extra and brings many long-term benefits.

Yet in other areas new spending undoubtedly will be required. And here it is a question of priorities. The response that "we just don't have the money" is an excuse for inaction. After all, the Rio Earth Summit Conference funding of the entire Agenda 21 program was expected to cost $128 billion, only one-tenth of the global arms budget.[15] Those of us in industrialized countries live in some of the richest societies the world has ever known. If we can't find the resources to make our public realm livable, to protect and restore our environment, and to care for our least well-off neighbors, it is because we choose not to. Confronting the reality of the social and political situation, in which various elite, politically powerful groups divert our collective resources for their own purposes and promote overly materialistic values, will be one of the main challenges ahead.

AN AGENDA FOR THE FUTURE

This whole book has essentially been about "how we get there from here." We do so through personal awareness, compassion, and action. But we also do so through strategic thinking, collective activity in professional fields such as urban planning, and through the development of institutions and social capital that can create a context for even more significant actions in the future.

Professionals involved in areas such as city and regional planning play a crucial role in the process of sustainable development in that they address the material conditions of our human environment. To a substantial extent, such as by helping create a more vibrant public realm, planners can help create community and literally bring people together. Planners can also play a more explicit role in promoting equity, public health, alternative local economies, and a less automobile-dependent society. Last but not least, they can both protect natural ecosystems, and lead the twenty-first-century process of restoring many of those elements of the environment that have been damaged in the past. These are some of the core planks of the new agenda of planning.

Other fields such as architecture, landscape architecture, economics, public health, and public administration also relate directly to questions of how to have more sustainable and livable communities. In the past all these professions have too often been sidetracked into narrow, conventional ways of doing business, or have been disempowered politically, institutionally, or economically. But it is possible both to be professional and to play more proactive roles in planning for sustainability.

In a broader sense, planning—systematically preparing for a better future—involves all of us. Our actions may simply involve consciousness raising and shared education about the challenges of the current world, but they are still significant. Within planning and design fields, a long line of pioneers has called for action on behalf of the human and ecological needs we can observe all around us. Now, more than ever, all of us can figure out ways big and small to take this.

NOTES

PREFACE

1   See Berry (1999).

CHAPTER 1   INTRODUCTION

1   See for example Met Office Hadley Center (2010).
2   Kiehl (2011).
3   See for example Lynas (2008); Ward (2007).
4   For background on the process of suburban sprawl, see Jackson (1985); Weiss (1987); Schuyler (1986); Daniels (1999); and Hayden (2003).
5   This process of land conversion is well described in the planning literature. See for example Mark Gottdiener's (1977) classic description of the collusion of development interests on Long Island. The seminal notion of "growth coalitions" was developed by Harvey Molotch and others, for example in Logan and Molotch (1987).
6   Lewis (1922).
7   See Farmland Information Center (2012).
8   For further background on problems related to sprawl, see Kunstler (1993); Transportation Research Board (1998); Benfield et al. (1999); US Environmental Protection Agency (2001, p. 6); Duany et al. (2000); and Lindstrom and Bartling (2003).
9   Benfield et al. (1999, p. 5).
10 Kotkin (2011).
11 US Environmental Protection Agency (2001, p. 6).
12 US Bureau of the Census (2001).
13 US Department of Energy (2012).
14 See Stephen Marshall, "The Challenge of Sustainable Transport," in Layard et al. (2001, p. 133).
15 Millard-Ball and Schipper (2001).
16 Cited in Pucher et al. (2007).
17 Gersmann, Hanna, "Recycling Rates in England top 40% for the First Time," *The Guardian*, 3 November 2011; Simin Davoudi, "Planning and Sustainable Waste Management," in Layard et al. (2001, p. 194).
18 Yardley, William, "Cities Get So Close to Recycling Ideal, They Can Smell It," *The New York Times*, 28 June 2012, p. A14.
19 Smartpackaging, "Recycling Rates: A Global Comparison," 2012, available at http://www.smartpackaging.org.nz/global-trends-in-packaging/.
20 McGinn (2001, p. 7).
21 Cited in McGinn (2001, p. 31).

22   Economic Policy Institute, "The State of Working America," 2012, available at http://stateof-workingamerica.org/; Lowrey, Annie, "For Two Economists, the Buffett Rule is Just a Start," *The New York Times*, 16 April 2012.
23   Ortiz and Cummins (2011, 19).
24   Kunstler (1993).
25   Putnam (2000).
26   Hall (1996, pp. 14ff.).
27   Isard (1975, p. 2).
28   See for example Krizek and Power (1996); Feiden and Hamin (2011).
29   See for example Congress for the New Urbanism (1999); Calthorpe (1993); Duany et al. (2000).
30   Downs (2005).
31   See for example Wagner and Caves (2012); Lennard et al. (1997); Lennard and Lennard (1995); Langdon (1994); Corbett and Zykofsky (1996); Kunstler (1993, 1996).
32   See for example Dannenberg et al. (2011).
33   See for example Bullard (1990, 2005); Agyeman (2005).
34   See for example Forman and Godron (1986); Dramstad et al. (1996); Farina (2000).
35   See for example Spirn (1984); Hough (1984, 1990); Lyle (1999); Alberti (2009).
36   See for example Riley (1998); Thompson and Sorvig (2007); Todd and Todd (1993); Peck (1998); Birch and Wachter (2008).
37   See for example Forester (1989); Innes (1995); Innes and Booher (2010).
38   See for example Hester (2006).
39   Davidoff (1968).

CHAPTER 2   SUSTAINABLE DEVELOPMENT

1   Good historical discussions of the emergence of the concept are provided by Kidd (1992, pp. 1–26); Mitlin (1992); and Beatley (1995).
2   See for example Ostrom (1990); National Research Council (2002).
3   Marsh (1864).
4   Wurster (1993, pp. 144–5).
5   See particularly Mumford (1934).
6   See for example Mazmanian and Kraft (2009); Gottlieb (1993).
7   Carson (1962); Commoner (1971); Roszak (1972).
8   Ward and Dubos (1972, pp. 192–3).
9   Ibid, p. 126.
10   See for example Teilhard de Chardin (1959).
11   See for example Gilligan (1982); Belenky et al. (1986).
12   Robert Gottlieb's 1993 book *Forcing the Spring* provides an excellent history of the evolution of environmentalism during this time.
13   Meadows et al. (1972).
14   Simon (1981).
15   Meadows et al. (1992); Meadows et al. (2004).
16   Goldsmith and Allen (1972).
17   Ibid, pp. 3–4.
18   Ibid, pp. 21–2.
19   See especially Hayes (1987); Brown (1981); and Lowe (1990, 1991).
20   Callenbach (1975, 1981).
21   World Commission on Environment and Development (1987, p. 8).
22   See for example Adams (1990).
23   Pearce et al. (1990, p. 1).
24   Trainer (1986, p. 5).

25 In Norberg-Hodge and Goering (1986, p. 127).

26 Ibid, p. 20.

27 Mander (1991).

28 For a more detailed discussion of key characteristics of the modernist worldview, see Chapter 6 of Norgaard (1994).

29 See for example Le Corbusier (1987 [1929]).

30 Ellin (1999, pp. 154ff.).

31 Dear (2000).

32 Harvey (1990).

33 See for example Capra (1996).

34 Layard et al. (2001, p. 14).

35 Giddens (1984); Ostrom (2005).

CHAPTER 3   THEORY OF SUSTAINABILITY PLANNING

1 Owens and Cowell (2002, p. 7).

2 Cronon (1991).

3 The preceeding discussion is indebted to Paul Jones, David Selby, and Stephen Sterling's discussion of interdisciplinarity in their book *Sustainability Education* (Jones et al., 2010).

4 The AICP Code of Ethics and Professional Conduct is available online at http://www.planning. org/ethics/ethicscode.htm.

5 See Webber et al. (1964).

6 Friedmann and Weaver (1979).

7 See for example Castells (1997).

8 Lefebvre (1991); Harvey (2009); Soja (1989).

9 Congress for the New Urbanism (1999); Duany et al. (2000); Beatley and Manning (1997); Beatley (2011).

10 London 2012 (2010, p. 3).

11 Blanco (1994, p. 3).

12 See Charles J. Hoch, "Making Plans," in Hoch et al. (2000, p. 23). Blanco (1994, p. 68) demonstrates that this process is similar to John Dewey's five-stage process of inquiry within his philosophy of instrumentalism.

13 For classic descriptions of this model, see Kent (1964) and Kaiser et al. (1995).

14 Lindblom (1959).

15 Etzioni (1967, 1971).

16 See for example Harvey (2009, 1990); Massey (1984); Lefebvre (1974); Castells (1977, 1978); Friedmann (1987); Fainstein and Fainstein (1979); Beauregard (1989); Forester (1989, 1993).

17 Logan and Molotch (1987).

18 Stone (1989).

19 Mollenkopf (1983).

20 Friedmann (1987).

21 Judge et al. (1995).

22 Sen (1982, 1999); Escobar (1995); Korten (1995, 2006); Khor (2007); and Shiva (2010).

23 Healey (1997, pp. 28–9).

24 Arnstein (1969).

25 See for example Innes (1995).

26 See for example Habermas (1995, pp. 43ff.).

27 Warburton (1998, p. 2).

28 Helling (1998).

29 Warburton (1998, pp. 28–34); Freire (1970).

30 Davidoff (1968).

31 Zukin (1991).
32 See for example Bartholomew (1999); Keene (2003); Edwards (2009); Tarrow (2005).
33 See for example Castells (1983); Susan S. Fainstein and Clifford Hirst, "Urban Social Movements," in Judge et al. (1995); Hess (2009); Pastor et al. (2009).
34 Castells (1997).
35 Miraftab (2009).
36 Especially Giddens (1984).
37 Berger and Luckmann (1966).
38 Friedmann (1987, pp. 181ff.).
39 Healey (1997, p. 61).
40 Putnam, (2000, pp. 183–284).
41 Ibid, p. 403.
42 Ibid, pp. 404–14.
43 Lindblom (1959).
44 A number of authors have written on the subject of how planners actively facilitate inequitable or unsustainable forms of development. See for example Harvey (2009); Beauregard (1989); Gottdiener (1977); Kunstler (1993); Benfield et al. (1999); and Marshall (2000).
45 See for example Howe (1994).

CHAPTER 4   SUSTAINABILITY PLANNING AND THE "THREE Es"

1 See Mazmanian and Kraft (2009). Leslie Paul Thiele (1999) takes a somewhat longer viewpoint, discerning four waves of environmentalism, the first starting in the late nineteenth century with the rise of preservationists and resource conservationists, the second in the 1960s and 1970s with the rapid growth of the modern movement, the third in the 1980s with a period of cooptation and mainstreaming, and the fourth beginning in the 1990s with growing understanding of global interdependence and the rise of widespread concern about sustainable development.
2 See Mazmanian and Kraft (2009); Gottlieb (1993); and Shabecoff (2000).
3 See Simon (1981).
4 See for example Rees (1992).
5 We have seen how capital subverts democratic institutions particularly in the United States, which according to observers such as Kevin Phillips has teetered historically between democracy and plutocracy, with the latter being in ascendance at the beginning of the twenty-first century. See Phillips (2002, 2008).
6 See for example Barbier and Markandya (2012); Pearce et al. (1989); Pearce and Warford (1993); and Turner (1988).
7 Costanza et al. (1991, p. 3); see also Daly and Farley (2010); Repetto (1985).
8 Hawken (2010); Hawken et al. (1999).
9 See for example Morris (1983); Sale (1985).
10 Nader et al. (1976).
11 Letter to George Logan, 12 November 1816, available at http://babel.hathitrust.org/cgi/pt?id=uc2.ark:/13960/t3pv6bn68;seq=93;view=1up;num=69.
12 Redclift (1987, p. 4).
13 See for example Massey (1994); Harvey (2009, 1990); Martin Khor, "Development, Trade, and the Environment: A Third World Perspective," in Goldsmith et al. (1992); Escobar (1995); Stavrianos (1981).
14 Frank (1969).
15 Orfield (1997).
16 Wilson (1987, 1996).
17 Gottlieb (1993, pp. 235 ff.).
18 US General Accounting Office (1983).
19 United Church of Christ/Commission for Racial Justice (1987).
20 Howard (1902).

CHAPTER 5 SUSTAINABILITY PLANNING IN PRACTICE

1 Various past texts in planning strategy are useful in relation to this chapter, including Hack et al. (2009); Schön (1983); Friedmann (1987); Forester (1989, 1993); and Barrett (2001).

CHAPTER 6 TOOLS FOR SUSTAINABILITY PLANNING

1 See Charles J. Hoch, "Making Plans," in Hoch et al. (2000, p. 23).
2 Phil Berke and Maria Manta Conroy (Berke and Conroy, 2000) published an interesting study comparing high-quality local plans with and without an explicit sustainability focus. They found little difference in whether these plans actually met sustainability goals, implying that an overt orientation toward "sustainability" may not be essential. However, this could simply mean that much progressive planning is headed in these directions these days anyway, whether under the guise of "sustainability," "livability," "environmental protection," "quality of life," or other catch-phrases.
3 Neuman (1998).
4 The most recent of these follows the Three Es format common to much sustainability planning. See Yaro and Hiss (1996).
5 Register (1987).
6 See www.iaia.org.
7 Counsell and Haughton (2002).
8 See for example Benfield et al. (2001); Ewing (1996); Bragado et al. (1995); Corbett and Corbett (2000); Corbett (1996); Fader (2000); Jacobs (1993); Lennard and Lennard (1995); Condon and Moriarty (1999); Roseland (2012); Wheeler (2002a), available at www.greenbelt.org. See also websites of the UN-related Best Practices database (www. bestpractices.org or www. sustainabledevelopment.org); the International Council for Local Environmental Initiatives (www.iclei.org); the Sustainable Communities Network (www.sustainable.org); the Resource Renewal Institute (www.rri.org); the Local Government Commission (www.lgc.org); the Rocky Mountain Institute (www.rmi.org); the International City/County Management Association (www.icma.org); the American Planning Association (www.planning.org); and the US Department of Energy's Center of Excellence for Sustainable Development (www.sustainable.doe.gov).
9 See http://dubai-award.dm.gov.ae/web/default.aspx.
10 Portnoy (2003, p. ix).
11 See for example Maclaren (1996).
12 Sustainable Seattle Coalition (1995).
13 http://www.sustainableseattle.org.
14 One good source on the subject of indicators is the Sustainable Measures website, at www. sustainablemeasures.com.
15 Keough (2003). See also www.sustainablecalgary.ca.
16 See Southworth and Ben-Joseph (1997).
17 More information on LEED is also available from the US Green Building Council, www.usgbc.org.
18 See for example Duany and Plater-Zyberk (1991); Talen (2009); http://www.smartcodecentral.org/.
19 Wackernagel and Rees (1996); Wackernagel and Yount (1998).
20 For one version, see www.rprogress.org.
21 Girardet (1999, p. 29).
22 Roy and Caird (2001, p. 277).
23 Shen (2001).
24 Moudon and Hess (2000).
25 See Kent and Klosterman (2000).
26 For further information on this field, see the International Association for Impact Assessment at www.iaia.org; the European Community's EIA home page at http://ec.europa.eu/environment/eia/home.htm; and a helpful resource list from the Canadian International Development Agency at www.iaia.org/eialist.html.

27   See http://www.epa.gov/ncea/ and http://www.environment-agency.gov.uk/.
28   See http://www.unece.org/env/eia/welcome.html.
29   Wheeler (2000).
30   See Abbott (1997); Abbott et al. (1994).
31   Friedmann (1987, pp. 181ff.).
32   Putnam et al. (1993, p. 167).
33   See for example Gittell and Vidal (1998); Serageldin (1996).

CHAPTER 7   CLIMATE CHANGE PLANNING

1   See for example Christianson (1999); Romm (2005); Lynas (2008); Weart (2008).
2   See European Environment Agency (2012).
3   International Energy Agency (2011).
4   See http://ec.europa.eu/clima/policies/ets/index_en.htm.
5   Granberg and Elander (2007).
6   See Boswell et al. (2012).
7   See http://www.iclei.org/index.php?id=10828.
8   Wheeler (2008a); Boswell et al. (2010); Wheeler (2012).
9   Wheeler (2012).

CHAPTER 8   ENERGY AND MATERIALS USE

1   Worrell and Galitsky (2005).
2   Ravetz (2000, p. 165).
3   Earth Policy Institute, "World Wind Power Climbs to New Record in 2011," available at http://www.earth-policy.org/indicators/C49/wind_power_2012.
4   Cushman, John, "In Drought: A Debate over Quota for Ethanol," *The New York Times*, 16 August 2012.
5   *The Economist*, "Energiewende: Germany's Energy Transformation," 28 July 2012.
6   See www.freecycle.org.
7   US Environmental Protection Agency, "Municipal Solid Waste," available at http://www.epa.gov/epawaste/nonhaz/municipal/index.htm.
8   Eurostat, "Waste Statistics," Table 2, available at http://epp.eurostat.ec.europa.eu/statistics_explained/index.php/Waste_statistics#Recovery.
9   US Environmental Protection Agency, "Plastics," available at http://www.epa.gov/osw/conserve/materials/plastics.htm.
10   CalRecycle, "Calendar Year 2011 Report of Beverage Container Sales, Returns, Redemption, and Recycling Rates," available at http://www.calrecycle.ca.gov/Publications/Documents/BevContainer/2012013.pdf.
11   Friedman (2003, pp. 5–9)
12   *The Guardian*, "Recycling Rates in England: How Does Your Town Compare?" 11 April 2011.
13   See for example Ehrenfeld and Gertler (1997); IISD, "Kalundborg," at http://www.iisd.org/business/viewcasestudy.aspx?id=77.
14   See Lyle (1994), Chapter 8.
15   Peck and Associates (2000, p. 52).
16   Wikipedia, "Motor vehicle," available at http://en.wikipedia.org/wiki/Motor_vehicle.

CHAPTER 9   ENVIRONMENTAL PLANNING

1   See for example Faber and O'Connor (1988).
2   See http://www.ideastransformlandscapes.org/.

3 The Nature Conservancy, "Modernizing Water Management: A National Sustainable Rivers Program," available at http://www.nature.org/ourinitiatives/habitats/riverslakes/srp-successes-and-nationwide-benefitsxml.null.

4 Subak (2000).

5 See   http://www.govisitcostarica.com/travelInfo/government-programs/government-helps-promote-reforestation.asp.

6 Carpenter, Murray, "Dam Removal to Help Restore Spawning Grounds," *The New York Times*, 11 June 2012.

7 See http://www.penobscotriver.org/content/4030/unprecedented-collaboration.

8 See for example Kehslinger (2008).

9 See http://www.nyc.gov/html/dep/html/stormwater/nyc_green_infrastructure_plan.shtml.

10 The Tbilisi Declaration is available at http://www.gdrc.org/uem/ee/tbilisi.html.

11 See for example http://www.classroomearth.org/.

12 See for example Danks (2010).

13 See http://edibleschoolyard.org/.

CHAPTER 10   LAND USE AND URBAN GROWTH

1 See for example Gordon and Richardson (1997).

2 See Yaro and Hiss (1996, p. 197).

3 Cox (2009).

4 Cullingworth and Nadin (2002, p. 1).

5 Elkin et al. (1991).

6 See Hall (1992, pp. 223ff.). Urban renewal and redevelopment programs in the United States have indeed represented a proactive public role, but unfortunately have often had negative results in that older urban neighborhoods have been bulldozed and replaced with sterile, modernist urban environments. Partly as a result, cities have been cautious about activist planning ever since.

7 US Environmental Protection Agency (2007, iii).

8 Ewing et al. (2007).

9 See Transportation Research Board (1998, p. 49).

10 Speir and Stephenson (2002, p. 67).

11 Lynch (1981)

12 Warner (1962)

13 Juri Pill, quoted in Cervero (1998, p. 89).

14 See Southworth and Owens (1993); Wheeler (2008b); Wheeler and Beebe (2011).

15 See Beatley (1994).

16 Cullingworth and Nadin (2002, p. 172).

17 Jackson (1985); Hayden (2003).

18 See for example Beatley (2000a).

19 See for example Duany and Plater-Zyberk (1991); Duany et al. (2000); Duany and Talen (2002).

20 California eliminated traditional redevelopment agencies in 2011, but as of 2012 it appeared likely that the same powers would re-emerge in a different form.

21 Winter (2001a, p. 27).

22 See Dye and England (2009).

23 Strong (1979).

CHAPTER 11   URBAN DESIGN

1 McDonough and Braungart (2002).

2 Lynch (1981, p. 235).

3 Brand (1994).

4 Alexander et al. (1977).

5 Alexander (1979).
6 Lyle (1994). See also Van der Ryn and Cowan (1996).
7 Chase et al. (2008).

CHAPTER 12   TRANSPORTATION

1 It is inherently difficult to evaluate synergistic effects of multiple types of actions within societies, due to the difficulty statistical methods have in evaluating the effect of multiple, sometimes
   hypothetical actions within complex systems. Studies such as Ewing et al. (2008) have shown
   that compact land use development alone can produce a 20–40 percent decrease in VMT for
   each increment of new development.
2 See for example American Association of State Highway and Transportation Officials (2004).
3 See Newman and Kenworthy (1999).
4 Cullingworth and Nadin (2002, p. 1); Rio2016, "Rio Bets on Bikes," 2012, available at http://www.
   rio2016.org/en/news/news/rio-de-janeiro-bets-on-bikes;   EMBARQ,   "Bringing   Bicycles   to
   Favelas," 2012, available at http://www.embarq.org/en/news/12/07/12/bringing-bicycles-favelas.
5 For more information on this and other BRT projects, see US Department of Transportation
   (2005) or http://fta.dot.gov/about/12351_4240.html.
6 See for example Cervero (1998).
7 Adapted from Stephen Marshall, "The Challenge of Sustainable Transport," in Layard et al.
   (2001).
8 Warner (1962).
9 Calthorpe (2011).

CHAPTER 13   HOUSING, FOOD, AND HEALTH

1 See the Vancouver Declaration on Human Settlements, Section III (8), available at http://
   habitat.igc.org/vancouver/van-decl.htm.
2 See Marcus and Sarkissian (1986); Marcus and Francis (1990); Francis (2003); Buchwald
   (2003).
3 Livable Housing Australia (2012).
4 Philpott (2006).
5 Food and Water Watch (2010).
6 Coleman-Jensen et al. (2012).
7 http://en.wikipedia.org/wiki/Decline_of_the_Roman_Empire#Lead_poisoning.
8 Centers for Disease Control, "Adult Obesity Facts," 2012, available at http://www.cdc.gov/
   obesity/data/adult.html.
9 See http://www.healthycityinitiative.com/.
10 City of Philadelphia (2010).
11 World Health Organization, "Making cities healthy for everyone: European mayors share experiences," 2012, available at http://www.euro.who.int/en/what-we-do/health-topics/environment-
   and-health/urban-health/news/news/2012/06/making-cities-healthy-for-everyone-european-
   mayors-share-experiences.
12 See http://www.belfasthealthycities.com/PDFs/Healthier%20Neighbourhoods%20Report.pdf.

CHAPTER 14   GREEN ARCHITECTURE AND BUILDING

1 Randolph and Masters (2008).
2 Birkeland (2002, p. 13).
3 Winter (2001b, p. 52).

4 For BedZED see http://www.oneplanetcommunities.org/communities/bedzed/. It is debatable whether BedZED is in fact zero net energy; the wood-fired combined heat and power plant was not operating as of this writing. For West Village, see http://westvillage.ucdavis.edu/ and Wheeler and Segar (2012).

5 See http://www.toronto.ca/environment/initiatives/cooling.htm.

6 See http://www.terramai.com/.

7 "Paving Without Asphalt or Concrete," *Environmental Building News*, 8 (11), 1999, available at www.buildinggreen.com/products/road_oyl.html.

8 Easton (1996).

9 Van der Ryn and Cowan (1996); see also Lyle (1994, 1999).

10 Kelbaugh (1997, pp. 52ff.).

11 Bacchus (2012).

12 See McCamant and Durrett (1994).

13 See Norwood and Smith (1995).

CHAPTER 15   SOCIAL EQUITY AND ENVIRONMENTAL JUSTICE

1 See Phillips (2002, 2008).

2 Agyeman et al. (2003); Agyeman (2005).

3 See for example Norberg-Hodge and Goering (1986); Khor (2007); and Shiva (2010).

4 See http://laborcenter.berkeley.edu/livingwage/resources.shtml.

5 US Bureau of the Census, *Statistical Abstract of the United States 2012*, Washington, DC available at http://www.census.gov/compendia/statab/.

6 See http://www.sentencingproject.org/template/page.cfm?id=122.

7 Uggen et al. (2012).

8 Bullard (1990, 2005).

9 Available at http://www.indiaresource.org/issues/energycc/2003/baliprinciples.html.

CHAPTER 16   ECONOMIC DEVELOPMENT

1 See for example Barnes (2006); Hawken (2010); Hawken et al. (1999); Daly (2007); and Barbier and Markandya (2012).

2 See Logan and Molotch (1987).

3 Daly and Cobb (1989).

4 Redefining Progress, "Genuine Progress Indicator," available at http://rprogress.org/ sustainability_indicators/genuine_progress_indicator.htm.

5 Waring (1988)

6 Hawken (2010); Hawken et al. (1999); Barbier and Markandya (2012).

7 Lovins (1977); Hawken et al. (1999).

8 Henderson (1995, 2006).

9 Schumacher (1973).

10 Daly (1973).

11 Logan and Molotch (1987).

12 Jacobs (1969, 1984); Shuman (1998, 2006, 2012).

13 C. Wright Mills, cited in Green (1973).

14 See http://jccu.coop/eng/aboutus/index.php.

CHAPTER 17   POPULATION

1 Ehrlich and Holdren (1971).

2 Devall and Sessions (1985, p. 70).

3   UN News Center, "Global Population to Pass 10 Billion by 2100, UN Projections Indicate," 2011, available at http://www.un.org/apps/news/story.asp?NewsID=38253#.UIXBU04tWDM.

4   Extrapolated from per capita figures in Baumert et al. (2005, p. 22).

CHAPTER 18   GOVERNANCE AND SOCIAL ECOLOGY

1   Barber (2004).
2   Prugh et al. (2000).
3   Library of Congress, "Campaign Finance: Germany," available at http://www.loc.gov/law/help/campaign-finance/germany.php.
4   US Bureau of the Census, "Voting and Registration," available at http://www.census.gov/hhes/www/socdemo/voting/.
5   UK political info, "General Election Turnout 1945–2010," available at http://www.ukpolitical.info/Turnout45.htm
6   http://en.wikipedia.org/wiki/Compulsory_voting.
7   Marquetti et al. (2012).

CHAPTER 19   INTERNATIONAL PLANNING

1   Morris (1994, pp. 305–6).
2   Gertler (1978).
3   Section 28.3. Text of Agenda 21 available at http://sustainable development.un.org/content/documents/Agenda21.pdf.
4   ICLEI (1997).
5   Ashley (2002, p. 460).
6   UN Human Settlements Programme (1996).
7   UN (2012).
8   For more information see http://whc.unesco.org/nwhc/pages/sites/main.htm.
9   Frost (2001, p. 213).
10 FERN (1998).
11 Colwell (2001, pp. 220–1).
12 Available at http://ec.europa.eu/environment/newprg/final.htm.
13 For further information, see http://europa.eu/about-eu/basic-information/decision-making/treaties/index_en.htm.
14 Cullingworth and Nadin (2002, p. 34).
15 Rydin (1998, pp. 3, 129ff.).
16 Available at http://www.infocooperare.ro/Files/Sustainable%20Spatial%20Development_2009 319345259.pdf.
17 http://epp.eurostat.ec.europa.eu/portal/page/portal/sdi/indicators
18 http://sustainable-cities.eu.
19 Chaibva (2000).
20 For further information, see www.earthcharter.org.
21 See for example Klein (2007); Korten (1995); Gross (1980).
22 Scientific Assessment Panel of the Montreal Protocol on Substances that Deplete the Ozone Layer, Scientific Assessment of Ozone Depletion, 2002, available at http://www.esrl.noaa.gov/csd/assessments/ozone/2010/report.html.
23 For more information, see www.biodiv.org.
24 See Barry G. Rabe, "Sustainability in a Regional Context: The Case of the Great Lakes Basin," in Mazmanian and Kraft (2009).
25 UN Department of Economic and Social Affairs (2011).
26 Rabinovich and Leitman (1996).
27 See for example Braidotti et al. (1994).

CHAPTER 20   NATIONAL PLANNING

1 See for example Jackson (1985).
2 See for example Markusen et al. (1991).
3 Macey (2001).
4 *Southern Burlington County NAACP v. Mount Laurel*, 67 N.J. 151,336 A.2d 713, appeal dismissed and cert. Denied, 423 US 808 (1975).
5 Hoch et al. (2000, pp. 250–1); Fair Share Housing Center, "Mount Laurel Doctrine," 2012, available at http://fairsharehousing.org/mount-laurel-doctrine/.
6 Parkinson and Roseland (2002). See http://fcm.ca/home/awards/fcm-sustainable-communities-awards.htm.
7 Urban Task Force (1999).
8 Stern et al. (2006).
9 National Round Table (2012). For further information, see http://nrtee-trnee.ca/.
10 President's Council on Sustainable Development (1996).
11 See http://www.hrlc.org.au/about-us/impacts/.
12 Information about national green plans is available from the Resource Renewal Institute in San Francisco, an NGO devoted to promoting these national environmental planning policies, at www.rri.org.
13 Resource Renewal Institute, "Case Study: The Netherlands' National Environmental Policy Plan," San Francisco, 1999.
14 See, for example, Van Eeten and Roe (2000).
15 See Alexander (2000, pp. 91–2); Beatley (2000b, pp. 98–100).
16 Rowe and Fudge (2003).
17 See http://www.miljomal.se/Environmental-Objectives-Portal/.
18 See Keil (1999), available at www.rri.org/envatlas/europe/germany/de-index.html.
19 Ravetz (2000, p. 146).
20 See Peter Hall, "The View from London Centre: London Docklands Development Corporation," in Blowers and Evans (1997, p. 131).
21 Cullingworth and Nadin (2002, p. 208).
22 DTLR (2001b).
23 Ibid, Section 4.7.
24 Warburton (2002, pp. 82–4).
25 DTLR (2000a).
26 DTLR (2001a).
27 ODPM (2003).
28 Resource Renewal Institute, "Case Study: New Zealand's Resource Management Act," San Francisco, 2000, available at www.rri.org/envatlas/oceania/new-zealand/nz-index.html.
29 For more information, see www.deh.gov.au/esd/.
30 Barrett and Usui (2002).
31 Resource Renewal Institute, "Development of Japan's Environmental Policy," San Francisco, 2000, available at www.rri.org/envatlas/asia/japanljp-dev.html.
32 See http://www.japanfs.org/en/pages/028625.html.
33 See http://www.japanfs.org/en/pages/009397.html.
34 President's Council on Sustainable Development (1996).
35 Hall (1992, p. 223).
36 Rydin (1998, p. 110).

CHAPTER 21   STATE AND PROVINCIAL PLANNING

1 Happaerts (2008).
2 See http://www.msbg.umn.edu/overview.html.
3 Bollens (1992).

4 Abbott (1997).
5 See for example http://smartgrowth.umd.edu/smartgrowthinmaryland.html; http://www.mdp. state.md.us/OurWork/smartGrowth.shtml.
6 For a detailed description of the Oregon land use planning system, see Abbott et al. (1994); Walker and Hurley (2011); Oregon DLCD (1997).
7 Available at http://www.nj.gov/state/planning/plan.html.
8 See http://www.njfuture.org/.
9 Available at www.mnplan.state.mn.us/pdf/2000/eqb/ModelOrdWhole.pdf.
10 See http://www.toolkit.bc.ca/icsp.
11 See http://www.mddep.gouv.qc.ca/developpement/strategie_gouvernementale/index_en.htm.
12 California Energy Commission, "Building Energy Efficiency Standards: Frequently Asked Questions," available at http://www.energy.ca.gov/title24/2013standards/rulemaking/documents/ 2013_Building_Energy_Efficiency_Standards_FAQ.pdf.
13 See the American Council for an Energy-Efficient Economy's 2011 State Energy Efficiency Scorecard, available at http://aceee.org/sector/state-policy/scorecard.
14 California Integrated Waste Management Board, http://www.calrecycle.ca.gov/lgcentral/rates/ diversion/2000.htm.
15 Florida (2005).
16 See http://www.scotland.gov.uk/About/Performance/scotPerforms/performance.
17 See http://www.nrg4sd.org/content/saluto-sino.
18 Strategy for Sustainable Development of Catalonia, available in English through http://www. gencat.cat/.
19 Parayil (2000).
20 Véron (2001).
21 Oommen (2008).

CHAPTER 22 REGIONAL PLANNING

1 Calthorpe and Fulton (2001).
2 Wheeler (2002b); Pastor et al. (2009).
3 See Greenstein and Wiewel (2000); Altshuler et al. (1999); Katz (2000); Pastor et al. (2001); Bullard (2007); Benner and Pastor (2012).
4 See Calthorpe and Fulton (2001); Daniels (1999); Calthorpe (2011).
5 See Peirce (1993); Ohmae (1995); Christopherson and Clark (2007).
6 Geddes (1915, p. 18).
7 Ibid, p. 16.
8 Howard (1902 [1898], pp. 51–2).
9 See Luccarelli (1995); Susman (1976).
10 Odum and Estill Moore (1938).
11 Friedmann and Alonso (1964, p. 1).
12 Ibid, p. 497.
13 See Webber et al. (1964).
14 For more information, including an updated Chesapeake 2000 agreement, see www. chesapeakebay.net.
15 Dramstad et al. (1996); Peck (1998).
16 Cartier (2005). See also Friedmann (2005).
17 Bai et al. (2011).
18 For more information, see the websites of the Metro Council (http://www.metro-region.org/) and 1000 Friends of Oregon (http://www.friends.org/); or Abbott et al. (1994); Abbott (1997); and Abbott (2011).
19 On Plan Bay Area see http://www.onebayarea.org/; on SB 375 see http://www.arb.ca.gov/cc/ sb375/sb375.htm.
20 Bartholomew (1999); Wheeler (2002b); Pastor et al. (2009).

21 For more information, see www.cbf.org/.
22 For more information, see www.keeptahoeblue.com/.
23 See http://www.london21.org/.
24 Tzung (2001); Lee (2012).
25 For more information, see www.atlantaregional.com/.
26 For more information, see www.mtc.ca.gov/.
27 Bae and Jun (2003).
28 Neuman (2000).
29 See Calthorpe (1993); Calthorpe and Fulton (2001); Downs (1994).
30 See http://www.sierranevada.ca.gov/our-work.
31 See http://www.evergladesplan.org/index.aspx.
32 Yaro and Hiss (1996); RPA (2012).
33 Box et al. (2001, pp. 210–12); http://www.wildlifetrusts.org/living-landscape/schemes/telford-and-wrekin-forest.
34 For a good overview of LA's air quality planning struggles, see Daniel A. Mazmanian, "Los Angeles' Transition from Command-and-Control to Market-based Clear Air Policy Strategies and Implementation," in Mazmanian and Kraft (2009).
35 For more information, see www.chesapeakebay.net/ or http://www.chesapeakebay.net/content/publications/cbp_12081.pdf.
36 For a history of Great Lakes cleanup efforts, see Barry G. Rabe, "Sustainability in a Regional Context: The Case of the Great Lakes Basin," in Mazmanian and Kraft (2009). On the St Lawrence Plan, see http://www.cglg.org/projects/water/legal.asp.
37 For more information, see www.calfed.water.ca.gov, http://deltacouncil.ca.gov/.
38 See Rabe, op. cit (Note 36).
39 See http://healthylakes.org.
40 See for example Orfield (1997, 2002); Orfield and Luce (2010); and Powell (2000, 2012).
41 Orfield (1997, p. 87).
42 See http://www.thestrategycenter.org/project/bus-riders-union/.
43 See Robert D. Bullard and coauthors. "Dismantling Transportation Apartheid: The Quest for Equity," in Bullard et al. (2000, pp. 54–5).
44 See for example Renner (1991).
45 Gottlieb (2002).

CHAPTER 23   LOCAL PLANNING

1 See for example Hom (2002).
2 Danielson and Doig (1983).
3 Eisenberg and Yost (2001).
4 See for example Rabinovitch and Leitman (1996); Irazábel (2005); and Moore (2007).
5 Berke and Conroy (2000).
6 Brown and Dohr (2000); Nadin et al. (2000).
7 Mittler (2001).
8 Portnoy (2003, pp. 64–71).
9 Livingstone (2002).
10 Munt (2001, pp. 292–3).
11 City of London (2010).

CHAPTER 24   NEIGHBORHOOD PLANNING

1 See Perry (1939).
2 Riesman (1953).
3 Bellah et al. (1985).

4   Putnam (2000).
5   See for example Moudon and Hess 2000.
6   Fogelsong (1986).
7   Bucher (2008).
8   Schuyler (1996).
9   Schuyler (1986, p. 41).
10 In Nolan (1916, pp. 94–5).
11 See for example Mumford (1938, 1961).
12 Whyte (1968).
13 Jacobs (1961).
14 See for example Marcus and Sarkissian (1986); Marcus and Francis (1990).
15 Former Toronto Mayor John Sewell (1993, 2009) provides a fascinating description of how different eras of neighborhood design philosophy have shaped one city in his books *The Shape of the City: Toronto Struggles with Modern Planning* and *The Shape of the Suburbs: Understanding Toronto's Sprawl*.
16 Nelessen (1994, pp. 83ff.).
17 See Peter Hall, "Sustainable Cities or Town Cramming?" in Layard et al. (2001, pp. 101–13).
18 Rapoport (1975).
19 Rudlin and Falk (1999, p. 125).
20 Wheeler et al. (forthcoming).
21 These and other strategies are detailed in a guidebook I wrote entitled *Smart Infill: Creating More Livable Communities in the Bay Area*, published in 2002 by the San Francisco-based Greenbelt Alliance (Wheeler, 2002a), and revised in a second edition in 2008.
22 See for example Grant (2002).
23 A study carried out in 2000 by the Natural Resources Defense Council and the US Environmental Protection Agency of an infill subdivision in Sacramento vs. a greenfield counterpart found that the infill neighborhood substantially reduced driving and travel distances (NRDC/US Environmental Protection Agency, 2000).
24 Natural Resources Defense Council and US Environmental Protection Agency. 2000. *Environmental Characteristics of Smart Growth Neighborhoods: An Exploratory Case Study.* Washington, DC.
25 See   http://www.princes-foundation.org/what-we-do/projects/uk/poundbury-dorchester-teaching-poundbury.
26 See Jacobs (1993); Jacobs et al. (2001).
27 Jacobs (1961, pp. 363ff.).
28 Rudofsky (1969).
29 Appleyard (1981); Appleyard and Appleyard (2013).
30 Engwicht (1993).
31 See Riley (1998).

CHAPTER 25   SITE PLANNING

1   For a historical description of do-it-yourself housing in one metropolitan area, see Harris (1996).
2   For an excellent description of the rise of large-scale homebuilding, see Weiss (1987).
3   For examples of well-designed, context-sensitive affordable housing, see Jones et al. (1997).
4   Eisenberg and Yost (2001).
5   McHarg (1969).
6   See for example Van der Ryn and Cowan (1996); Lyle (1994, 1999); Todd and Todd (1993); and architect William McDonough's Hannover Principles (McDonagh, 1992).
7   Condon and Moriarty (1999).
8   Alexander et al. (1977); Alexander (1979).

CHAPTER 26 HOW DO WE GET THERE FROM HERE?

1 Owens and Cowell (2001, p. 170).
2 Tuchman (1984).
3 See Jacobs (1985).
4 See, for example, Hiss (1990); Seamon (1993).
5 Seamon (1993, p. 16).
6 Norgaard (1994).
7 Inglehart (1977, 1990, 1997).
8 A substantial literature relates to the various ways we structure our own realities, either internally within ourselves through cognitive processes or externally within society through institutions and culture. Particularly useful works include Berger and Luckmann (1966); Giddens (1984, 1994); and Castells (1996, 1997, 1998).
9 See among others Phillips (2002, 2008).
10 Giddens (1998).
11 Etzioni (1993).
12 Dryzek (1990, 2000).
13 Lerner (1996).
14 Giddens (1998, p. 65).
15 Ravetz (2000, p. 6).

BIBLIOGRAPHY

Abbott, Carl, "The Portland Region: Where Cities and Suburbs Talk to Each Other – and Often Agree," *Housing Policy Debate*, Winter 11–51, 1997.

——, *Portland in Three Centuries: The Place and the People*, Corvallis, OR: Oregon State University Press, 2011.

Abbott, Carl, Deborah Howe, and Sy Adler, *Planning the Oregon Way: A Twenty-Year Evaluation*, Corvallis, OR: Oregon State University Press, 1994.

Abramson, Alan U., Mitchell S. Tobin, and Matthew R. VanderGoot, "The Changing Geography of Metropolitan Opportunity: The Segregation of the Poor in U.S. Metropolitan Areas, 1970 to 1990," *Housing Policy Debate*, 6 (1), 45–72, 1995.

Adams, W.M., *Green Development: Environment and Sustainability in the Third World*, New York: Routledge, 1990.

Agyeman, Julian, *Sustainable Communities and the Challenge of Environmental Justice*, New York: New York University Press, 2005.

Agyeman, Julian, Robert D. Bullard, and Bob Evans, *Just Sustainabilities: Development in an Uneven World*, Cambridge, MA: MIT Press, 2003.

Alberti, Marina, *Advances in Urban Ecology: Integrating Humans and Ecological Processes in Urban Ecosystems*, New York: Springer, 2009.

Alexander, Christopher, *The Timeless Way of Building*, New York: Oxford University Press, 1979.

Alexander, Christopher, Sara Ishikawa, Murray Silverstein, with Max Jacobson, Ingrid Fiksdahl-King, and Shlomo Angel, *A Pattern Language*, New York: Oxford University Press, 1977.

Alexander, Ernest R., "Netherlands' Planning: The Higher Truth," *Journal of the American Planning Association*, 67 (1), 91–2, 2000.

Altshuler, Alan, Harold Wolman (eds), with Faith Mitchell and William Morrill, *Governance and Opportunity in Metropolitan America*, Washington, DC: National Academy Press, 1999.

American Association of State Highway and Transportation Officials, *Guide for the Planning, Design, and Operation of Pedestrian Facilities*, Washington, DC: American Association of State Highway and Transportation Officials, 2004.

Appleyard, Donald, *Livable Streets*, Berkeley, CA: University of California Press, 1981.

Appleyard, Donald, and Bruce Appleyard, *Livable Streets 2.0.*, Washington, DC: Island Press, 2013.

Arnstein, Sherry, "A Ladder of Citizen Participation," *Journal of the American Institute of Planners*, 8(3), 1969.

Ashley, Mike, "Local Government and the WSSD," *Local Environment* (7) 4, 459–63, 2002.

Atkinson, Anthony B., Thomas Piketty, and Emmanuel Saez, "Top Incomes in the Long Run of History," *Journal of Economic Literature*, 49 (1), 3–71, 2011.

Audirac, Ivonne, *Rural Sustainable Development in America*, New York: John Wiley & Sons, 1997.

Bacchus, Jamy, "Title 24: Saving Homeowners Money, Reducing Pollution, and Creating Jobs: What's Not to Like?" 2012, blog posting at http://switchboard.nrdc.org/blogs/jbacchus/title_24_saving_homeowners_mon.html.

Bae, Chang-Hee Christine, and Myung-Jin Jun, "Counterfactual Planning: What if There Had Been No Greenbelt in Seoul?" *Journal of Planning Education and Research*, 22, 374–83, 2003.

Bai, Y., Huang, Y., Wang, M., Huang, S., Sha, C., and Ruan, J., "The Progress of Ecological Civilization Construction and Its Indicator System in China," *Shengtai Xuebao/Acta Ecologica Sinica* 31 (20) 6295–304, 2011.

Barber, Benjamin R., *Strong Democracy: Participatory Politics for a New Age*, Berkeley, CA: University of California Press, 2004 [1984].

Barbier, Edward B., and Anil Markandya, *A New Blueprint for a Green Economy*, London: Earthscan, 2012.

Barnes, Peter, *Capitalism 3.0: A Guide to Reclaiming the Commons*, San Francisco, CA: Berrett-Koehler, 2006.

Barrett, Brendan, and Mikoto Usui, "Local Agenda 21 in Japan: Transforming Local Environmental Governance," *Local Environment*, 7 (1), 49–67, 2002.

Barrett, Carol D., *Everyday Ethics for Practicing Planners*, Washington, DC: American Institute of Certified Planners, 2001.

Bartholomew, Keith, "The Evolution of American Nongovernmental Land Use Planning Organizations," *Journal of the American Planning Association*, 65 (4), 357–63, 1999.

Baumert, Kevin A., Timothy Herzog, and Jonathan Pershing, *Navigating the Numbers: Greenhouse Gas Data and International Climate Policy*, Washington, DC: World Resources Institute, 2005, available at http://pdf.wri.org/navigating_numbers.pdf.

Beatley, Timothy, *Habitat Conservation Planning*, Austin, TX: University of Texas Press, 1994.

——, "The Many Meanings of Sustainability" and "Planning and Sustainability: The Elements of a New (Improved?) Paradigm," *Journal of Planning Literature*, 9 (4), 339–42, 383–95, 1995.

——, *Green Urbanism: Learning from European Cities*, Washington, DC: Island Press, 2000a.

——, "Dutch Green Planning More Reality Than Fiction," *Journal of the American Planning Association*, 67 (1), 98–100, 2000b.

——, *Native to Nowhere: Sustaining Home and Community in a Global Age*, Washington, DC: Island Press, 2004.

——, *Biophilic Cities: Integrating Nature into Urban Design and Planning*. Washington, DC: Island Press, 2011.

Beatley, Timothy, and Kristy Manning, *The Ecology of Place: Planning for Environment, Economy, and Community*, Washington, DC: Island Press, 1997.

Beauregard, Robert, "Between Modernity and Postmodernity: The Ambiguous Position of U.S. Planning," *Environment and Planning D: Society and Space*, 7 (4), 381–95, 1989.

Becker, Egon, and Thomas Jahn (eds), *Sustainability and the Social Sciences: A Cross-disciplinary Approach to Integrating Environmental Considerations into Theoretical Reorientation*, London: Zed Books, 1999.

Belenky, Mary Field, Blythe McVicker Clinchy, Nancy Rule Goldberger, and Jill Mattuck Tarule, *Women's Ways of Knowing: The Development of Self, Voice, and Mind*, New York: Basic Books, 1986.

Bellah, Robert, Richard Madsen, William M. Sullivan, and Steven M. Tipton, *Habits of the Heart: Individualism and Commitment in American Life*, Berkeley, CA: University of California Press, 1985.

Benfield, F. Kaid, Matthew D. Raimi, and Donald D.T. Chen, *Once There Were Greenfields*, Washington, DC: Natural Resources Defense Council, 1999.

Benfield, F. Kaid, Jutka Terris, and Nancy Vorsanger, *Solving Sprawl: Models of Smart Growth in Communities Across America*, New York: Natural Resources Defense Council, 2001.

Benner, Chris, and Manuel Pastor, *Just Growth: Inclusion and Prosperity in America's Metropolitan Regions*, New York: Routledge, 2012.

Berger, Peter L., and Thomas Luckmann, *The Social Construction of Reality*, New York: Doubleday, 1966.

Berke, Philip R., and Maria Manta Conroy, "Are We Planning for Sustainable Development?" *Journal of the American Planning Association*, 66 (1), 21–34, 2000.

Berry, Thomas, *The Great Work: Our Way Into the Future*, New York: Bell Tower, 1999.

Birch, Eugenie, and Susan Wachter, *Growing Greener Cities: Urban Sustainability in the Twenty-First Century*, Philadelphia, PA: University of Pennsylvania Press, 2008.

Birkeland, Janis, *Design for Sustainability, A Sourcebook of Integrated Eco-logical Solutions*, London: Earthscan, 2002.

Blakely, Edward J., and Mary Gail Snyder, *Fortress America: Gated and Walled Communities in the United States*, Washington, DC: Brookings Institution Press, 1997.

Blanco, Hilda, *How to Think about Social Problems: American Pragmatism and the Idea of Planning*, Westport, CT: Greenwood Press, 1994.

Blowers, Andrew (ed.), *Planning for a Sustainable Environment: A Report by the Town and County Planning Association*, London: Earthscan, 1993.

Blowers, Andrew, and Bob Evans (eds.), *Town Planning into the 21st Century*, London: Routledge 1997.

Bollens, Scott A., "State Growth Management: Intergovernmental Frameworks and Policy Objectives," *Journal of the American Planning Association*, 58 (4), 454–66, 1992.

Boswell, Michael R., Adrienne I. Greve, and Tammy L. Seale, "An Assessment of the Link Between Greenhouse Gas Emissions Inventories and Climate Action Plans," *Journal of the American Planning Association*, 76 (4), 451–62, 2010.

Boswell, Michael R., Adrienne I. Greve, and Tammy L. Seale, *Local Climate Action Planning*, Washington, DC: Island Press, 2012.

Boulding, Kenneth, *The Meaning of the Twentieth Century: The Great Transition*, New York: Harper & Row, 1964.

Box, John, Veronica Cossons, and Jan McKelvey, "Sustainability, Biodiversity, and Land Use Planning," *Town & Country Planning*, July/August, 210–12, 2001.

Bragado, Nancy, Judy Corbett, and Sharon Sprowls, *Building Livable Communities: A Policymaker's Guide to Infill Development*, Sacramento, CA: Local Government Commission, 1995.

Braidotti, Rosi, Ewa Charkiewicz, Sabine Hausler, and Saskia Wieringa, *Women, the Environment and Sustainable Development: Towards a Theoretical Synthesis*, London: Zed Books, 1994.

Brand, Stewart, *How Buildings Learn: What Happens After They're Built*, New York: Viking, 1994.

Brown, Caroline, and Stefanie Dohr, "Understanding Sustainability and Planning in England: An Exploration of the Sustainability Content of Planning Policy at the National, Regional and Local Levels," in Rydin, Yvonne, and Andy Thomley (eds), *Planning in the UK: Agendas for the New Millennium*, Aldershot: Ashgate, 2000.

Brown, Lester R., *Building a Sustainable Society*, New York: Norton, 1981.

Brown, Lester R., Christopher Flavin, and Sandra Postel, *Saving the Planet: How to Shape an Environmentally Sustainable Global Economy*, New York: Norton, 1991.

Bucher, Genevieve Susanne, *Implementing Sustainability in Surrey: Amending the East Clayton Neighborhood Concept Plan*, Masters Thesis, Simon Fraser University, 2008.

Buchwald, Emilie, *Toward the Livable City*, Minneapolis, MN: Milkweed Editions, 2003.

Bullard, Robert D., *Dumping in Dixie: Race, Class, and Environmental Quality*, Boulder, CO: Westview Press, 1990.

— (ed.), *The Quest for Environmental Justice: Human Rights and the Politics of Pollution*. San Francisco, CA: Sierra Club Books, 2005.

——, *Growing Smarter: Achieving Livable Communities, Environmental Justice, and Regional Equity*. Cambridge, MA: MIT Press, 2007.

Bullard, Robert D., Glenn S. Johnson, and Angel O. Torres (eds) *Sprawl City: Race, Politics, and Planning in Atlanta*, Washington, DC: Island Press, 2000.

Burnham, Daniel, *Plan of Chicago*, Chicago, IL: Commercial Club, 1909.

Callenbach, Ernest, *Ecotopia*, New York: Bantam Books, 1975.

——, *Ectopia Emerging*, Berkeley, CA: Banyan Tree Books, 1981.

Calthorpe, Peter, *The Next American Metropolis: Ecology, Community and the American Dream*, New York: Princeton Architectural Press, 1993.

——, *Urbanism in the Age of Climate Change*, Washington, DC: Island Press, 2011.

Calthorpe, Peter, and William Fulton, *The Regional City: Planning for the End of Sprawl*, Washington, DC: Island Press, 2001.

Capra, Fritjof, *The Web of Life: A New Scientific Understanding of Living Systems*, New York: Anchor Books, 1996.

Carson, Rachel, *Silent Spring*, New York: Fawcett Crest, 1962.

Cartier, Carolyn, "Scale Relations and Administrative Hierarchy," in Ma, Laurence J.C., and Fulong Wu, *Restructuring the Chinese City: Changing Society, Economy, and Space*. London: Routledge, 2005.

Castells, Manuel, *The Urban Question: A Marxist Approach*, London: Edward Arnold, 1977.

——, *City, Class and Power*, New York: St Martin's Press, 1978.

——, *The City and the Grassroots*, Berkeley, CA: University of California Press, 1983.

——, *The Information Age, Vol. 1: The Rise of the Network Society*, Cambridge, MA: Blackwell, 1996.

——, *The Information Age, Vol. 2: The Power of Identity*, Cambridge, MA: Blackwell, 1997.

——, *The Information Age, Vol. 3: End of Millennium*, Cambridge, MA: Blackwell, 1998.

Cervero, Robert, *The Transit Metropolis: A Global Inquiry*, Washington, DC: Island Press, 1998.

Chaibva, Shem, "EIA Training for Urban Local Authorities in Africa," *Local Environment* (5) 1, 97–106, 2000.

Chase, John, Margaret Crawford, and John Kaliski (eds), *Everyday Urbanism*, New York: Random House, 2008.

Christianson, Gale B., *Greenhouse: The 200-Year Story of Global Warming*, New York: Walker and Company, 1999.

Christopherson, Susan, and Jennifer Clark, *Remaking Regional Economies: Power, Labor, and Firm Strategies in the Knowledge Economy*. London: Routledge, 2007.

City of London, *London 2012 Sustainability Plan Summary*, 2010, available at http://www.london2012.com/about-us/publications/publication=london-2012-sustainability-plan-summary/.

City of Philadelphia, *Philadelphia 2035: Planning and Zoning for a Healthier City*, 2010, available at http://phila2035.org/wp-content/uploads/2011/02/Phila2035_Healthier_City_Report.pdf.

Coleman-Jensen, Alisha, Mark Nord, Margaret Andrews, and Steven Carlson, *Household Food Security in the US 2011*, Washington, DC: USDA, 2012, available at http://www.ers.usda.gov/publications/err-economic-research-report/err141.aspx.

Colwell, Adrian, "Setting the Direction for EU Environmental Policy," *Town & Country Planning*, July/August, 220–1, 2001.

Commoner, Barry, *The Closing Circle: Nature, Man and Technology*, New York: Knopf, 1971.

Condon, Patrick, and Stacy Moriarty (eds), *Second Nature: Adapting LA's Landscape for Sustainable Living*, Los Angeles, CA: Treepeople, 1999.

Congress for the New Urbanism, *Charter of the New Urbanism*, New York: McGraw-Hill, 1999.

Corbett, Judy, *A Policymaker's Guide to Transit-oriented Development*, Sacramento, CA: Local Government Commission, 1996.

Corbett, Judy, and Michael Corbett, *Designing Sustainable Communities: Learning from Village Homes*, Washington, DC: Island Press, 2000.

Corbett, Judy, and Paul Zykofsky, *Building Livable Communities: A Policymaker's Guide to Transit-Oriented Development*, Sacramento, CA: Center for Livable Communities/Local Government Commission, 1996.

Costanza, Robert, Herman E. Daly, and Joy A. Bartholomew, "Goals, Agenda, and Policy Recommendations for Ecological Economics," in Costanza, Robert (ed.), *Ecological Economics: The Science and Management of Sustainability*, New York: Columbia University Press, 1991.

Counsell, David, and Graham Haughton, "Sustainability Appraisal – Delivering More Sustainable Regional Planning Guidance?" *Town & Country Planning*, January, 14–17, 2002.

Cox, Wendell, "Special Report: Infill in US Urban Areas," posting on newgeography.com, 2009, available at http://www.newgeography.com/content/00852-special-report-infill-us-urban-areas.

Cronon, William, *Nature's Metropolis: Chicago and the Great West*, New York: Norton, 1991.

Cullingworth, J.B., and Vincent Nadin, *Town and Country Planning in the UK*, 13th edn, London and New York: Routledge, 2002.

Daly, Herman E. (ed.), *Toward a Steady-State Economy*, San Francisco, CA: W.H. Freeman, 1973.

——, *Ecological Economics and Sustainable Development: Selected Essays of Herman Daly*, Northampton, MA: Edward Elgar, 2007.

Daly, Herman E., and John B. Cobb, Jr, *For the Common Good: Redirecting the Economy Toward Community, the Environment, and a Sustainable Future*, Boston, MA: Beacon Press, 1989.

Daly, Herman E., and Joshua Farley, *Ecological Economics: Principles and Applications*, Washington, DC: Island Press, 2010.

Daniels, Thomas L., *When City and Country Collide: Managing Growth in the Metropolitan Fringe*, Washington, DC: Island Press, 1999.

Danielson, Michael N., and Jameson W. Doig, *New York: The Politics of Urban Regional Development*, Berkeley, CA: University of California Press, 1983.

Danks, Sharon, *Asphalt to Ecosystems: Design Ideas for Schoolyard Transformation*, Oakland: New Village Press, 2010.

Dannenberg, Andrew, Howard Frumkin, and Richard Jackson, *Making Healthy Places: Designing and Building for Health, Well-being, and Sustainability*, Washington, DC: Island Press, 2011.

Davidoff, Paul, "Advocacy Planning," *Journal of the American Institute of Planners*, 21 (4), 1968.

Dear, Michael J., *The Postmodern Urban Condition*, Malden, MA: Blackwell, 2000.

Devall, Bill, and George Sessions, *Deep Ecology: Living as if Nature Mattered*, Salt Lake City, UT: Peregrine Smith Books, 1985.

Downs, Anthony, *New Visions for Metropolitan America*, Washington, DC: Brookings Institution, 1994.

——, "Smart Growth: Why We Discuss It More than We Do It," *Journal of the American Planning Association*, 71 (4), 367–80, 2005.

Dramstad, Wenche E., James D. Olson, and Richard T.T. Forman, *Landscape Ecology Principles in Landscape Architecture and Land-use Planning*, Cambridge, MA: Harvard University GSD and Washington, DC: Island Press, 1996.

Dryzek, John S., *Discursive Democracy: Politics, Policy, and Political Science*, New York: Cambridge University Press, 1990.

——, *Deliberative Democracy and Beyond: Liberals, Critics, Contestations*, New York: Oxford University Press, 2000.

DTLR, *Planning Policy Guidance 3: Housing*, (Department of Transport, Local Government and the Regions), London: The Stationery Office, 2000a.

——, *Good Practice Guide on Sustainability Appraisal of Regional Planning Guidance*, (Department of Transport, Local Government and the Regions), London: The Stationery Office, 2000b.

——, *Planning Policy Guidance 13: Transport*, (Department of Transport, Local Government and the Regions), London: The Stationery Office, 2001a.

——, *Planning Green Paper Planning: Delivering a Fundamental Change*, (Department of Transport, Local Government and the Regions), London: The Stationery Office, 2001b.

Duany, Andres, and Elizabeth Plater-Zyberk, *City and Placemaking Principles*, New York: Rizzoli, 1991.

Duany, Andres, and Emily Talen, "Transect Planning," *Journal of the American Planning Association*, 68 (3), 245, 2002.

Duany, Andres, Elizabeth Plater-Zyberk, and Jeff Speck, *Suburban Nation: The Rise of Sprawl and the Decline of the American Dream*, New York: North Point, 2000.

Dye, Richard F., and Richard W. England (eds), *Land Value Taxation: Theory, Evidence, and Practice*, Cambridge, MA: Lincoln Institute of Land Policy, 2009, available at http://www.lincolninst.edu/pubs/1568_Land-Value-Taxation.

Easton, David, *The Rammed Earth House*, White River Junction, VT: Chelsea Green Publishers, 1996.

Eckerberg, Katerina, and Bjorn Forsberg, "Implementing Agenda 21 in Local Government: The Swedish Experience," *Local Environment*, 3 (1), 333–47, 1998.

Edwards, Michael, *Civil Society*, 2nd edn, Cambridge, MA: Polity Press, 2009.

Ehrenfeld, John, and Nicholas Gertler, "Industrial Ecology in Practice: The Evolution of Interdependence," *Journal of Industrial Ecology*, 1 (1), 67–79, 1997, available at http://www. johnehrenfeld.com/Kalundborg.pdf.

Ehrlich, Paul R., and Ann Ehrlich, *The Population Bomb*, New York: Sierra Club, 1969.

Ehrlich, Paul R., and John P. Holdren, "Impact of Population Growth," *Science* 171, 1212–17, 1971.

Eisenberg, David, and Peter Yost, "Sustainability and Building Codes", *Environmental Building News*, 10 (9), 1, 8–15, 2001.

Elkin, Tim, Duncan McLaren, and Mayer Hillman, *Reviving the City: Towards Sustainable Urban Development*, London: Continuum, 1991.

Ellin, Nan, *Postmodern Urbanism*, rev. edn, New York: Princeton Architectural Press, 1999 [1996].

Engwicht, David, *Reclaiming Our Cities and Towns: Better Living with Less Traffic*, Philadelphia, PA: New Society Publishers, 1993.

Escobar, Arturo, *Encountering Development: The Making and Unmaking of the Third World*, Princeton, NJ: Princeton University Press, 1995.

Etzioni, Amitai, "Mixed Scanning: A 'Third' Approach to Decision-making," *Public Administration Review*, 27, 153–69, 1967.

——, *The Active Society: A Theory of Societal and Political Processes*, New York: Free Press, 1971.

——, *The Spirit of Community: Rights, Responsibilities, and the Communitarian Agenda*, New York: Crown Publishers, 1993.

European Environment Agency, *Greenhouse Gas Emission Trends and Projections in Europe 2012*, Copenhagen, 2012, available at http://www.eea.europa.eu/publications/ghg-trends-and-projections-2012/.

Evans, Peter (ed.), *Livable Cities? Urban Struggles for Livelihood and Sustainability*, Berkeley, CA: University of California Press, 2002.

Ewing, Reid, *Best Development Practices*, Chicago, IL: American Planning Association, 1996.

——, "Is Los Angeles-style Sprawl Desirable?" *Journal of the American Planning Association*, 63 (1), 107–27, 1997.

Ewing, Reid, Keith Bartholomew, Steve Winkelman, Jerry Walters, and Don Chen, *Growing Cooler: The Evidence on Urban Development and Climate Change*, Washington, DC: Urban Land Institute, 2007.

Ewing, Reid, Keith Bartholomew, Steve Winkelman, Jerry Walters, and Geoffrey Anderson, "Urban Development and Climate Change," *The Journal of Urbanism*, 1 (3), 201–16, 2008.

Faber, Daniel, and James O'Connor, "The Struggle for Nature: Environmental Crises and the Crisis of Environmentalism in the United States," *Capitalism, Nature, Socialism*, 1 (2), 12–39, 1988.

Fader, Steven, *Density by Design: New Directions in Residential Development*, Washington, DC: Urban Land Institute, 2000.

Fainstein, N.I., and S.S. Fainstein, "New Debates in Urban Planning: The Impact of Marxist Theory," *International Journal of Urban and Regional Research*, 3, 381–401, 1979.

Farina, Alma, *Landscape Ecology in Action*, Boston, MA: Kluwer Academic Publishers, 2000.

Farmland Information Center, "Natural Resources Inventory," available at http://www.farmlandinfo. org/agricultural_statistics/, 2012.

Feiden, Wayne, and Elizabeth Hamin, *Assessing Sustainability: A Guide for Local Governments*. Chicago, IL: APA Planning Advisory Service, 2011.

FERN, "The EU's Fifth Environmental Action Plan, September 1998," 1998, available at www.fern. org/pubs/archive/5eap.html.

Fernandez, Edesio (ed.), *Environmental Strategies for Sustainable Development in Urban Areas: Lessons from Africa and Latin America*, Brookfield, VT, and Aldershot: Ashgate, 1998.

Florida, Richard, *Cities and the Creative Class*, New York: Routledge, 2005.

Fogelsong, Richard, *Planning the Capitalist City: The Colonial Era to the 1920s*, Princeton, NJ: Princeton University Press, 1986.

Food and Water Watch, "Consolidation and Buyer Power in the Grocery Industry," available at http://documents.foodandwaterwatch.org/doc/RetailConcentration-web.pdf, 2010.

Forester, John, *Planning in the Face of Power*, Berkeley, CA: University of California Press, 1989.

——, *Critical Theory, Public Policy, and Planning Practice: Toward a Critical Pragmatism*, Albany, NY: SUNY Press, 1993.

Forman, Richard T.T., and Michel Godron, *Landscape Ecology*, New York: John Wiley & Sons, 1986.

Francis, Mark, *Urban Open Space: Designing for User Needs*, Washington, DC: Island Press, 2003.

Frank, Andre Gunder, *Capitalism and Underdevelopment in Latin America*, New York: Monthly Review Press, 1969.

Friedman, Andrew, "Recycling Redux: The Ups and Downs of the Waste-not Movement," *Planning*, 69 (10), 5–9, 2003.

Friedmann, John, *Planning in the Public Domain: From Knowledge to Action*, Princeton, NJ: Princeton University Press, 1987.

——, *China's Urban Transition*, Minneapolis, MN: University of Minnesota Press, 2005.

Friedmann, John, and William Alonso (eds), *Regional Development and Planning: A Reader*, Cambridge, MA: MIT Press, 1964.

Friedmann, John, and Clyde Weaver, *Territory and Function: The Evolution of Regional Planning*, London: Edward Arnold, 1979.

Freire, Paolo, *Pedagogy of the Oppressed*, trans. Myra Bergman Ramos, New York: Seabury Press, 1970.

Frost, Peter, "Urban Biosphere Reserves: Reintegrating People with the Natural Environment," *Town & Country Planning*, July/August, 213–15, 2001.

Geddes, Patrick, *Cities In Evolution: An Introduction to the Town Planning Movement and to the Study of Civics*, London: Williams & Norgate, 1915.

Gertler, Len, *Habitat and Land*, Vancouver: University of British Columbia Press, 1978.

Giddens, Anthony, *The Constitution of Society: Outline of the Theory of Structuration*, Berkeley, CA: University of California Press, 1984.

——, *Beyond Left and Right: The Future of Radical Politics*, Palo Alto, CA: Stanford University Press, 1994.

——, *The Third Way: The Renewal of Social Democracy*, Cambridge, UK: Polity Press, 1998.

Gilliard, Geoff, "Surrey Shifts Sustainability Status Quo with Neighborhood Concept Plan," *Plan*, 43 (1), 13–14, 2003.

Gilligan, Carol, *In a Different Voice: Psychological Theory and Women's Development*, Cambridge, MA: Harvard University Press, 1982.

Girardet, Herbert, *Creating Sustainable Cities*, Totnes: Green Books, 1999.

Gittell, Ross, and Avis Vidal, *Community Organizing: Building Social Capital as a Development Strategy*, Thousand Oaks, CA: Sage Publications, 1998.

Global Cities Project, *Building Sustainable Communities: An Environmental Guide for Local Government*, San Francisco, CA: The Center for the Study of Law and Politics, 1991.

Goldsmith, Edward, *The Way: An Ecological World View*, Boston, MA: Shambhala, 1993.

Goldsmith, Edward, and Robert Allen, *Blueprint for Survival*, Boston, MA: Houghton Mifflin, 1972.

Goldsmith, Edward, Martin Khor, Helena Norberg-Hodge, and Vandana Shiva, *The Future of Progress: Reflections on Environment & Development*, Bristol: The International Society for Ecology and Culture, 1992.

Gordon, Peter, and Harry Richardson, "Are Compact Cities a Desirable Planning Goal?" *Journal of the American Planning Association*, 63 (1), 95–107, 1997.

Gottdiener, Mark, *Planned Sprawl: Private and Public Interests in Suburbia*, Beverly Hills, CA: Sage Publications, 1977.

Gottlieb, Paul. D., *Growth Without Growth: An Alternative Economic Development Goal for Metropolitan Areas*, Washington, DC: The Brookings Institution, 2002, available at http://www.brookings.edu/research/reports/2002/02/useconomics-gottlieb.

Gottlieb, Robert, *Forcing the Spring: The Transformation of the American Environmental Movement*, Washington, DC: Island Press, 1993.

Granberg, Michael, and Ingemar Elander, "Local Governance and Climate Change: Reflections on the Swedish Experience," *Local Environment*, 12 (5), 537–48, 2007.

Grant, Jill, "Mixed Use in Theory and Practice: Canadian Experience with Implementing a Planning Principle," *Journal of the American Planning Association*, 68 (1), 71–84, 2002.

Green, Mark, "The Corporation and the Community," in Nader, Ralph, and Mark J. Green (eds), *Corporate Power in America*, New York: Grossman Publishers, 1973.

Greenbelt Alliance, *Reviving the Sustainable Metropolis: Guiding Bay Area Conservation and Development into the 21st Century*, San Francisco, CA: Greenbelt Alliance, 1989.

Greenstein, Rosalind, and Wim Wiewel (eds), *Urban-Suburban Interdependencies*, Cambridge, MA: Lincoln Institute of Land Policy, 2000.

Gross, Bertram, *Friendly Fascism: The New Face of Power in America*, New York: M. Evans, 1980.

Gruen, Victor, *The Heart of Our Cities: The Urban Crisis, Diagnosis and Cure*, New York: Simon & Schuster, 1964.

Habermas, Jurgen, *Moral Consciousness and Communicative Action*, trans. Christian Lenhardt and Shierry Weber Nicholsen, Cambridge, MA: MIT Press, 1995 [1983].

Hack, Gary, Eugenie Birch, Paul Sedway, and Mitchell Silver, (eds), *Local Planning: Contemporary Principles and Practices*, Washington, DC: International City/County Management Association, 2009.

Hall, Peter, *City and Regional Planning*, 3rd edn, New York: Routledge, 1992.

——, *Cities of Tomorrow: An Intellectual History of Urban Planning and Design in the Twentieth Century*, Cambridge, MA: Blackwell, 1996.

Hamm, Bernd, and Pandurang K. Muttagi (eds), *Sustainable Development and the Future of Cities*, London: Centre for European Studies, 1998.

Happaerts, Sander, "Governance for Sustainable Development at the Inter-Subnational Level," Katholieke Universiteit Leuven, 2008, available at http://steunpuntdo.be/papers/Paper_Sander_Governance_februari%202008.pdf.

Hardoy, Jorge E., Diana Mitlin, and David Satterthwaite, *Environmental Problems in Third World Cities*, London: Earthscan, 1992.

Harris, Jonathan M., Timothy A. Wise, Kevin P. Gallagher, and Neva R. Goodwin, *A Survey of Sustainable Development: Social and Economic Dimensions*, Washington, DC: Island Press, 2001.

Harris, Richard, *Unplanned Suburbs: Toronto's American Tragedy 1900–1950*, Baltimore, MD: Johns Hopkins University Press, 1996.

Hart, Maureen, *Guide to Sustainable Community Indicators*, Ipswich, MA: QLF/Atlantic Center for the Environment, 1995.

——, *Guide to Sustainable Community Indicators*, 2nd edn, North Andover, MA: Sustainable Measures, 1999.

Harvey, David, *The Condition of Postmodernity: An Enquiry into the Origins of Cultural Change*, Cambridge, MA: Blackwell, 1990.

——, *Social Justice and the City*, rev. edn, Athens, GA: University of Georgia Press, 2009.

Hawken, Paul, *The Ecology of Commerce: A Declaration of Sustainability*, rev. edn, New York: HarperCollins, 2010.

Hawken, Paul, Amory Lovins, and L. Hunter Lovins, *Natural Capitalism: Creating the Next Industrial Revolution*, London: Earthscan, 1999.

Hayden, Dolores, *Redesigning the American Dream: The Future of Housing, Work, and Family Life*, New York: W.W. Norton & Company, 1984.

——, *Building Suburbia: Green Fields and Urban Growth, 1820–2000*, New York: Pantheon, 2003.

Hayes, Denis, *Repairs, Reuse, Recycling: First Steps Toward a Sustainable Society*, Washington, DC: Worldwatch Institute, 1987.

Healey, Patsy, *Collaborative Planning: Shaping Places in Fragmented Societies*, Vancouver: University of British Columbia Press, 1997.

Helling, Amy, "Collaborative Visioning: Proceed with Caution!: Results from Evaluating Atlanta's Vision 2020 Project," *Journal of the American Planning Association*, 64 (3), 335–49, 1998.

Henderson, Hazel, *Paradigms in Progress: Life Beyond Economics*, Indianapolis, IN: Knowledge Systems, Inc, 1995.

——, *Ethical Markets: Growing the Green Economy*, White River Junction, VT: Chelsea Green, 2006.

Hess, Daniel J., *Localist Movements in a Global Economy: Sustainability, Justice, and Urban Development in the United States*, Cambridge, MA: MIT Press, 2009.

Hester, Randolph T., *Design for Ecological Democracy*, Cambridge, MA: MIT Press, 2006.

Hiss, Tony, *The Experience of Place*, New York: Vintage Books, 1990.

Hoch, Charles J., Linda C. Dalton, and Frank S. So (eds), *The Practice of Local Government Planning*, 3rd edn, Washington, DC: International City/County Management Association, 2000.

Holmberg, Johan (ed.), *Making Development Sustainable: Redefining Institutions, Policy, and Economics*, Washington, DC: Island Press, 1992.

Hom, Leslie, "The Making of Local Agenda 21: An Interview with Jeb Brugmann," *Local Environment*, (7) 3, 251–6, 2002.

Hough, Michael, *City Form and Natural Process: Towards a New Urban Vernacular*, New York: Routledge, 1984.

——, *Out of Place: Restoring Identity to the Regional Landscape*, New Haven, CT: Yale University Press, 1990.

Howard, Ebenezer, *Garden Cities of Tomorrow*, London: S. Sonnenschein & Co, 1902 [1898].

Howe, Elizabeth, *Acting on Ethics in Planning*, New Brunswick, NJ: Centre for Urban Policy Research, 1994.

ICLEI, "Local Agenda 21 Survey," Toronto, International Council on Local Environmental Initiatives, 1997, available at http://www.iclei.org/index.php?id=798.

Inglehart, Ronald, *The Silent Revolution: Changing Values and Political Styles among Western Publics*, Princeton, NJ: Princeton University Press, 1977.

——, *Culture Shift in Advanced Industrial Society*, Princeton, NJ: Princeton University Press, 1990.

——, *Modernization and Postmodernization: Cultural, Economic, and Political Change in 43 Societies*, Princeton, NJ: Princeton University Press, 1997.

Innes, Judith, "Planning Theory's Emerging Paradigm: Communicative Action and Interactive Practice," *Journal of Planning Education and Research*, 14 (3), 128–35, 1995.

Innes, Judith, and David E. Booher, *Planning With Complexity: An Introduction to Collaborative Rationality for Public Policy*, New York: Routledge, 2010.

International Energy Agency, "CO_2 Emissions from Fuel Combustion – Highlights," Paris, 2011, available at http://www.iea.org/co2highlights/co2highlights.pdf.

International Forum on Globalization, *Alternatives to Economic Globalization: A Better World is Possible*, San Francisco, CA: Berrett-Koehler, 2002.

Irazábel, Clara, *City-Making and Urban Governance in the Americas: Curitiba and Portland*, Burlington, VT: Ashgate, 2005.

Isard, Walter, *Introduction to Regional Science*, Englewood Cliffs, NJ: Prentice Hall, 1975.

IUCN, *World Conservation Strategy*, International Union for the Conservation of Nature, 1980.

Jackson, Kenneth T., *Crabgrass Frontier: The Suburbanization of the United States*, New York: Oxford University Press, 1985.

Jacobs, Allan, *Looking at Cities*, Cambridge, MA: Harvard University Press, 1985.

——, *Great Streets*, Cambridge, MA: MIT Press, 1993.

Jacobs, Allan, Elizabeth Macdonald, and Yodan Roffe, *The Boulevard Book*, Cambridge, MA: MIT Press, 2001.

Jacobs, Jane, *The Death and Life of Great American Cities*, New York: Random House, 1961.

——, *The Economy of Cities*, New York: Random House, 1969.

——, *Cities and the Wealth of Nations: Principles of Economic Life*, New York: Random House, 1984.

Jones, Paula, David Selby, and Stephen Sterling (eds), *Sustainability Education: Perspectives and Practice Across Higher Education*, London: Earthscan, 2010.

Jones, Tom, William Pettus, and Michael Pyatok, *Good Neighbors: Affordable Family Housing*, Melbourne: The Images Publishing Group, 1997.

JSIC, "Tomamae Is 'Japan's Denmark' and a Mecca for Wind Farms," (Japan for Sustainability Information Center), 2002, available at www.japanfs.org.

Judge, David, Gerry Stoker, and Harold Wolman (eds.), *Theories of Urban Politics*, Thousand Oaks, CA: Sage Publications, 1995.

Kaiser, Edward L., David R. Godschalk, and F. Stuart Chapin Jr, *Urban Land Use Planning*, 4th edn, Urbana, IL: University of Illinois, 1995.

Katz, Bruce (ed.), *Reflections on Regionalism*, Washington, DC: Brookings Institution Press, 2000.

Keene, John, *Global Civil Society?* Cambridge, UK: Cambridge University Press, 2003.

Kehslinger, Rebecca, "Success of Wetland Mitigation Projects," *National Wetlands Newsletter*, 30 (2), 2008, available online at http://www.eli.org/pdf/research/nwn.30.2.kihslinger.pdf.

Keil, Claudia, *Environmental Policy in Germany*, San Francisco, CA: Resource Renewal Institute, 1999.

Kelbaugh, Douglas, *Common Place: Toward Neighborhood and Regional Design*, Seattle, WA: University of Washington Press, 1997.

Kent, Robert B., and Richard E. Klosterman, "GIS and Mapping: Pitfalls for Planners," *Journal of the American Planning Association*, 66 (2), 189–98, 2000.

Kent, T.J., *The Urban General Plan*, San Francisco, CA: Chandler Publishing Co, 1964.

Keough, Noel, "Calgary's Citizen-led Community Sustainability Indicators Project," *Plan*, 43 (1), 35–6, 2003.

Khor, Martin, *Globalisation, Liberalisation, and Protectionism: The Global Framework Affecting Rural Producers in Developing Countries*, Penang, Malaysia: Third World Network, 2007.

Kidd, Charles V., The Evolution of Sustainability, *Journal of Agricultural and Environmental Ethics*, 5 (1), 1–26, 1992.

Kiehl, Jeffrey, "Lessons From the Earth's Past," *Science*, 331 (6014), 158–9, 2011.

Kinsley, Michael J., and L. Hunter Lovins, *Paying for Growth, Prospering from Development*, Snowmass, CO: Rocky Mountain Institute, 1997, available at http://www.natcapsolutions. org/publications_files/PayingForGrowth_ChronPilot_Sep1997.pdf.

Klein, Naomi, *The Shock Doctrine: The Rise of Disaster Capitalism*, Toronto: Knopf, 2007.

Korten, David C., *When Corporations Rule the World*, West Hartford, CT: Kumarian Press, 1995.

——, *The Great Turning: From Empire to Earth Community*, San Francisco, CA: Berrett-Koehler, 2006.

Kotkin, Joel, "Cities and the Census," 2011, available at http://www.joelkotkin.com/content/00406-cities-and-census-cities-neither-booming-nor-withering.

Krizek, Kevin J., and Joe Power, *A Planners Guide to Sustainable Development*, Chicago, IL: American Planning Association, 1996.

Kunstler, James Howard, *The Geography of Nowhere: The Rise and Decline of America's Manmade Landscape*, New York: Simon & Schuster, 1993.

——, *Home from Nowhere: Remaking Our Everyday World for the 21st Century*, New York: Simon & Schuster, 1996.

Langdon, Philip, *A Better Place to Live: Reshaping the American Suburb*, New York: HarperCollins, 1994.

Layard, Antonia, Simin Davoudi, and Susan Batty (eds), *Planning for a Sustainable Future*, New York: Spon Press, 2001.

Le Corbusier, *The City of Tomorrow and its Planning*, New York: Dover Publications, 1987 [1929].

Lee, Wei-chin, "Diplomatic Impetus and Altruistic Impulse: NGOs and the Expansion of Taiwan's International Space," Washington, DC: Brookings Institution, 2012, available at http://www.brookings.edu/research/opinions/2012/07/16-taiwan-ngo-lee.

Lefebvre, Henri, *The Production of Space*, trans. Donald Nicholson-Smith, Cambridge, MA: Blackwell, 1991 [1974].

Lennard, Suzanne H., and Henry L. Lennard, *Livable Communities Observed*, Carmel, CA: Gondolier Press, 1995.

Lennard, Suzanne H., Henry L. Lennard, and Sven von Ungern-Stemberg (eds), *Making Cities Livable*, Carmel, CA: Gondolier Press, 1997.

Leopold, Aldo, *A Sand County Almanac*, New York: Ballantine Books, 1949 [1970].

Lerner, Michael, *The Politics of Meaning: Restoring Hope and Possibility in an Age of Cynicism*, New York: Addison-Wesley, 1996.

Lewis, Sinclair, *Babbitt*, New York: Harcourt, Brace & Co, 1922.

Lindblom, Charles E., "The Science of Muddling-through," *Public Administration Review*, 19, 79–88, 1959.

Lindstrom, Matthew J., and Hugh Bartling (eds), *Suburban Sprawl: Culture, Theory, and Politics*, Lanham MD: Rowman & Littlefield, 2003.

Lipke, David J., "Green Homes: Eco-friendly Building Trends," *American Demographics*, January 2001.

Livable Housing Australia, *Livable Housing Guidelines*, 2nd edn, Sidney, 2012, available at http://www.fahcsia.gov.au/sites/default/files/documents/09_2012/lhd_guidelines_2012_secondedition1.pdf.

Livingstone, Ken, *The Draft London Plan: A Summary*, London: City Hall, 2002.

Logan, John R., and Harvey L. Molotch, *Urban Fortunes: The Political Economy of Place*, Berkeley, CA: University of California Press, 1987.

London 2012, "Towards One Planet 2012: Sustainability Plan Summary," London: London 2012, 2010, available at http://www.london2012.com.

Lovins, Amory, *Soft Energy Paths: Toward a Durable Peace*, San Francisco, CA: Friends of the Earth, 1977.

Lowe, Marcia D., *Alternatives to the Automobile: Transport for Livable Cities*, Worldwatch Paper 98, Washington, DC: The Worldwatch Institute, 1990.

———, *Shaping Cities: The Environmental and Human Dimensions*, Worldwatch Paper 105, Washington, DC: The Worldwatch Institute, 1991.

Luccarelli, Mark, *Lewis Mumford and the Ecological Region*, New York: The Guilford Press, 1995.

Lusk, Ann, "Designing the Active City: The Case for Multi-use Paths," *Progressive Planning: The Magazine of Planners Network*, 157 (Fall), 18–21, 2003, available at http://www.plannersnetwork.org/2003/10/designing-the-active-city-the-case-for-multi-use-paths/.

Lyle, John Tillman, *Regenerative Design for Sustainable Development*, New York: John Wiley & Sons, 1994.

———, *Design for Human Ecosystems: Landscape, Land Use, and Natural Resources*, Washington, DC: Island Press, 1999.

Lynas, Mark, *Six Degrees: Our Future on a Hotter Planet*, Washington, DC: National Geographic, 2008.

Lynch, Kevin, *A Theory of Good City Form*, Cambridge, MA: MIT Press, 1981.

McCamant, Kathryn, and Charles Durrett, *Co-housing: A Contemporary Approach to Housing Ourselves*, 2nd edn, Berkeley, CA: Ten Speed Press, 1994.

McDonough, William, and Partners, "The Hannover Principles: Design for Sustainability," Charlottesville, VA: William McDonough Architects, 1992, available at http://www.mcdonough.com/principles.pdf.

———, and Michael Braungart, *Cradle to Cradle: Remaking the Way We Make Things*, New York: North Point Press, 2002.

Macey, Jonathan R., *Property Rights in Sweden: A Law and Economics Perspective*, Stockholm: Swedish Center for Business and Policy Studies, SNS Occasional Paper No. 85, January 2001.

McGinn, Anne Platt, *Why Poison Ourselves? A Precautionary Approach to Synthetic Chemicals*, Worldwatch Paper 153, Washington, DC: Worldwatch Institute, 2001.

McHarg, Ian L., *Design With Nature*, Garden City, NY: The Natural History Press, 1969.

McHarg, Ian L., and Frederick Steiner, *To Heal the Earth: Selected Writings of Jan L. McHarg*, Washington, DC: Island Press, 1998.

Maclaren, Virginia, "Urban Sustainability Reporting," *Journal of the American Planning Association*, 62 (2), 184–202, 1996.

Mander, Jerry, *In the Absence of the Sacred: The Failure of Technology and the Survival of the Indian Nations*, San Francisco, CA: Sierra Club Books, 1991.

Marcus, Clare Cooper, and Wendy Sarkissian, *Housing as if People Mattered: Site Design Guidelines for Medium-density Family Housing*, Berkeley, CA: University of California Press, 1986.

Marcus, Clare Cooper, and Carolyn Francis (eds), *People Places: Design Guidelines for Urban Open Space*, New York: Van Nostrand Reinhold, 1990.

Markusen, Ann, Peter Hall, Scott Campbell, and Sabina Deitrick, *The Rise of the Gunbelt: The Military Remapping of Industrial America*, New York: Oxford University Press, 1991.

Marquetti, Adalmir, Carlos E. Schonerwald da Silva, and Al Campbell, "Participatory Economic Democracy in Action: Participatory Budgeting in Porto Alegre, 1989–2004," *Review of Radical Political Economics*, 44 (1), 62–81, 2012.

Marsh, George Perkins, *Man and Nature, or Physical Geography as Modified by Human Action*, New York: C. Scribner, 1864.

Marshall, Alex, *How Cities Work*, Austin, TX: University of Texas Press, 2000.

Massey, Doreen, *Spatial Divisions of Labour: Social Structures and the Geography of Production*, London: Macmillan, 1984.

——, *Space, Place and Gender*, Cambridge, UK: Polity Press, 1994.

Mazmanian, Daniel A., and Michael E. Kraft (eds), *Toward Sustainable Communities: Transition and Transformations in Environmental Policy*, 2nd edn, Cambridge, MA: MIT Press, 2009.

Meadows, Donella, Dennis L. Meadows, Jörgen Randers, and William W. Behrens III, *The Limits to Growth*, New York: Universe Books, 1972.

Meadows, Donella, Dennis L. Meadows, and Jörgen Randers, *Beyond the Limits: Confronting Global Collapse, Envisioning a Sustainable Future*, Post Mills, VT: Chelsea Green, 1992.

Meadows, Donella, Jörgen Randers, and Dennis Meadows, *The Limits to Growth: The 30-Year Update*. White River Junction, VT: Chelsea Green, 2004.

Merchant, Carolyn, *Radical Ecology*, New York: Routledge, 1992.

Met Office Hadley Center, *Advance: Improved Science for Mitigation Policy Advice*, 2010, available at http://www.metoffice.gov.uk/media/pdf/n/c/advance.pdf.

Millard-Ball, Adam, and Lee Schipper, "Are We Reaching Peak Travel? Trends in Passenger Transport in Eight Industrialized Countries," *Transport Reviews*, 31 (3), 357–78, 2001.

Miraftab, Faranak, "Insurgent Planning: Situating Radical Planning in the Global South," *Planning Theory*, 9 (1), 32–50, 2009.

Mitlin, Diana, "Sustainable Development: A Guide to the Literature," *Environment and Urbanization*, 4 (1), 111–24, 1992.

Mittler, Daniel, "Hijacking Sustainability? Planners and the Promise and Failure of Agenda 21," in Layard, Antonia, Simin Davoudi, Susan Batty (eds), *Planning for a Sustainable Future*, New York: Spon Press, 2001.

Mollenkopf, John H., *The Contested City*, Princeton, NJ: Princeton University Press, 1983.

Moore, Steven A., *Alternative Routes to the Sustainable City: Austin, Curitiba, and Frankfurt*, Plymouth UK: Lexington Books, 2007.

Morris, A.E.J., *History of Urban Form Before the Industrial Revolutions*, 3rd edn, New York: John Wiley & Sons, 1994.

Morris, David, *Self-reliant Cities: Energy and the Transformation of Urban America*, San Francisco, CA: Sierra Club Books, 1983.

Moudon, Anne Vernez, and Paul Hess, "Suburban Clusters: The Nucleation of Multifamily Housing in Suburban Areas of the Central Puget Sound," *Journal of the American Planning Association*, 66 (3), 243–64, 2000.

Mumford, Lewis, *Technics and Civilization*, New York: Harcourt, Brace & Co, 1934.

——, *The Culture of Cities*, New York: Harcourt, Brace & Co, 1938.

——, *The City in History: Its Origins, Its Transformations, and Its Prospects*, New York: Harcourt, Brace & World, 1961.

——, *The Urban Prospect*, New York: Harcourt, Brace & World, 1968.

Munt, Ian, "An Exemplary Sustainable World City?" *Town & Country Planning*, November 2001, 292–3.

Nadel, Steve, and Howard Geller, *Smart Energy Policies: Saving Money and Reducing Pollutant Emissions through Greater Energy Efficiency*, Washington, DC: Tellus Institute, 2001.

Nader, Ralph, and Mark J. Green (eds), *Corporate Power in America*, New York: Grossman Publishers, 1973.

Nader, Ralph, Mark Green, and Joel Seligman, *Taming the Giant Corporation*, New York: W.W. Norton & Co., 1976.

Nadin, Vincent, Stefanie Dohr, and Caroline Brown (eds), *Sustainability, Development, and Spatial Planning in Europe*, London: Routledge, 2000.

National Research Council, *The Drama of the Commons*, Washington, DC: National Academy Press, 2002.

National Round Table, *Reality Check: The State of Climate Progress in Canada*, Ottawa: National Round Table on the Environment and the Economy, 2012.

Nelessen, Anton, *Visions for a New American Dream*, Chicago, IL: Planners Press, 1994.

Neuman, Michael, "Does Planning Need the Plan?" *Journal of the American Planning Association*, 64 (2), 215ff., 1998.

——, "Regional Design: Recovering a Great Landscape Architecture and Urban Planning Tradition," *Landscape and Urban Planning*, 47, 115–28, 2000.

Newman, Peter, and Jeffrey Kenworthy, *Cities and Automobile Dependence: An International Sourcebook*, Aldershot, UK: Gower, 1989.

——, *Sustainability and Cities: Overcoming Automobile Dependence*, Washington, DC: Island Press, 1999.

Nicholson-Lord, David, "Myopia," *Town & Country Planning*, March/April 2003, 100–103.

Nolan, John (ed.), *City Planning: A Series of Papers Presenting the Essential Elements of a City Plan*, New York: D. Appleton and Co, 1916.

Norberg-Hodge, Helena, and Peter Goering (eds), *The Future of Progress: Reflections on Environment and Development*, Berkeley, CA: International Society for Ecology and Culture, 1986.

Norgaard, Richard, *Development Betrayed: The End of Progress and a Coevolutionary Revisioning of the Future*, New York: Routledge, 1994.

Norwood, Ken, and Kathleen Smith, *Rebuilding Community in America: Housing for Ecological Living, Personal Empowerment, and the New Extended Family*, Berkeley, CA: Shared Living Resource Center, 1995.

NRDC/US Environmental Protection Agency, *Environmental Characteristics of Smart Growth Neighborhoods: An Exploratory Case Study*, Washington, DC: Natural Resources Defense Council and US Environmental Protection Agency, 2000.

O'Connor, Martin (ed.), *Is Capitalism Sustainable? Political Economy and the Politics of Ecology*, New York: The Guilford Press, 1994.

ODPM, *Sustainable Communities: Building for the Future*, London: Office of the Deputy Prime Minister, 2003.

Odum, Howard W., and Harry Estill Moore, *American Regionalism: A Cultural-historical Approach to National Integration*, New York: Henry Holt & Co, 1938.

Ohmae, Kenichi, *The End of the Nation-State: The Rise of Regional Economies*, New York: The Free Press, 1995.

Oommen, M.A., "Reforms and the Kerala Model," *Economic and Political Weekly*, 43 (2), 22–5, 2008.

Oregon DLCD, "Oregon Statewide Planning Program," Oregon Department of Land Conservation and Development, 1997, available at http://www.oregon.gov/LCD/docs/publications/brochure.pdf.

Orfield, Myron, *Metropolitics: A Regional Agenda for Community and Stability*, Washington, DC, and Cambridge, MA: Brookings Institution Press and Lincoln Institute of Land Policy, 1997.

——, *American Metropolitics: The New Suburban Reality*, Washington, DC: Brookings Institution Press, 2002.

Orfield, Myron, and Thomas F. Luce, Jr., *Region: Planning the Future of the Twin Cities*, Minneapolis, MN: University of Minnesota Press, 2010.

Ortiz, Isabel, and Matthew Cummins, *Global Inequality: Beyond the Bottom Billion: A Rapid Review of Income Distribution in 141 Countries*, New York: UNICEF, 2011, available at http://www.unicef.org/socialpolicy/index_58230.html.

Osborn, Fairfield, *Our Plundered Planet*, New York: Grosset & Dunlap, 1948.

Ostrom, Elinor, *Governing the Commons: The Evolution of Institutions for Collective Action*, New York: Cambridge University Press, 1990.

——, *Understanding Institutional Diversity*, Princeton, NJ: Princeton University Press, 2005.

Owens, Susan, and Richard Cowell, "Planning for Sustainability: New Orthodoxy or Radical Challenge," *Town & Country Planning*, June, 170–72, 2001.

——, *Land and Limits: Interpreting Sustainability in the Planning Process*, London: Routledge, 2002.

Parayil, Govindan (ed.), *Kerala: The Development Experience: Reflections on Sustainability and Replicability*, London: Zed Books, 2000.

Parkinson, Sarah, and Mark Roseland, "Leaders of the Pack: An Analysis of the Canadian 'Sustainable Communities' 2000 Municipal Competition," *Local Environment*, 7 (4), 422–9, 2002.

Pastor, Manuel Jr., Marta Lopez-Garza, Peter Dreier, and J. Eugene Grigsby III, *Regions that Work: How Cities and Suburbs Can Grow Together*, Minneapolis, MN: University of Minnesota Press, 2001.

Pastor, Manuel Jr., Chris Benner, and Martha Matsuoka, *This Could Be the Start of Something Big: How Social Movements for Regional Equity are Reshaping Metropolitan America*, Ithaca, NY: Cornell University Press, 2009.

Pearce, David, and Edward B. Barbier, *Blueprint for a Sustainable Economy*, London: Earthscan, 2000.

Pearce, David, and Jeremy J. Warford, *World Without End: Economics, Environment and Sustainable Development*, New York: Oxford University Press, 1993.

Pearce, David, Edward Barbier, and Anil Markandya, *Blueprint for a Green Economy*, London: Earthscan, 1989.

Pearce, David, Edward Barbier, and Anil Markandya, *Sustainable Development: Economics and Environment in the Third World*, London: Edward Elgar, 1990.

Peck, Sheila, *Planning for Biodiversity: Issues and Examples*, Washington, DC: Island Press, 1998.

Peck and Associates, *Implementing Sustainable Community Development: Charting a Federal Role for the 21st Century*, Toronto: Canada Mortgage and Housing Corporation, 2000, available at http://cmhc.ca/odpub/pdf/62484.pdf?lang=en.

Peirce, Neal R., *Citistates: How Urban America Can Prosper in a Competitive World*, Washington, DC: Seven Locks Press, 1993.

Perry, Clarence Arthur, *Housing for the Machine Age*, New York: Russell Sage Foundation, 1939.

Phillips, Kevin, *Wealth and Democracy: A Political History of the American Rich*, New York: Broadway Books, 2002.

——, *Bad Money: Reckless Finance, Failed Politics, and the Global Crisis of American Capitalism*, New York: Viking, 2008.

Philpott, Tom, "How Food Processing Got into the Hands of a Few Giant Companies," *GRIST*, 2006, available at http://grist.org/article/giants/.

Porter, Douglas R., *Managing Growth in America's Communities*, Washington, DC: Island Press, 1997.

Portnoy, Kent E., *Taking Sustainable Cities Seriously: Economic Development, the Environment, and Quality of Life in American Cities*, Cambridge, MA: MIT Press, 2003.

Powell, John A., "Addressing Regional Dilemmas for Minority Communities," in Katz, Bruce (ed.), *Reflections on Regionalism*, Washington, DC: Brookings Institution Press, 2000.

——, *Racing to Justice: Transforming our Conceptions of Self and Other to Build an Inclusive Society*, Bloomington, IN: Indiana University Press, 2012.

President's Council on Environmental Quality, *Global 2000 Report*, Washington, DC: US Government Printing Office, 1981.

President's Council on Sustainable Development, *Sustainable America: A New Consensus for Prosperity, Opportunity, and a Healthy Environment for the Future*, Washington, DC: US Government Printing Office, 1996.

Prugh, Thomas, Robert Costanza, and Herman Daly, *The Local Politics of Global Sustainability*, Washington, DC: Island Press, 2000.

Pucher, John, Zhong-Ren Peng, Neha Mittal, Yi Zhu, and Nisha Korattyswaroopam, "Urban Transport Trends and Policies in China and India: Impacts of Rapid Economic Growth," *Transport Reviews*, 27 (4), 279–410, 2007.

Putnam, Robert D., *Bowling Alone: The Collapse and Revival of American Community*, New York: Simon & Schuster, 2000.

Putnam, Robert D., Robert Leonardi, and Raffaella Y. Nanetti, *Making Democracy Work: Civic Traditions in Modern Italy*, Princeton, NJ: Princeton University Press, 1993.

Rabinovitch, Jonas, and Josef Leitman, "Urban Planning in Curitiba," *Scientific American*, March 1996, 46–53.

Randolph, John, and Gilbert M. Masters, *Energy for Sustainability: Technology, Planning, Policy*, Washington, DC: Island Press, 2008.

Rapoport, Amos, "Toward a Redefinition of Density," *Environment and Behavior*, 7 (2), 133–58, 1975.

Ravetz, Joe, *City-Region 2020: Integrated Planning for a Sustainable Environment*, London: Earthscan, 2000.

Redclift, Michael, *Sustainable Development: Exploring the Contradictions*, London: Methuen, 1987.

Rees, William E., "Ecological Footprints and Appropriated Carrying Capacity: What Urban Economics Leaves Out," *Environment and Urbanization*, 4 (2), 121–30, 1992.

Register, Richard, *Ecocity Berkeley: Building Cities for a Healthy Future*, Berkeley, CA: North Atlantic Books, 1987.

——, *Ecocities: Building Cities in Balance with Nature*, Berkeley, CA: Berkeley Hills Books, 2002.

Renner, Michael, *Jobs in a Sustainable Economy*, Worldwatch Paper 104, Washington, DC: The Worldwatch Institute, 1991.

Repetto, Robert (ed.), *The Global Possible: Resources, Development, and the New Century*, New Haven, CT: Yale University Press, 1985.

Resource Renewal Institute, "Case Study: The Netherlands' Environmental Policy Plan," San Francisco, 1999.

Riesman, David, *The Lonely Crowd*, Garden City, NY: Doubleday, 1953.

Riley, Ann L., *Restoring Streams in Cities*, Washington, DC: Island Press, 1998.

Romm, Joseph, *Hell and High Water*, New York: HarperCollins, 2005.

Roseland, Mark, *Toward Sustainable Communities: Resources for Citizens and their Governments*, 4th edn, Stony Creek, CT: New Society Publishers, 2012.

Roszak, Theodore, *Where the Wasteland Ends: Politics and Transcendence in Post-industrial Society*, Garden City, NY: Doubleday, 1972.

Rowe, Janet, and Colin Fudge, "Linking National Sustainable Development Strategy and Local Implementation: A Case Study in Sweden," *Local Environment*, 8 (2), 125–40, 2003.

Roy, Robin, and Sally Caird, "Household Ecological Footprints: Moving Towards Sustainability?" *Town & Country Planning*, October, 277–9, 2001.

RPA, *Landscapes: Improving Conservation Practice in the Northeast Megaregion*, New York: Regional Plan Association, 2012, available at http://www.rpa.org/library/pdf/RPA-Northeast-Landscapes.pdf.

Rudlin, David, and Nicholas Falk, *Building the 21st Century Home: The Sustainable Urban Neighborhood*, Oxford: Architectural Press, 1999.

Rudofsky, Bernard, *Streets for People: A Primer for Americans*, Garden City, NY: Doubleday, 1969.

Rydin, Yvonne, *Urban and Environmental Planning in the UK*, London: Macmillan, 1998.

Rydin, Yvonne, and Andy Thomley (eds), *Planning in the UK: Agendas for the New Millennium*, Aldershot: Ashgate, 2000.

Sale, Kirkpatrick, *Dwellers in the Land: The Bioregional Vision*, San Francisco, CA: Sierra Club Books, 1985.

SAP/MPSDOL, *Scientific Assessment of Ozone Depletion: 2002*, New York: United Nations Environment Programme, Scientific Assessment Panel of the Montreal Protocol on Substances that Deplete the Ozone Layer, 2002, available at http://www.unep.org/ozone/pdf/execsumm-sap2002.pdf.

Satterthwaite, David (ed.), *The Earthscan Reader in Sustainable Cities*, London: Earthscan, 1999.

Schön, Donald A., *The Reflective Practitioner: How Professionals Think in Action*, New York: Basic Books, 1983.

Schumacher, E.F., *Small Is Beautiful: A Study of Economics as if People Mattered*, London: Blond and Briggs, 1973.

Schuyler, David, *The New Urban Landscape: The Redefinition of City Form in Nineteenth-century America*, Baltimore: Johns Hopkins University Press, 1986.

——, *Apostle of Taste: Andrew Jackson Downing, 1815–1852*, Baltimore: Johns Hopkins University Press, 1996.

Seamon, David (ed.), *Dwelling, Seeing, and Designing: Toward a Phenomenological Ecology*, Albany, NY: SUNY Press, 1993.

Sen, Amartya, *Poverty and Famines: An Essay on Entitlements and Deprivation*, Oxford: Clarendon Press, 1982.

——, *Development as Freedom*, Oxford: Oxford University Press, 1999.

Serageldin, Ismail, "Sustainability as Opportunity and the Problem of Social Capital," *Brown Journal of World Affairs*, 3 (2), 187–203, 1996.

Sewell, John, *The Shape of the City: Toronto Struggles with Modern Planning*, Toronto: University of Toronto Press, 1993.

——, *The Shape of the Suburbs: Understanding Toronto's Sprawl*, Toronto: University of Toronto Press, 2009.

Shabecoff, Philip, *Earth Rising: American Environmentalism in the 21st Century*, Washington, DC: Island Press, 2000.

Shen, Quin, "A Spatial Analysis of Job Openings and Access in a US Metropolitan Area," *Journal of the American Planning Association*, 67 (1), 53–68, 2001.

Shiva, Vandana, *Staying Alive: Women, Ecology, and Development*, Brooklyn NY: South End Press, 2010.

Shuman, Michael, *Going Local: Creating Self-Reliant Communities in a Global Age*, New York: Free Press, 1998.

——, *The Small-Mart Revolution: How Local Businesses are Beating the Global Competition*, San Francisco, CA: Berrett-Koehler, 2006.

——, *Local Dollars, Local Sense: How to Shift Your Money from Wall Street to Main Street and Achieve Real Prosperity*, White River Junction, VT: Chelsea Green, 2012.

Simon, Julian L., *The Ultimate Resource*, Princeton, NJ: Princeton University Press, 1981.

Sirkis, Alfredo, "Bike Networking in Rio: The Challenges for Non-motorised Transport in an Automobile-dominated Government Culture," *Local Environment*, 5 (1), 83–95, 2000.

Soja, Edward W., *Postmodern Geographies: The Reassertion of Space in Critical Social Theory*, New York: Verso, 1989.

Southworth, Michael, and Eran Ben-Joseph, *Streets and the Shaping of Towns and Cities*, New York: McGraw-Hill, 1997.

Southworth, Michael, and Peter M. Owens, "The Evolving Metropolis: Studies of Community, Neighborhood, and Street Form at the Urban Edge," *Journal of the American Planning Association*, 59 (3), 271–87, 1993.

Speir, Cameron, and Kurt Stephenson, "Does Sprawl Cost Us All? Isolating the Effects of Housing Patterns on Public Water and Sewer Costs," *Journal of the American Planning Association*, 68 (1), 56–70, 2002.

Spirn, Ann Whiston, *The Granite Garden: Urban Nature and Human Design*, New York: Basic Books, 1984.

Stavrianos, L.S., *Global Rift: The Third World Comes of Age*, New York: Morrow, 1981.

Stern, Sir Nicholas, et al., *Stern Review on the Economics of Climate Change*, London: Her Majesty's Treasury, 2006.

Stivers, Robert L., *The Sustainable Society: Ethics and Economic Growth*, Philadelphia, PA: Westminster Press, 1976.

Stone, Clarence, *Regime Politics: Governing Atlanta, 1946–1988*, Lawrence, KS: University Press of Kansas, 1989.

Stren, Richard, Rodney R. White, and Joseph B. Whitney (eds), *Sustainable Cities: Urbanization and the Environment in International Perspective*, Boulder, CO: Westview Press, 1992.

Strong, Ann L., *Land Banking: European Reality, American Prospect*, Baltimore, MD: Johns Hopkins University Press, 1979.

Subak, S, "Forest Protection and Reforestation in Costa Rica: Evaluation of a Clean Development Mechanism Prototype," *Environmental Management*, 26 (3), 283–97, 2000.

Susman, Carl (ed.), *Planning the Fourth Migration: The Neglected Vision of the Regional Planning Association of America*, Cambridge, MA: MIT Press, 1976.

Sustainable Seattle Coalition, *Indicators of Sustainable Community*, Seattle, WA: Sustainable Seattle Coalition, 1995.

Talen, Emily, "Design by the Rules: The Historical Underpinnings of Form-Based Codes," *Journal of the American Planning Association*, 75 (2), 144–60, 2009.

Tarrow, Sidney, *The New Transnational Activism*, Cambridge, UK: Cambridge University Press, 2005.

Teilhard de Chardin, Pierre, *The Phenomenon of Man*, New York: Harper, 1959.

Thayer, Robert, *Gray World, Green Heart: Technology, Nature, and the Sustainable Landscape*, New York: Wiley, 1994.

Thiele, Leslie Paul, *Environmentalism for a New Millennium: The Challenge of Coevolution*, New York: Oxford University Press, 1999.

Thompson, J. William, and Kim Sorvig, *Sustainable Landscape Construction: A Guide to Green Building Outdoors*, 2nd edn, Washington, DC: Island Press, 2007.

Todd, Nancy Jack, and John Todd, *From Eco-cities to Living Machines*, Berkeley, CA: North Atlantic Books, 1993.

Tomalty, Ray, "Growth Management in the Vancouver Region," *Local Environment*, 7 (4), 431–45, 2002.

Trainer, Ted, *Abandon Affluence!*, London: Zed Books, 1986.

Transportation Research Board, *Costs of Sprawl Revisited: The Evidence of Sprawl's Negative and Positive Impacts*, Washington, DC: National Academy Press, 1998, available at http://onlinepubs.trb.org/onlinepubs/tcrp/tcrp_rpt_39-a.pdf.

Tuchman, Barbara, *The March of Folly: From Troy to Vietnam*, New York: Knopf, 1984.

Turner, R. Kerry (ed.), *Sustainable Environmental Management: Principles and Practice*, Boulder, CO: Westview Press, 1988.

Tzung, Lin Shen, "The Evolution of Taiwan's Environmental Movement," in Turner, Jennifer L., *Hong Kong Conference Report,* Washington, DC: The Wilson Center, 2001, available at http://www.wilsoncenter.org/publication/hong-kong-conference-report-section-1-english.

Uggen, Christopher, Sarah Shannon, and Jeff Manza, *State-Level Estimates of Felon Disenfranchisement in the United States, 2010,* Washington, DC: The Sentencing Project, 2012.

UN, *The Millennium Development Goals Report 2012,* New York: United Nations, 2012, available at http://www.slideshare.net/undesa/millennium-development-goals-report-2012.

UN Department of Economic and Social Affairs, *World Urbanization Prospects: The 2011 Revision,* New York: United Nations, 2011, available at http://esa.un.org/unup/pdf/WUP2011_Highlights.pdf

UN Framework Convention on Climate Change, *Report on the In-depth Review of the Third National Communication of the United Kingdom,* New York: United Nations, 2003.

UN Human Settlements Programme, "Habitat Declaration," Nairobi: United Nations Center on Human Settlements, 1996, available at www.unchs.org.

United Church of Christ/Commission for Racial Justice, *Toxic Wastes and Race in the United States: A National Report on the Racial and Socioeconomic Characteristics of Communities with Hazardous Waste Sites,* Washington, DC: United Church of Christ/Commission for Racial Justice, 1987.

Urban Ecology Inc., *Blueprint for a Sustainable Bay Area,* Oakland, CA: Urban Ecology, Inc., 1996.

Urban Task Force, *Towards an Urban Renaissance,* (Lord Rogers, Chair) London: Department of the Environment, Transport and the Regions, 1999.

US Bureau of the Census, *Statistical Abstract of the United States,* Washington, DC: US Government Printing Office, 2001.

US Department of Energy, "Data, Analysis & Trends," 2012, available at http://www.afdc.energy.gov/afdc/data/vehicles.html.

US Department of Transportation, *Bus Rapid Transit Ridership Analysis,* Washington, DC: US Government Printing Office, 2005, available at http://www.nbrti.org/docs/pdf/WestStart_BRT_Ridership_Analysis_Final.pdf.

US Environmental Protection Agency, *Our Built and Natural Environments: A Technical Review of the Interactions between Land Use, Transportation, and Environmental Quality,* Washington, DC: US Environmental Protection Agency, 2001.

——, *Measuring the Air Quality and Transportation Impacts of Infill Development,* Washington, DC: US Environmental Protection Agency, 2007, available at http://www.epa.gov/dced/pdf/transp_impacts_infill.pdf.

——, *Executive Summary of the Inventory of US Greenhouse Gas Emissions, and Sinks, 1990–2000,* Washington, DC: US Environmental Protection Agency, 2012, available at http://www.epa.gov/climatechange/ghgemissions/usinventoryreport.html.

US General Accounting Office, *Siting of Hazardous Waste Landfills and their Correlation with Racial and Economic Status of Surrounding Communities,* Washington, DC: General Accounting Office, GAO/RCED-83-168, 8–211461, 1983.

Vale, Brenda, and Robert Vale, *The Autonomous House: Design and Planning for Self-sufficiency,* London: Thames and Hudson, 1975.

Van der Ryn, Sim, and Peter Calthorpe (eds), *Sustainable Communities,* San Francisco, CA: Sierra Club Books, 1984.

Van der Ryn, Sim, and Stuart Cowan, *Ecological Design,* Washington, DC: Island Press, 1996.

Van Eeten, Michel, and Emery Roe, "When Fiction Conveys Truth and Authority: The Netherlands Green Heart Planning Controversy," *Journal of the American Planning Association,* 66 (1), 58–67, 2000.

Véron, René, "The 'New' Kerala Model: Lessons for Sustainable Development," *World Development,* 29 (4), 601–17, 2001.

Vogt, William, *Road to Survival,* New York: W. Sloane Associates, 1948.

Wackernagel, Mathis, and William Rees, *Our Ecological Footprint: Reducing Human Impact on the Earth,* Gabriola Island, BC: New Society Publishers, 1996.

Wackernagel, Mathis, and J. David Yount, "The Ecological Footprint: An Indicator of Progress Toward Regional Sustainability," *Environmental Monitoring and Assessment*, 51, 511–29, 1998.

Wagner, Fritz, and Roger Caves (eds), *Community Livability: Issues and Approaches to Sustaining the Well-being of People and Communities*, New York: Routledge. 2012.

Walker, Peter A., and Patrick T. Hurley, *Planning Paradise: Politics and Visioning of Land Use in Oregon*, Tucson, AZ: University of Arizona Press, 2011.

Warburton, Diane (ed.), *Community and Sustainable Development: Participation in the Future*, London: Earthscan, 1998.

——, "Participation: Delivering Fundamental Change," *Town & Country Planning*, March 2002, 82–4.

Ward, Barbara, and Rene Dubos, *Only One Earth: The Care and Maintenance of a Small Planet*, New York: W.W. Norton & Co, 1972.

Ward, Peter Douglas, *Under a Green Sky: The Mass Extinctions of the Past, and What They Mean for Our Future*, New York: Smithsonian Books, 2007.

Waring, Marilyn, *If Women Counted: A New Feminist Economics*, San Francisco, CA: Harper & Row, 1988.

Warner, Sam Bass, *Streetcar Suburbs: The Process of Growth in Boston 1870–1900*, Cambridge, MA: Harvard University Press, 1962.

Weart, Spencer B., "The Discovery of Global Warming," American Institute of Physics, website supplement to Weart, Spencer B., *The Discovery of Global Warming*, 2nd edn, Cambridge, MA: Harvard University Press, 2008, available at http://www.aip.org/history/climate/index.html.

Webber, Melvin M., John W. Dyckman, Donald L. Foley, Albert Z. Guttenberg, William L. C. Wheaton, and Catherine Bauer Wurster (eds), *Explorations into Urban Structure*, Philadelphia, PA: University of Pennsylvania Press, 1964.

Weiss, Marc, *The Rise of the Community Builders: The American Real Estate Industry and Urban Land Planning*, New York: Columbia University Press, 1987.

Wheeler, Stephen M., "Planning for Metropolitan Sustainability," *Journal of Planning Education and Research*, 20, 133–45, 2000.

——, *Smart Infill: Creating More Livable Communities in the Bay Area*, San Francisco, CA: Greenbelt Alliance, 2002a.

——, "The New Regionalism: Key Characteristics of an Emerging Movement," *Journal of the American Planning Association*, 68 (3), 267–78, 2002b.

——, "State and Municipal Climate Change Plans: The First Generation," *Journal of the American Planning Association*, 74 (4), 481–96, 2008a.

——, "The Evolution of Built Landscapes in Metropolitan Regions," *Journal of Planning Education and Research*, 27 (4), 400–416, 2008b.

——, *Climate Change and Social Ecology*, Abingdon and New York: Routledge, 2012.

Wheeler, Stephen M., and Timothy Beatley (eds), *The Sustainable Urban Development Reader*, 2nd edn, London and New York: Routledge, 2009.

Wheeler, Stephen M., and Craig Beebe, "The Rise of the Postmodern Metropolis: Spatial Evolution of the Sacramento Metropolitan Region," *Journal of Urban Design*, 16 (3), 307–32, 2011.

Wheeler, Stephen M., and Robert B. Segar, "West Village: Development of an Ecological Neighborhood in Davis, CA," *Planning Theory and Practice*, 13 (1), 145–56, 2012.

Wheeler, Stephen M., Mihaela Tomuta, Ryan Van Hayden, and Louise Jackson, "The Impacts of Alternative Patterns of Urbanization on Greenhouse Gas Emissions in an Agricultural County," *Journal of Urbanism*, forthcoming.

Whyte, William H., *The Organization Man*, Garden City, NY: Doubleday, 1956.

——, *The Last Landscape*, Garden City, NY: Doubleday, 1968.

Wilson, William Julius, *The Truly Disadvantaged: The Inner City, the Underclass, and Public Policy*, Chicago, IL: University of Chicago Press, 1987.

——, *When Work Disappears: The World of the New Urban Poor*, New York: Knopf, 1996.

Winter, Paul, "Sustainable Housing: The Legal Context, Part I," *Town & Country Planning*, January, 26–8, 2001a.

——, "Sustainable Housing: The Legal Context, Part II," *Town & Country Planning*, February, 50–3, 2001b.

World Commission on Environment and Development, (The Brundtland Commission), *Our Common Future*, New York: Oxford University Press, 1987.

Worrell, Ernst, and Christina Galitsky, *Energy Efficiency Improvement and Cost Saving Opportunities for Petroleum Refineries*, Berkeley, CA: Lawrence Berkeley National Laboratory, 2005.

Wurster, Donald, *The Wealth of Nature: Environmental History and the Ecological Imagination*, New York: Oxford University Press, 1993.

Yaro, Robert D., and Tony Hiss, *A Region at Risk: The Third Regional Plan for the New York–New Jersey–Connecticut Metropolitan Area*, Washington, DC: Island Press, 1996.

Zukin, Sharon, *Landscapes of Power: From Detroit to Disney World*, Berkeley, CA: University of California Press, 1991.

INDEX

Note: page numbers in italic type refer to Figures.